CORN Publication Series 7

D/2005/0095/110
ISBN 2-503-51780-3

Land, shops and kitchens

Technology and the food chain in twentieth-century Europe

Edited by Carmen Sarasúa, Peter Scholliers

& Leen Van Molle

BREPOLS

CONTENTS

List of contributors 7

List of figures 8

List of tables 9

Preface 11

1. The rise of a food market in European history
   Carmen SARASÚA and Peter SCHOLLIERS 13

2. Conflict and environmental tension in the adoption of technological innovation in the agrarian sector
   Ramon GARRABOU SEGURA 30

3. Environment conditions and biological innovations in European agrarian growth
   Josep PUJOL-ANDREU 42

4. The role of EU policies in technological innovation in agriculture
   Konstandinos MATTAS and Efstratios LOIZOU 68

5. Recent innovations in the horticultural production system in the Southeast of Spain
   Carmen NAVARRO DEL AGUILA, José LÓPEZ-GÁLVEZ and José SALAZAR MATO 85

6. Nitrogen in modern European agriculture
   Vaclav SMIL 110

7. Modernisation and the international food system: re-articulation or resistance?
   David GOODMAN and Michael REDCLIFT 120

8. *Kulturkampf* in the countryside. Agricultural education, 1800–1940: a multifaceted offensive
   Leen VAN MOLLE 139

9. Changing tastes. The role of scientific and medical discoveries in changing the modern diet
   Rayna GAVRILOVA 170

10. The rise of supermarkets in twentieth-century Britain and France
    Isabelle LESCENT-GILES 188

11. Milky ways. Dairy, landscape and nation building until 1930 212
Barbara ORLAND

12. Fast food and slow food. The fastening food chain and recurrent countertrends in Europe and the Netherlands (1890–1990)
Anneke H. VAN OTTERLOO 255

13. Industrialising catering. Technological developments and its effects in the twentieth century 278
Ulrike THOMS

LIST OF CONTRIBUTORS

RAMON GARRABOU SEGURA	Dept. of Economic History Universitat Autónoma de Barcelona, ES
RAYNA GAVRILOVA	Dept. of Cultural Studies University of Sofia, BG
DAVID GOODMAN	Dept. of Environmental Studies University of California, Santa Cruz, US
ISABELLE LESCENT-GILES	Dept. of History University of Paris – Sorbonne, FR
EFSTRATIOS LOIZOU	Dept. of Agricultural Products Marketing and Quality Control Technological Educational Institute of Western Macedonia, Florina, GR
JOSÉ LÓPEZ-GÁLVEZ	Dept. of Economics Universidad de Almería, ES
KONSTANDINOS MATTAS	School of Agriculture Aristotle University of Thessaloniki, GR
LEEN VAN MOLLE	Dept. of History University of Leuven, BE
CARMEN NAVARRO DEL AGUILA	Dept. of Applied Economics Universidad de Almería, ES
BARBARA ORLAND	History of Knowledge Center Swiss Federal Institute of Technology Zurich, CH
ANNEKE H. VAN OTTERLOO	Dept. of Sociology University of Amsterdam, NL
JOSEP PUJOL-ANDREU	Dept. of Economic History Universitat Autónoma de Barcelona, ES
MICHAEL REDCLIFT	Dept. of Geography Kings College, London, UK
JOSÉ SALAZAR MATO	Dept. of Applied Economics Universidad de Almería, ES
CARMEN SARASÚA	Dept. of Economic History Universitat Autónoma de Barcelona, ES
PETER SCHOLLIERS	Dept. of History Free University of Brussels (VUB), BE
VACLAV SMIL	Dept. of Geography University of Manitoba, CA
ULRIKE THOMS	Dept. of History of Medicine Freie Universität Berlin, DE

LIST OF FIGURES

3.1 Main flows of wheat seeds between 1830 and 1914
3.2 Price indexes in Barcelona (Spain) in constant pesetas, 1865–1935
4.1 Simplified form of the development, transfer and use of technology in agriculture
5.1 Evolution of production, home market and export in the Almería agrarian sector
5.2 Evolution of fruit and horticultural production in Almería, season 1996/97
5.3 Distribution of working hours in the cucumber and green bean season 1987/88
5.4 Estimation of labour force needs in one ha of tomato crop grown in a greenhouse without heating, season 1998/99
5.5 Sketch representation of the nursery fertigation system
5.6 Plant of a seeding tray with irrigation by flooding
5.7 The NFT system
5.8. Arrangement of the double-paired lines in the cultivation channels
8.1 Kilograms of cereals per head and per year
8.2 The knowledge chain
8.3 The knowledge network
11.1 *Railmilk*-flow to Berlin, 1903 and 1927
11.2 Development of the Carl Bolle Dairy
11.3 Development of producer prices for fresh milk in Switzerland, 1800–1914
11.4 Development of the milk production in Switzerland, 1837–1930
11.5 Development of export/import of cheese in Switzerland. Export 1851–1939 / Import 1851–1913
11.6 Export rates Swiss cattle, 1851–1913
12.1 The food chain
12.2 Private consumption in the Netherlands, 1925–1985
12.3 Attributes of 'regional food' from consumers' point of view in 1998–1999
12.4 Reasons for acceptance and disapproval of higher prices of regional food in 1998–1999
13.1 Expenditure for wages of the kitchen staff per patient's day of stay in the Hospital St. Jacobs, Leipzig 1875–1908
13.2 Number of canteens in West-Germany and their turnover per employee, 1959–1999

LIST OF TABLES

3.1 European Experimental Centres, members of the International Association of Seed Testing, 1931
3.2 Institutions of wheat improvement in Europe, 1880–1938
3.3 New varieties of wheat between 1880 and 1938
3.4 Pedigrees of wheat hybrids obtained between 1880 and 1938
3.5 Wheat improvement activities in Spain, 1880–1935
3.6 Consumption of N, P_2O_5 and K_2O from mineral and chemical fertilizers between 1880–1936
3.7 Agronomic conditioning factors and technical change
5.1 Cost and destination of the investment
5.2 Production prices and income from the main cultivation alternatives (1990/91, 1993/94 and 1998/99 seasons)
5.3 Evolution of internal yield taxes
5.4 Work required by cucumber crops in greenhouses in two exploitations, one with family labour and the other with hired labour
5.5 Residues generated yearly by tomato crops grown on the NFT system and on rockwool substrate
5.6 Efficiency of the use of water and fertilisers, and contamination relations between tomato crops grown on the NFT system and on rockwool substrate
5.7 Plant densities within the nursery and in the greenhouse, relation between nursery surface and greenhouse surface for different crop conditions
5.8 Composition and characteristics of the nutrient solutions used for the different crops grown on NFT
5.9 Main results obtained with tomato, green bean and cucumber crops grown on NFT
8.1 Lectures organised annually by the *Belgische Boerenbond*, 1905–1939
10.1 Household consumption of canned and frozen food in France in the early 1970s
10.2 Numbers of supermarket outlets in the UK
10.3 Supermarket penetration in Europe in January 1972
10.4 Hypermarket penetration in Europe at 1/1/1972
10.5 Share of top five food retailers in Europe in 2000
10.6 American acquisitions by European food retailers 1973–1984
10.7 International network of major European retailers in 1990
12.1 The ten leading world-wide food companies in 1997
12.2 Average availability of meat by type (g/person/day) in 1998
13.1 The composition of cost of meals from the canteen of the Emser Plumb and Silverworks in Ems, Germany, 1875–1891
13.2 Per capita consumption of potatoes, West-Germany, 1948/49–1975
13.3 Increase in the consumption of frozen food in private and collective households in Switzerland, 1964–1970
13.4 Sales figures of frozen food 1991 and 2001 in Germany
13.5 Costs per menu in a defrosting and in a normal kitchen in 1964
13.6 Food sold to the catering industry as percentage of total food sales in 1986

Preface

Bringing together historians from chiefly the North Sea countries, the CORN network conducts comparative research into rural developments in the North Sea Area from the Middle Ages to the twentieth century. Particular attention has been given by the research group to such key themes as land productivity, agro-systems, price development, the rural labour market, marriage patterns, common rights, tenure and land ownership.

Agriculture and the countryside have no sharply defined geographic or intrinsic boundaries: town and country, agriculture and industry, the land, the cowshed, the market and the kitchen are all closely intertwined. In consequence, agrarian and rural history is not confined to the farm or does not stop at the borders of the village. Historical research follows the logic of increasing connections between agriculture and the surrounding world, as is clearly illustrated by the present CORN book *Land, Shops and Kitchens* and as will be further demonstrated by the succeeding volume *Exploring the Food Chain*. Both publications show the extent to which the inclusion of technological innovations has become a crucial factor in the proper understanding of the entire nutritional process.

Land, Shops and Kitchens links the nineteenth century with the twenty-first and the North Sea Area with Central Europe and the Atlantic and Mediterranean regions. It traces the line from the land to what comprises our daily nourishment, and in doing so draws on the expertise of historians, economists, sociologists, geographers, natural scientists and engineers – a very enriching combination. Production, retailing and consumption are inextricable parts of the same reality of feeding the people. In this respect, European farmers are producing steadily less food for their own needs, concentrating rather on supplying local, regional, national and international markets. They are producing more and better, according to the demands of the customers in the food chain: the expanding food industry, the powerful wholesale business and the ever more numerous, wealthy and demanding public. Moreover, the development of the means of transport and new preservation techniques signify that there are nearly no limitations on transporting food; food processing indeed is transforming the notion of perishable food. Agriculture is no longer the pivotal and incontestable food supplier it once was, but has become just another of the many elements in the extensive food chain, together with fundamental and applied research, political interference, the industrialisation of food processing, the standardisation of packaging, the commercialisation of food in department stores and shifting eating patterns.

The contributions of eighteen authors, coupled – it must be said – with the stimulating efforts of its initiators and editors, Carmen Sarasúa and Peter Scholliers, have enabled this book to offer an exploration of numerous facets of the kaleidoscopic food chain. It is thus very gratifying that this publication has been accorded an excellent place in the CORN series.

Leuven, June 2005

Leen Van Molle

1 The rise of a food market in European history

Carmen SARASÚA, Universitat Autónoma de Barcelona
Peter SCHOLLIERS, Free University of Brussels (VUB)

The history of contemporary Europe is the history of the effort of European peasants and farmers to produce enough food for themselves, and to provide the market. It is the history of technicians, scientists, and individual inventors who, together with the peasants, developed new tools, machines, seeds and breeds in order to produce more wheat and more meat, and to produce a better quality and taste. It is also the history of governments to put an end to starvation crises which fuelled the masses' opposition to the *Ancient Régime* across Europe in the second half of the 18th century, and to guarantee lower prices for food, so that workers could subsist with very poor wages.

Since the 18th century, the ever-growing European populations accounted for an increasing demand for food. Furthermore, an increasing percentage of this population was urban, dependent upon food markets, and of which the middle and higher classes, given their higher incomes and social aspirations, aimed at an increasingly diversified diet. To what extent did these massive changes in the *demand* for food transform its *supply*, i.e., agricultural productivity, techniques, and total output? They did transform it in the first place because of the central importance of agriculture in 18th-century economic thought and policies. Growing agricultural output meant the increase of the nation's wealth, to 'perpetuate wealth in the form of corn, drink, wood, livestock, raw materials for manufactured goods', as Quesnay wrote in his 1764 *Tableau Économique* (1972: i). The strategic importance of agricultural development for European governments explains the central role of agricultural *policies* and *institutions* that have regulated European agriculture in the last three centuries, developing a new revolutionary framework of property rights, and investing public money in water and transport infrastructures. The *political* importance of food was also apparent in the central role of agricultural price policies within economic policy. In the 18th and 19th centuries, when food shortages were common, European governments intervened to guarantee the supply of food at low prices for consumers and, at the same time, the profits of farmers (with protectionist policies). In the second half of the 20th century, with overproduction and increasing pressure from non-European producers, public intervention (EU agricultural policies) has consisted of subsidising agricultural prices in order to maintain farmers' profits.

To a large extent, the fact that European governments shifted from keeping low food prices to subsidising agricultural prices, reflects the success of the impressive ensemble of technological innovations known as the *agricultural revolution*. The pressures from the demand not only stimulated the rapid diffusion of new machines and tools, but what was more important, also the expansion of an international market for food. This in turn was possible because of the major transformation of the transatlantic transport system, achieved thanks to technological innovations such as steam power applied to ships, which allowed for a much cheaper and regular connection between the American highly productive fields and the European markets. The transport revolution meant cheaper food and an increasingly integrated international food market, for wheat in the first place, which was a powerful incentive to further adoption of technological innovations by European farmers (a process, nevertheless, delayed or ignored thanks to the protectionist reaction of European agriculture).

In this sequence, changes in the demand for food appear as the origin of the transformation of European agricultural production, and particularly of the central role that technological innovations had in the increase of agricultural productivity. But this new, competitive, and capital-intensive agriculture had in turn major impacts on the demand for food, that is, on the Europeans' diet: cheaper and better food, with a more diversified diet, was possible for more people now, particularly in the 20th century. Thus, the impact of technological innovations on relative prices was a key factor of this transformation, with strong consequences on consumption and on the social redistribution of income.

To account for this story, this volume brings together three different research traditions: the historiography of agriculture, that of food, and that of retailing, and connects them to the history of technology. Although these fields have undergone distinct developments during the last decade, we strongly believe in their integration when it comes to the explanation of the history of food. Land tenure, soil characteristics, labour organisation, and climate or technical innovations affect directly the type, quality, significance, and quantity of the food. In turn, through consumers' preferences, snob effects, retailers' sales methods, and marketing, the retailers and the consumers have always influenced production. The history of food is, in fact, a perfect example of how the stories of production and consumption must be told together in order to explain the economic growth of the last three centuries. How the complex relationship between food production, distribution, and consumption did operate, and to what views and outcomes this has led and will lead, is a central concern of this volume.

I. The role of technology in the food chain

Such integral approach is common among food sociologists and food geographers. Particularly the notion of 'chains of provision' or 'food systems' has been applied (Fine and Leopold, 1991; Atkins and Bowler, 2001). So far, historians have hardly considered chains of food provision. One but crucial exception, though: Dutch historians have adapted the 'chain of food products' that includes primary production (agriculture), secondary production (agribusiness), distribution (transport, packaging, wholesale and retail), food preparing (shopping, cooking and serving), actual consumption (eating, conviviality, identification), and waste disposal. Each phase is mirrored by a temporal-spatial step, namely the farm, the factory, the market, the kitchen, the table and the garbage can (Van Otterloo, 2000; see also her chapter in this volume). Such approach has great merits with regard to the 20th century, when the so-called middle field is gaining weight (lobby organisations, marketing, household and cooking schools, food regulation by the state, medical counsellors et cetera). However, historians have directed their attention primarily to this middle field, and they neglected the whole chain of production, consumption and their relations.

True, historians have not waited for theoretical insights to develop some kind of integrated view. We may refer, for example, to the large literature on past hunger crises, where attention is paid to crop failure, price increases, failing wholesale and retail trade, private and public relief initiatives, nutritional information and cooking advice, shrivelling calorie intake, rising social inequality, emigration, and sharply growing mortality

(Rotberg and Rabb, 1985). There are of course other examples that advocate a direct link between food production, retailing and consumption (a connection that is not new in economics, that has traditionally analysed markets' supply and demand). But despite this strive for 'integral' consumption history, historians of agriculture, those of food retailing and food consumption seem to live in separate worlds with own journals, conferences, celebrities and bibliographical references. This is linked to different traditions and epistemological developments. Most agricultural historians have a background of economic history, devoting themselves to the study of employment, yield ratios, productivity and output, price developments, market structure, land tenure, technological innovation, labour and wages. Food historians have a very similar background when they study consumed quantities, price developments, institutions' diets, household spending on food, business histories, safety regulations and technological innovations. The 'culturalisation' of historiography, however, has taken many food historians away from the economics toward matters of taste, preference, sociability, representation, gender, identity, classifications, and other issues that became popular in the 1990s (Flandrin, 1999). Historians of retailing have largely followed the same path as food historians. Often, they started with the study of a big corporation (Wilson, 1970), or a spectacular innovation such as the department store (Miller, 1981). The cultural turn is presently shifting the attention to the significance and perception of shopping, fashion, gender, and especially to modernity. So, although the three approaches do have common grounds, they develop largely within own traditions and questions.

It takes a very broad outlook to tie together the land, the kitchen, the supermarket, medical counselling, agribusiness, the table, health concerns, and taste. Technology may be saluted as a common ground that brings together the *time-spaces* of the food chain. We use a functional definition of technology here, as 'the sum of the methods by which a social group provides themselves with the material objects of their civilization' (Long and Post, 2003: viii). Historians of technology have described a multistage process, from invention to information, diffusion, and final adoption, with many factors intervening in it: educational and credit institutions, firms, groups of interest et cetera. Choices made among competing techniques should be taken into account, as well as stories of failure and success.

Technology has been and is very present in the study of production, distribution and consumption of food. Agricultural historians have written extensively on innovations in work organisation, crop rotation, tools, fertilisation, machines, new seeds and breeds, in relation particularly to increasing productivity and output (van Zanden, 1991). Food and retailing historiography has focused on technological innovations transforming manufacturing and introduction of new produces, conservation, packaging, health, food distrust, and grocers' rationalisation and reorganisation (den Hartog, 1995). A definition of technology as inducing economic superiority (that can be traced back to Adam Smith and his description of a technologically superior capitalist division of labour) has been uncontested in economic and agricultural history for a long time. In this vision, technological superiority is defined as the method that costs less, with labour-saving technologies (with their massive implications on work organisation) as best indicators of economic efficiency. But technological innovations depend as much on economic and social institutions (that are in control of agricultural production).

We think that one of the major contributions of this volume is to show the strength of the critical approaches to the technological box used in the last decades in Western agriculture and food industry. Interestingly, these criticisms do not stem from academic circles, but from consumers themselves. The critical reception (or direct refusal) of certain agriculture and food technologies shows the very active role of consumers in shaping and modelling technological production (Oldenziel et al, 2005). In fact, although not a new development, agriculture and food have become more of a battlefield in recent years, with heavy economic but also political interests at stake. And the fact that social scientists are being increasingly critical with the technological models adopted after World War II is only a reflection of consumers' growing discontent and a practical evidence that, in any given market, production cannot be understood without a deep comprehension of how consumption works.

II. Risk society, McDonaldization and other concerns about our food

The historical study of technology in relation to the food chain contributes to important, past and present-day societal issues. Two of them, food scares and hyper-rationalisation, have gained large attention in recent years and ought to be put in historical context.

Food scares form part of the recent concept of a society of risks, with fears, doubts, mistrust and uncertainties, which may be linked to neo-liberal, post-industrial developments (Beck, 1992). The role of technology is double, here. On the one hand, technology offers possibilities to control the whole chain of food provisioning more than ever, and to reassure consumers by boosting feelings of trust and security. Technological advancements have allowed the improvement of such control in the course of time. But on the other hand, the same technological innovations which have been the basis of the productivistic model of agricultural growth dominant in the 20th century have had side effects on the world's natural resources and supply of raw materials. Genetic manipulation of plants and animals has contributed highly to the already important distrust of food among an increasing number of people. They (we!) have clear feelings of *food alienation*, which may lead to a loss of identity, tormented social relations, growing anxiety and augmenting tensions, and to the search for food authenticity and food safety (Fischler, 2001). In this process, governments are expected to intervene, and this since far-remote times (Bruegel and Stanziani, 2004). Surely, a society of risk is of all times, and the relation between technology and food insecurity should thus be investigated in a long-term perspective.

Feelings of *food alienation* and food dangers lead to the second important phenomenon that links up with the first one: the so-called McDonaldization of social life, which is considered by sociologists as one of the three trends that affect present-day food consumption (together with social differentiation and self-rationalisation) (Germov and Williams, 1999, 6–8). At the *production* level, McDonaldization is a model of enduring rationalisation (Ritzer, 1996), fordism applied to the preparation of food. It refers to the ensemble of techniques that have transformed the food sector, seeking lower costs: most prominently the 'assembling line', but also standardisation of the product, exchangeability of components, and dequalification of the 'cooks', workers reduced to the repetition of a pre-fixed and limited number of movements. McDonaldization has transformed as well the *consumption* of food, implying a totally uniformed offer all over the world: same product, same taste, same process of preparation, same presentation.

Again, technology has played a central role with a double effect. The massive success of a globalised model of fast-food eating-places has been possible thanks to technical innovations such as frozen foods, diffusion of standard types (in many cases genetically modified) of seeds and breeds (the same meat everywhere), industrial products such as ketchup or mayonnaise based on colouring and preserver chemical products, and transport and marketing innovations such as airlifting, packaging (take-away boxes) and retailing techniques. In this sense, technological developments appear as making possible new business opportunities and low cost food.

But on the other hand, McDonaldization involves invariable taste, form and quality of the food, atomisation of meal patterns, disruption of 'traditional' social life, ever-increasing distance between producer and consumer, et cetera. It replaces local products, tastes and traditional ways of preparing and consuming food. It ignores culinary culture of both cooks and eaters. The uniformed fast-food system has, thus, consequences that are undesirable for part of the population. The social and political struggle against these effects (and against the very technologies that have made this food system possible, like the demand for fresh, locally grown vs. frozen, imported foods, shows) has taken the form of a defence of 'traditional' or 'authentic' food. This reaction must not only be seen within the broader context of a defence (headed by France) of European food as an essential part of European culture, but also of European firms and markets. In making choices about their food, the new conscious consumers are also acting as protectionist barriers to the EU market as much as tariffs used to do. When José Bové and his farmer colleagues boycott a fast-food store, they are doing much more than advocating the return to the 'traditional' cheese, bread and wine diet.

III. Asking questions, seeking *integral* answers

The present volume is based on a workshop organised in Barcelona, March 2003, as part of the European Science Foundation funded project 'Tensions of Europe: Technology and the Making of 20th century Europe'.[1] Our main question to the participants was about the economic, social and cultural role of technology in the food chain, with special interest in tensions that came along with the technology. With *tensions* we meant, for example, the confrontation of different food models (fast food vs. 'slow' food), the distrust with regard to the massive use of chemical fertilisers, the struggle between producers and big retailers about the control over the food chain, or the debate with regard to the role of the state. We proposed to focus on six themes that fit into the broader theoretical concern of the ToE-project (i.e. linking and de-linking of infrastructures, circulation of knowledge, and circulation of artefacts):

1. Models of agricultural technological innovations: the ecological restriction
2. The technical formation of the agricultural labour force
3. The role of EC policies in technological innovations in agriculture
4. Industrial catering, fast food and their rejection
5. Hygiene and technological innovations in food preservation
6. Modern forms of food retailing

[1] We would like to thank all the people and institutions that supported the organisation of the colloquium (particularly Johan Schot and Ruth Oldenziel from the *Tensions of Europe-project*, ToE), and the Universitat Autónoma de Barcelona for hosting our meeting. More on the ToE-project may be found at http://www.histech.nl/tensions.

The workshop brought together historians, sociologists, natural scientists, economists, geographers and engineers. This mix turned out to be extremely fruitful. We wished to create for each chapter a close connection between the historiography of food production, food distribution and food consumption around the question of technology. It is far from evident to assemble experts on chemical fertilisers and experts on identity construction through food, or to convince them to consider the whole food system. Yet, we believe that we generally succeeded in doing so: thus, agricultural historians and economists considered the consumers' reactions and resistance, while food historians linked the consumers' hopes and practices to changes in production. We consider this 'integration' of different traditions to be an important contribution to the interpretation of the development of food systems. The volume includes case studies from most of Europe, overcoming the restrictive Anglo-Saxon monopoly of the food system literature (Germany, France, Great Britain, The Netherlands, and Switzerland, but also Spain and Greece).

A *first* outcome of the workshop was the electronic publication of six annotated bibliographies dealing with technology and dairy products (Barbara Orland), fertilisers, and in particular nitrogen (Vaclav Smil), industrial catering (Ulrike Thoms), environment and biological innovations (Josep Pujol), retailing (Isabelle Lescent-Giles), and agricultural education (Leen Van Molle).[2] A *second* outcome is the publication of the present volume.

IV. The long-term perspective: which technology, how much food?

It seems useful to put the chain of food provisioning in a broader historical perspective. For our purpose, this may start in the 18th century. In pre-industrial Europe, the large majority of Europeans ate what the fields produced around them. Their food system and diet depended to a large extent upon the natural conditions of the land, soil, rain and work, that is upon a physical environment little shaped yet by the hands of women and men. European landscapes were modified *and* transformed, as Sereni (1974) proved when he described the Tuscan territory. Mountains were broken into terraces and cultivated, irrigation systems were developed, and by the 18th century new non-European crops had been adopted. But with the exception of the wealthy people who could afford non-local foods, the population had a diet mostly based upon local production. Yet, the concept of *local* food was always changing. In Spain, American plants such as potatoes, tomatoes or corn were incorporated since the beginning of the 18th century, and by the mid-19th they were regarded as local crops. On the other hand, soil and climate conditions never allowed for the adoption of plants such as coffee, cocoa and spices, which kept being imported and regarded as exotic (in the sense of non-European, not of expensive) food until today (with the exception of sugar cane in parts of Andalucía and bananas in the Canary Islands).

If food consumption was a basic way of class differentiation, the central role of religion in shaping food demand should be remembered as well (for example, the expansion of pork meat as a statement on the condition of the non-Jew consumer). Massive consumption of fish (and an array of techniques to cure, salt, smoke, preserve, or keep it fresh) in

[2] http://www.histechnl/tensions/defaultOud.html (theme "Agriculture and Food", see "Publications", and then "Annotated Bibliographies").

Spain (today the second largest world consumer of fish, only after Japan) was also rooted in the fasting practices of Catholicism (Sarasúa, 2001).

The 18th-century agricultural revolution meant the development of capital-intensive agricultural techniques, which had rising yields as a consequence. Not only new tools and machines emerged, but also new organisational methods, property rights, practices, and ideas about the land. The incentive to develop, finance and adopt all these novelties, was the increasing population with a growing number of mouths to be fed, but also rising urban incomes. Middle classes and the rise of, what has been called, a consumer society, meant more abundantly supplied urban food markets, more money devoted to food (in absolute terms), and a more diversified diet.

This increasing demand for food put new pressure on agricultural production: even if external markets were increasingly able to supply European demand, local farmers wanted of course take advantage and make a profit of it. And so they responded to the rising demand for food. The farmers' strategy was increasing investments, capitalisation, adoption of new techniques, diversification of output, and increasing productivity. After two centuries of capitalist agriculture, as new techniques have developed (agricultural, but also transportation, refrigeration, storage, and transformation techniques), the relation between agriculture (production) and food (consumption) has completely changed. Technology has allowed agriculture to overcome physical restrictions, and thus now we can eat what it pleases us at any moment: cheap tomatoes and strawberries arrive to Central and Northern Europe every day around the year.

In the last decades, new functions of food (for example, food as a medicine) and changes in taste (Gavrilova in this volume) are replacing the plain, traditional ones (to merely satisfy a physical need, to perform a ritual as part of being a member of a social group). Thus, the ways we eat, as well as the very nature of our food, are changing. As a result of this deeply changing diet, the demand for food is changing too. And given the increasing international competition, this new demand is deeply conditioning agricultural practices and productive systems (non-refined bread and sugar, replacement of red by white meats, transformation of the greases market, et cetera). Of course, technological change in agriculture, entailing new uses and meanings of food, did not start in the 18th century, but it may be argued that changes then were that huge that the whole process became irreversible. Capital intensive technologies seem to have liberated agricultural production from physical restrictions. Agriculture *can* produce increasing amounts of any product at any time.

V. (Agricultural) policies, (ecological) costs, and (food) antagonisms

Which role did policies have? As we mentioned above, agricultural production has had a strategic importance for European governments and institutions since the 18th century. It has been the goal of (particularly EU) agricultural policies after World War II to increase productivity by achieving economies of scale (by favouring large-scale production and concentration) and capitalisation. By the 1980s, however, decades of heavy subsidies had led to a number of serious problems in the EU, such as overproduction, increasing need for tariff protection – and subsequently problems with agricultural producers in the so-called Third World –, increasing regional disparities, heavy reduction in the number of farms and agricultural jobs, and growing political discontent for subsidies to the agricultural sector (which amounted to half the total EU expenditure). In an attempt to favour more productive units, subsidies have benefited more medium and large-scale units, and more developed regions. As a result, the costly and decades-long subsidies to the agricultural sector have actually *increased* economic differentiation among farmers and agricultural regions, and are being helpless in fighting the disappearance of agricultural activity.

These problems led to the 1992 reform of the EU agricultural policy, analysed in the chapters by Mattas and Loizou, and by Redclift and Goodman. The 1992 reform fostered a new model of 'rural development' to replace the old model of 'agricultural growth', a model that has been 'marketed' and presented to the public opinion as the solution to another problem not yet mentioned here: the damaged environment.

Focusing on the environmental problem, some of the chapters in the book put the question: at what cost? Very often, we understand costs in a very short-term way. A vision of long-term costs and externalities allows us to take into account the very high costs of producing while ignoring physical restrictions. In fact, failure or success in the adoption of a given technique also depends upon physical restrictions, as chapters by Pujol, Garrabou, and Navarro, López-Gálvez and Salazar show.

This is a crucial matter. To what extent recent changes in diet and food consumption are really induced and determined by the consumer? What has been the role of corporations in shaping the demand for food? We know that by using a given technological toolbox, food firms make choices that condition to a large extent agricultural production. Thoms shows in her chapter how the supply of potatoes was heavily conditioned by the technological choices made by canteens and other mass consumers. Furthermore, the rapidity with which agricultural producers are adapting themselves to these new developments (organically-grown agricultural production, et cetera) is explained by the fact that, in a context of increasing competition and 'open' markets, European producers are not competitive via prices. Ecologically produced, traceable, safe and controlled food is more expensive than food imports, and thus it is becoming the best possible strategy to compete in the EU market. Like in the case of the return to local food, the ever growing demand for quality and health controls has replaced (or better, reinforces) tariffs as 'protective' mechanisms.

The reaction against the type of standard mass-production that industrial food production entails, is another consequence of increasingly industrialised international food markets. This reaction, that may be labelled as 'slow food' in Van Otterloo's perspective, is taking two forms. One is the re-valuation of the 'traditional' way of producing food (i.e., non-industrial, home made food). This includes the opening of artisan-like shops to make cider, butter or jams 'the old way' (the use of pre-industrial technologies, including the restoration of old machines, tools and devices is an important mark), or the running of farms where hens are raised in open fields (vs. the chain-like industrial production). How food was produced 'traditionally' becomes the subject of a new narrative, as well as the target of food policies trying to preserve these 'traditions'. As Orland points out in her chapter, 'Discourses on *old* and *new* (...) can be read as a communication strategy which attempts to fill the gap between constant change in an industrialising world and the assumed character of unchanging and invariant peasants traditions'.

The second form of this reaction is the definition of locally produced food as quality food. This is entailing the return to old ('local') breeds and seeds, the development of narratives about the 'regional' or 'local' ways of cultivating the soil and breeding animals (in France, *le terroir*), as well as the reshaping of the industrial chains of distribution and marketing, well described in the chapter by Lescent-Gilles. *Authenticity* is the key word. 'Farmers markets', where farmers sell directly their production to the consumer, belong to this trend, which implies the return to seasonality and the rejection of exotic food. Finally, it seems obvious that one of the functions of these behaviours is to satisfy the need of wealthy consumers to distance themselves from mass consumers. Those who can pay more want to buy different products, and this is the new way to do it.

The 'national' definition of technologies has also much to do with the uneven results of their adoption. Every technology is inextricably linked with the place from which it has developed, and this is much truer for agricultural technologies, which aims at modifying the conditions of the soil, and the reproduction of plants and animals. The 19th-century agricultural technology developed mainly in England, the Netherlands, and Central Europe, and was rapidly marketed (in the form of new tools, machines, breeds or seeds) as the recipe for increasing yields. One century later, the results of the a-critical adoption of these new techniques are devastating in regions whose physical conditions differ greatly from the regions where these techniques were originally developed. Chapters by Navarro, López-Gálvez and Salazar on the one hand and Pujol on the other, show that the adoption of agricultural technologies developed in Northern European countries had a very negative impact on Mediterranean agriculture, where water scarcity and soil poverty demand a completely different technological set.

The questioning of intensive agricultural methods, high returns, high productivity, and industrialised technologies that have characterised the last two centuries, is taking different forms by Garrabou and Smil. The first is in favour of including the hidden costs of environmental destruction in the final balance of these productivistic models. Does it make sense to keep spending on technologies that have such a destructive impact on the environment? Should we abandon the tendency to ever increasing production? If Garrabou proposes to produce less (the abandon of intensive productive technologies), Smil proposes to eat less as the best way to lower the unnecessarily high demand for (and the waste of) food, and thus the non-sustainable pressure on the environment. In other words, to critically analyse consumers' habits and the political creation of *needs*.

VI. A further survey of the chapters...

It seems worthwhile to present very briefly each chapter separately, although some questions have already been raised. Each chapter shows very distinct points of interest, a specific approach or a particular conclusion, which will allow the emphasising of more questions.

Ramon Garrabou's "Conflict and environmental tension in the adoption of technological innovation in the agrarian sector" presents an overview of Europe's agricultural performance since 1800. He reminds of the rapid increase of agricultural productivity since the end of the 19th century, fuelled by technological change, stressing the different development between the Atlantic and the Mediterranean regions, and the growing tension caused by the ever-growing use of non-organic modes of agriculture. In terms of damage to the environment, Garrabou's view is quite pessimistic. He pleads to include fully the environment in the (historical) analysis of agricultural development, and to devote full attention to health, pollution, and destruction of traditional modes of production. Vaclav Smil ("Nitrogen in modern European agriculture") picks up arguments developed by Garrabou, to fiercely denounce the ever-growing use of chemical fertilisers, particularly nitrogen that has great difficulties to get totally absorbed. He points at two far-reaching consequences: Europe's changing diet following the American model (particularly, the much too high caloric intake of the average European), and the environmental impact (on soil, water, air, wild life). Smil very explicitly connects agriculture to food, when suggesting that a possible solution for the high nitrogen use would include consumers who eat less, balancing our consumption of food with our real physical needs of food, and avoiding waste and excessive consumption. Such change would require social engineering.

In "Modernisation and the international food system: re-articulation and resistance?", Michael Redclift and David Goodman place agricultural technological development in the broader context of the transformations of agri-food systems in 20th-century Western Europe and the region's insertion in global commodity markets. The chapter starts from the transition to capital intensive agriculture in late 19th century, then it discusses European agricultural policies since 1945 (fixing prices, absorbing surpluses...), and the problems related to this capital intensive model, such as 'food scares', to arrive to Common Agricultural Policy reform ('a massive decoupling of farm support from production'), and the new paradigm of an alternative agriculture, including new uses of land. This change, they argue, involves consumers' more active (political) role. They introduce a new project of social engineering, the so-called *repeasantization* (a renewed central role for farmers) and the *quality turn* in agriculture and food, with increasing importance of alternative agro-food networks.

In "Environment conditions and biological innovations in European agrarian growth", Josep Pujol develops the striking differences between the Atlantic and the Mediterranean regions, stressing the divergent relative cost on the one hand, and environmental conditions on the other since about 1820. He emphasises the way in which technological innovations have been introduced in relation to the environment in both regions. These were developed in northern Europe under particular environmental conditions, and so their adoption in the South has led to unexpected negative consequences (too intensive use of scarce resources). María del Carmen Navarro, José López-Gálvez and José Salazar develop this North – South tension further by concentrating on the adoption of the greenhouse agricultural technology in Southern Spain ("Recent innovations in the horticultural production system in the Southeast of Spain"). Surely, productivity rose manifold, but at the cost of high technological investments and massive environmental damage (in terms of resources, water and soil pollution, and irreversible harm due to irrigation techniques). They stress the urgent need for environmentally aware agricultural policies to reverse this trend.

Leen Van Molle's "*Kulturkampf* in the countryside. Agricultural education, 1800–1940: a multifaceted offensive" describes the evolution of agricultural education and the spread of technological know-how in Central Europe. In most countries the outcome of agricultural education was paradoxical: on the one hand, a strong discourse on *authentic* rural values emerged around the notion of the *Heimat* (homeland), while on the other a market-oriented, modern and scientifically based mode of production was promoted. What made this paradox possible was the 'hidden agenda' of agricultural technical education: not only improving the methods of production, but also maintaining and reinforcing class and gender differences. Education was used as a political instrument, and the transfer of technical knowledge was not neutral but aimed to maintaining the social, political and religious status quo.

In their chapter on "The role of EU policies in technological innovation in agriculture", Konstandinos Mattas and Efstratios Loizou identify six broad categories of technological innovation in agriculture (biotechnology, mechanical and electrical equipment, bio-energy and fossil fuels, environmentally friendly innovations, animal production innovations, and innovations in agro-industries). They discuss to what extent the price-support system characteristic of the UE Common Agricultural Policy induced farmers to adopt technical innovations. They conclude that CAP has led to new problems (pollution, soil mistreatment, water supply problems), which explains the ongoing reform of EU agricultural policy, not longer seeking an increase of productivity, but the supply of healthy food and sustained rural development.

These chapters, with an emphasis on the production side of the food system, address questions such as the measurement of the environmental costs, the chronology of innovation, the reaction of farmers to education, the role of mediators, and the responsibility of the state in education, the diffusion of technology and the environment. It appears that cultural elements need to be incorporated into the explanation of how agriculture has been conceived by rural workers, landowners, governments, lobbyists and consumers, and how, in turn, these cultural elements have influenced points of view, strategies and policies.

Anneke van Otterloo ("Fast food and slow food. The fastening food chain and recurrent countertrends in Europe and the Netherlands, 1890–1990") historically explores the initiatives countering the industrialisation and rationalisation of food production in Europe since the end of the 19th century. After describing fast food expansion, using as a case study the Netherlands, she analyses Slow Food (born in 1986 in Italy as a movement against fast food), and claims that 'the rise of the modern industrial food chain between 1890 and 1990 has been accompanied by recurrent manifestations of discontent and protest'. The documentation of this discontent allows her to point that the contemporary 'quality turn' described by Goodman and Redclift (in this volume) must be understood not so much as a novelty, but as 'part of a whole century of debates on food qualities'.

By focusing on mass consumption of food, Ulrike Thoms' chapter "Industrialising catering. Technological developments and its effects in the twentieth century", indirectly poses the question of defining who is the consumer of food: not only individuals or families but, as eating is increasingly done outside the homes, institutions (be these hospitals, schools, factories, other workplaces...). She looks at the cooking technology for 'the masses', considering the factors time, cost, scientific advice, endogenous elements (such as wars), the role of the state, and profit maximisation. Technical innovations, such as freezing and standardisation of the product, appear as preconditions for the increase in eating out. In turn, mass production of food implied the development of suitable (mass) cooking techniques and devices (friers, freezers), and the selection of a few varieties of each product (as the story of the potato shows). For Thoms, the most important innovations in food technology can be traced back to the specific demands of catering. Also, and importantly, the reciprocal influence between producers, scientists, the state and consumers is stressed, as well as the different evolution in East and West Germany.

Barbara Orland directly links cattle production and milk consumption to the construction of national identity ("Milky ways. Dairy, landscape and nation building until 1930"). She considers the case of Switzerland (especially between 1880–1930), investigating the way the milk market operated, how the image of the Swiss nation was built and how and when new, sophisticated technology was applied. She explores the invention of a model (and an image, *in casu* the 'brown cow'), the way old modes of production changed under the influence of new techniques, market-oriented thinking and nation building.

Rayna Gavrilova's chapter "Changing tastes. The role of scientific and medical discoveries in changing the modern diet" explores the influence of scientific and nutritional knowledge on taste and eating. She defines taste as a social construct that influences choices, and gives a survey of the nutritional research. She then considers science's points of interest related to food (energetic value, composition, vitamins, and food-linked diseases as obesity). Finally, she investigates how this knowledge affected the people's minds as well as their everyday eating practice (with health concerns and food scares). Particularly, Gavrilova studies how the scientific and medical knowledge has been diffused by means of texts, education, and policy. Overall, four interrelated phenomena may explain the growing importance of scientific knowledge for the general public: the trust in rational knowledge, the importance of the individual person, the role of the state, and the power of the bourgeoisie and middle-classes.

Isabelle Lescent-Giles presents a chronology of modern, large-scale retailing in relation to product, format and business innovation ("The rise of supermarkets in twentieth-century Britain and France"). After an overview of the innovations in food retailing up to the 1920s (advertising, promotions, price strategies, et cetera), she focuses on 20th-century changes, and in particular on the coming of the supermarket in France and the U.K. She deals with product innovation (convenience and healthy foods), sales innovations (self-service, price scanning, differentiation with various trends that include slow food produces), and business innovation (centralisation, efficiency, internationalisation, diversification), which gave birth to a power struggle between the retailer and the manufacturer, won by the big retailers. Lescent-Giles' chapter stresses the technological innovations behind these business developments.

These chapters, which emphasise the demand side of the food system (distribution, manufacturing and catering), address the questions of the role of kitchen technology (e.g. microwave) in the transformation of meal patterns (and family life), the reversing links between retailing and agriculture / manufacturing, the role of technology and science in identity formation, and the conflicts in relation to technological development (water and soil pollution, food quality, global markets, North – South and East – West tensions).

VII. …leads to new questions

Many blank spots need further attention, and among them the use of fertilisers throughout Europe, the diffusion of agricultural know-how or the image of agriculture and agricultural labour in past and present. As for food consumption, the agenda for future research would include the spread of fast-food restaurants, technical innovation on packaging, advertising, conservation techniques or consumer credit; and the changing relation between food consumption, work schedules and the now declining role of housewives as family cooks.

Yet, several conclusions may be drawn. The first one relates to a general observation: agriculture and food seem to offer a privileged way of looking at the contradictory effects of technology in the twentieth century. The second is a paradox: the technological innovations responsible for the dramatic increase in agricultural production in 20th-century Europe are now widely rejected by individual consumers and by different social and political groups. The third conclusion refers to persisting (old, new and future) problems in European production, distribution, manufacturing and consumption of food with the widening of markets. There will be at least three main consequences of the 2004 EU enlargement to the East: food markets in the Western countries have to make room for the production of new members; subsidies for agriculture and livestock in the Western countries will be drastically reduced; and agriculture and food industries and distribution in the Eastern countries will have to adapt to the EU stricter health and environmental requirements. Proper insight in these developments and problems require an historical dimension.

Soil pollution is one of the most serious environmental problems in Europe. Contaminants deposited in the soil pollute ground waters through leaking, and vice versa, polluted waters affect soil quality. This has obvious consequences for agricultural production, food safety, public health and the perception of food by consumers. The most direct source of water and soil pollution is the chemical products used in excess on agricultural production and lost to the environment. Yet, these agents of pollution were years ago the very sources of productivity increase, both in the West and in the East (Ardillier-Carras, forthcoming).

We will highlight one example of how technological innovations welcomed in the 20th century for their positive impact on production and productivity have come to be rejected due to their negative side effects on public health and the environment. What Vaclav Smil terms 'the 20th century most important agricultural revolution' was the synthesis of ammonia, the simplest of all stable inorganic nitrogen compounds, in the years around the Great War. It replaced traditional sources of nitrogen fertilisers (organic, such as guano or manure, and inorganic, such as nitrates). Dutch agriculture was the most intensive pre-1940 user of nitrogen fertilisers, with applications averaging above 40 kg N/ha during the late 1930s, and reaching 50–60 kg N/ha in parts of the country (the average US rate was less than 3 kg N/ha). By the early 1980s the Dutch application averaged 250 kg N/ha and was the highest in the world. European usage of nitrogenous fertilisers remained above half of the world's total. It has gone from 1.8 Mt N in 1950, to 9.6 Mt N two decades later (in 1970), and it has peaked in 1988 at 15.98 Mt N. Europe's intensive fertilisation (even before ammonia) whose post-1950 costs were increasingly supported by rising agricultural subsidies, was reflected in a steady growth of average crop yields. By 1900 wheat harvests were around 2 t/ha in the UK, the Netherlands, Denmark and Germany. Due to other innovations, Dutch, English and French wheat yields rose from 4.3 t/ha during the early 1960s to 7.3 t/ha two decades later.

But unwanted side effects of intensive fertilisation appeared: 'The most acute problems with nitrogen leaking from excessively fertilised agro-ecosystems is the leaching of highly soluble nitrates into surface and ground waters' (Smil, this volume). In 1991 the Council of the European Communities issued its nitrate directive (91/676/EEC) requiring the member states to reduce the nitrate load from the agricultural sector to acceptable levels. This pullback lowered the average European applications to 99 kg N/ha by the year 2000, a nearly 20 percent from the 1988 peak (2057 g). Despite this reduction, 'The price Europe pays for its surplus of food thus goes beyond the staggering cost of irrationally high agricultural subsidies and the health effects of changing diets. Environmental costs of excess food output are actually more worrisome as many of these impacts would persevere even if the subsidies were to be miraculously cut' (Smil, this volume).

Furthermore, there is a risk of over-production due to a fall in the demand for food. According to FAO's food balance sheets, Europe has by far the highest per capita food availability in the world. The EU mean was about 3,500 kcal/day in the year 2000, while actual daily food requirements are rarely above 2,000 kcal/person (a gap of 1,000 kcal/day). Food waste accounts for a large part of this gap which probably will lead to reduce food consumption (and thus, demand) in the near future. Changing diets and the continent's ageing population will further this tendency.

Not only does Europe have the highest per capita food availability, but the dominant model of protein consumption, based mainly on animal proteins, results extremely inefficient and costly. In the first half of the 20th century the 'climbing of the protein ladder' (Goodman and Redclift, this volume) was centered on cattle breeding: cows need to process massive amounts of pasture to produce every kg. of meat. The same process is today taking place with fish breeding in fish factories (after technological innovations have made it possible): to produce 1 kg of salmon, 4 kg of fish flour are needed. Fish flour is industrially produced using as raw material other less demanded (cheaper) types of fish, a process which incentives a systematic depredation of the seas. At the same time, consumption of vegetable proteins is falling dramatically.

The relation between agriculture and food is twofold, as stated above. Within the current system of the large agribusiness, agricultural production is dependant upon its conveniences and requirements, that have also changed cheap, standardised and industrially processed food in mass consumption, and has to a large extent dictate EU agricultural policies. This transformation has consigned agriculture to a subordinate role in modern food systems, while it shifted control from farmers to oligopolistic trans-national food industry conglomerates and retail multiples.

A more efficient use of technological innovations in the food chain implies, in the first place, a redefinition of *efficiency*: it is not about increasing yields and output at any cost or no matter how, but about sustainability and creating wealth without risking the environment or public health. Farmers must grow their crops without old chemical *friends* such as pesticides and fertilisers, and replace them with old traditional inputs such as manure. They must be very careful with, or completely avoid, their use of genetically modified seeds or breeds. They will be required, on the other hand, to produce under most strict hygienic and safety norms, recycling their packaging. Is a realistic option this return to traditional, labour intensive techniques, replacing the capital intensive, productivistic technological model that has been dominant for over two centuries? At least we know that for the first time in their history, current EU environmental and agricultural policies are in the process of convergence, which means that their goals are no longer in contradiction. This also means that institutions and firms involved in the production of technology will have to adapt to this new type of demand for technology, new legal and health requirements, a new vision of what the role of technology must be, a new place for technology in agricultural production, in the food industries, and in consumption in general.

VIII. Towards a new relationship between food and agriculture?

A redefinition of the functions of the agricultural sector is currently taking place in Europe and particularly debated in the European institutions. This has evolved from the exhaustion of the fifty-year old model of EU Common Agricultural Policy, and the new challenges from both the world food market and the European consumers' demands.

EU agriculture employs less than four percent of EU workers, but absorbs almost half the EU total budget. The reason for this dramatic unbalance is the old political decision by member states to subsidy a European agricultural sector, including a number of agricultural units of currently six millions (of which only one million would survive without subsidies). Heavy subsidies to agricultural production and exports have also led to other unsatisfactory unbalances. Within Europe, a double burden for the non-agricultural population arose, both in terms of how their taxes are being used and the prices they have to pay for their food. Outside Europe, the unfair competition with non-EU agricultural producers appears most particularly the non-industrialised, and thus the damaging consequences for food world markets.

There is also the question of who has benefited from this agricultural policy. Despite the general claims, the main beneficiaries have been by far the largest producers, particularly the multinational firms. Today, eighty percent of total beneficiaries of CAP subsidies receive around twenty percent of the money, whilst twenty percent of the beneficiaries receive the other eighty percent. This has led to serious social and regional inequalities in the distribution of the subsidies. According to the Commission (October 2002), half of all direct payments by CAP go to the largest beneficiaries in the more productive and competitive areas, such as the Paris basin, Low Saxony or East Anglia, most of which contain export-led, multinational firms.

Since 1 January 2005 the reformed CAP foresees the dis-entailment of subsidies from production: subsidies will be granted (and progressively reduced) to the agricultural unit, and no longer in connection with (as a percentage of) output. This is intended to end with overproduction and (what has been worst of all) with the fraud consisting in producing only to be granted the subsidy, no matter whether a demand existed for the product, and regardless any quality of environmental requirements.

Discussing overproduction does not mean that hunger and malnutrition are not a problem for Europeans anymore, given the growing numbers of poor people among the immigrant population, elder citizens – particularly women, and unemployed. Women and men collecting rejected meat or vegetables at night at the garbage cans, backdoors of supermarkets, are a common view in large cities, as people queuing at the entrance of private and public charity institutions for a simple, hot meal. Yet despite these new (or rather old) hungry poor, European agricultural policy is no longer concerned by food provision, but by increasingly sophisticated consumers' demands, such as quality and safety standards, traceability, protection of local denominations, environmental protection, and animal welfare.

These complex and new consumers' demands are shaping agricultural production in a wholly new way. Yet, this is not to say that the productive sector is in the hands of consumers. After a long century of innovations in organisation, marketing and distribution, and increasing vertical integration of firms in the food chain, favoured by CAP subsidies and policies, large agribusiness are able to market in their benefit every type of 'consumer demand', and increasing their profits through it. Not only that, but the high standards of quality and safety required by EU food legislation are acting, in practice, as powerful trade barriers to imports from non EU (and non Western) agricultural producers (and as such, defended by EU multinational firms).

Yet, even if agribusiness is profiting by the new changes in demands of food, it is not the only one in doing so. The new tendencies in food consumption include a strong flavour of 'back to the local', a re-valorisation of the local, non-industrial cultivation and food production systems. Even if main suppliers of ecological and biologically produced food are large firms, a small segment of small scale producers coexists with them, mainly distributing through local fairs and markets, or through the farms themselves. Although their dimension is very modest, their existence is significant of a new taste for non-mass, industrialised, food production, and has also interesting political implications. The role played by cow breeding and milk production in the Swiss process of nation building (Orland, this volume) echoes the role played by agriculture and food in the new regionalisms and nationalisms growing within the EU. The construction (and then the defence) of local 'traditions' in soil cultivation, animal breeding, and food producing and cooking, explains some of the latest developments in food consumption.

Increasingly, the old functions of agriculture are being replaced by a new vision of the rural regions. This implies a vision about what the rural population should devote themselves to, i.e. to protect the environment and landscapes, to receive tourists (and sports lovers, hunters and fishermen), to keep alive the old productive activities, the cultural heritage. Europeans are surely happier to pay for these social goods than for overproduction of unwanted or too costly produced food items, only to be later destroyed to keep a convenient price level. What can be envisioned for the future is a much smaller part of the European soil devoted to agricultural and cattle production (and of this, most outside the intensively technological model today increasingly rejected), together with a large part devoted to alternative, non agricultural production. This evolution means that the close connection made at the beginning of this Introduction between agriculture and food may be less close in a few decades. Not only the old 'agricultural sector' will be less and less connected with food production, and increasingly centered in providing services or alternative – not for food – products (such as wood). Part of our food may have no relation with agriculture anymore! It may come from industries producing at their own laboratories. Is it too shocking for Europeans to imagine ourselves buying our vitamins, minerals and proteins in the parapharmacies, like Americans already do, instead of in our old markets in the form of oranges and milk? A perfectly balanced and perfectly safe diet for the European of the future? Let us simply say that this is, already, *technically* possible.

Bibliography

Ardillier-Carras, F. (forthcoming), 'La descolectivización de la agricultura en Armenia o las dificultades de la transición post soviética', *Historia Agraria*, 36.

Atkins, P. and Bowler, I. (2001) *Food in society: economy, culture, geography*, London.

Beck, U. (1992) *Risk society. Towards a new modernity*, London.

Bruegel, M. and Stanziani, A. (2004) 'Pour une histoire de la "sécurité alimentaire"', *Revue d'Histoire Moderne et Contmporaine*, 51, 3, pp. 7–16.

Fine, B. and Leopold, E. (1991) *The world of consumption*, London.

Fischler, C. (2001) *L'Homnivore. Le goût, la cuisine et le corps*, Paris.

Flandrin, J.-L. (1999) 'Préface', in J.-L. Flandrin and J. Cobbi (eds), *Tables d'hier, tables d'ailleurs. Histoire et ethnologie du repas*, Paris, pp. 17–36.

Germov, J. and Williams, L. (eds) (1999) *A sociology of food and nutrition. The social appetite*, Oxford.

Hartog, A. den (ed.) (1995) *Food technology, science and marketing. The European diet in the twentieth century*, East Linton.

Long, P.O. and Post, R.C. (2003) 'Series Introduction', in R.C. Post (ed), *Technology, transport and travel in American history. Historical perspectives on technology, society and culture*, Washington, pp. vii–ix.

Miller, M.B. (1981) *The Bon Marché: bourgeois culture and the department store 1869–1920*, London.

Oldenziel, R., Albert de la Bruhèze, A. and Wit, O. de (2005) 'Europe's mediation junction: technology and consumer society in the 20th Century', *History and Technology*, 21, 1, pp. 107–139.

Otterloo, A. van (2000) 'Voeding in verandering', in J.W. Schot, [et al.] (eds), *Techniek in Nederland in de twintigste eeuw*, Zutphen, pp. 237–247.

Ritzer, G. (1996²) *The McDonaldization of society*, New York.

Rotberg, R. and Rabb, T.K. (eds) (1985) *Hunger and history*, Cambridge.

Quesnay's *Tableau Économique* (1972), edited and translated by Kuczynski, M. and Meek, R.L, London.

Sarasúa, C. (2001) 'Upholding status: the diet of a noble family in early 19th-century La Mancha', in P. Scholliers (ed.), *Food, drink and identity. Cooking, eating and drinking in Europe since the Middle Ages*, Oxford, pp. 37–61.

Sereni, E. (1974) *Storia del paesaggio agrario italiano*, Milan.

Wilson, C. (1970) *The history of Unilever: a study in economic growth and social change*, London, 1970.

Zanden, J.L. van (1991) 'The first green revolution: the growth of production and productivity in European agriculture, 1870–1914', *Economic History Review*, 44, 2, pp. 215–239.

2 Conflict and environmental tension in the adoption of technological innovation in the agrarian sector

Ramon GARRABOU SEGURA, Universitat Autónoma de Barcelona

There is little doubt today about the adverse effects that the technological complex behind the systems of agriculture in the world's industrialised nations had on the environment. Following several decades of use, serious problems of degradation have been observed to the extent that there are now growing doubts as to whether productive capacity can be maintained and food quality and safety be guaranteed in the future. Observation of the food production system of the end of the 20th century reveals the twin phenomenon of spectacular growth in production and productivity and, at the same time, alarming processes of environmental damage and worrying problems of food safety.

When we consider how this situation has emerged, there is little doubt about the decisive role technological innovation has had in the whole process. Analysis of technological change is thus evidently a significant factor in order to explain the contradictory nature of the evolution of agriculture and the environmental conflicts it has generated. It is my intention here not to systematically examine technological changes in agriculture but, rather, to show some of the significant differences between the technological innovations before and after the implementation of the technological complex behind the green revolution. Having characterised the innovations of advanced organic agriculture, I will show how a series of factors led, from the end of the 19th century onwards, to an intensification of technological change and how, above all, this moved off in new directions. Finally, I will look at the environmental conflicts that the application of the new technological model generated, concentrating on those caused by the spread of new fertilisation techniques and the emergence of a new stock-raising system.

I. Technological change in the context of advanced organic agriculture

Until the beginning of the 20th century, farming systems continued to be organically based. In Atlantic Europe from the 18th century on, the main line of innovation depended essentially on the diffusion of crop rotation that brought in leguminous forage plants and alternated cereals with turnips and beet.[1] Fallow was abandoned, which meant an intensification of land use and greater pressure than under previous systems without, however, destroying the equilibrium that guaranteed the maintenance of its productive capacity. Despite processes of specialisation, the succession of a very diverse series of plants on the land was the norm under these systems of agriculture, and the maintenance of vegetation on field boundaries meant the survival of levels of diversity that were of enormous value in agronomic terms. Farmers had observed that the technique of crop

---

[1] An excellent analysis of this fundamental aspect of technological change at this time is to be found in Chorley, 1981: 34.

alternating produced good results at harvest time and avoided the problem of infesting plants and, at the same time, was an efficient means of preventing or diminishing the effects of plagues and disease. The bringing in of leguminous forage plants, turnips and beet made possible the development of large-scale stock-raising, closely linked to the crops grown on the farms, which produced most of the fodder. Stock raising provided the driving force that accounted for a significant proportion of the energy inputs used on the farm, besides growing quantities of meat and milk supply.

As for the fundamental element of any system of agriculture, the question of replacing the nutrients extracted from the soil with the harvest, the innovations brought in with the new rotations were a major improvement, in particular as regarded the nitrogen cycle. With their ability to fix nitrogen from the atmosphere, leguminous forage plants, alfalfa and clover added to the nitrogen drawn from the soil free of charge, to be used for future crops. Besides that, livestock produced a significant proportion of the fertilising material in the form of manure, produced from farmyard waste, which meant the elimination of the waste that was later to cause such problems. Although over the course of the 19th century other fertilising materials began to be used, particularly guano as well as industrial fertilisers, their use was marginal and complementary to a system of fertilisation based on the reuse of waste matter of all kinds from the farms themselves.

The variety of animal breeds remained another characteristic of this technological model. Although there was development of the practice of selective stock breeding with a view to improving meat, dairy or work production, the degree of manipulation was moderate and the feeding system continued to be based on natural pasture and forage, and feed from the farms themselves. As for seeds, the process of ecological globalisation that occurred in the 19th century and the first attempts at hybridisation (in that they put species and plant varieties from around the world at the disposal of European farmers) guaranteed high levels of bio-diversity compatible with the growing homogeneousness of plantations.

In the context of systems of organically based agriculture in which the use of fossil fuels remained very limited, levels of mechanisation continued to be modest. With the exception of threshing machines, the use of the steam engine was very limited in other farming activities. The use of animal power was dominant, and the type of machinery that could be used with such traction had neither an adverse effect on the maintenance of the fertility of arable land nor caused significant levels of pollution.[2]

In the Mediterranean, the spread of this technological model and its principal innovation, the alternation of crops with the incorporation of forage plants in rotations, was less marked than in Atlantic Europe. Low rainfall was often an insurmountable obstacle, and extensive farming with the use of fallow remained the very generalised norm. The persistence of fallow has often been interpreted as proof of a reticence to technological change and of irrational behaviour on the part of Mediterranean farmers in not introducing a form of innovation that had spread throughout Atlantic Europe and that had brought marked improvements in production. Contemporary writers often made assessments of this nature, and historians have often repeated them. I believe this is a good example of the problems involved in the analysis of technological change in

---

[2] An overview of the main lines of innovation in European agriculture in the 18th and the 19th centuries can be found in Mazoyer and Roudart, 1997.

agriculture when environmental variables are overlooked.[3] As some agronomists and technicians began to argue in the early years of the 20th century, the arid conditions typical of much of the Mediterranean impeded the implementation of the type of rotation that had a major role in improvements in production in Atlantic Europe. In the climatic conditions of the Mediterranean it was difficult to replace fallow with leguminous forage plants as water levels and the cycle of rainfall did not guarantee the proper growth of such plants, while fallow served to retain in the soil some of the rain that could be used for cereal crops the following year. It was also an efficient means of eliminating weeds. Given frequent situations of hydraulic stress affecting most of the crops, having a little more water available, albeit in small amounts, was of enormous importance. Despite the persistence of fallow, the alternation of crops was commonplace, on occasion different varieties and species of cereals being planted and, on others, a small proportion of the land lying fallow was planted with leguminous plants. This diversification proved to be an efficient means of maintaining fertility and reducing the incidence of plagues, pests and competing plants.

Another aspect to be noted in Mediterranean agriculture was the trend towards specialisation in tree and bush crops more appropriate to the physical environment. Woody plants with lower water requirements adapt better to hydraulic cycles and allow better results than to be obtained with other crops. The expansion of the cultivation of vines, olives, almonds and carob beans, whether alone or in conjunction with herbaceous crops, meant that agricultural production increased, efficient advantage being taken of the particular natural resources the Mediterranean bio-region provided. There was also an expansion of the area under irrigation and, in that water was a major limiting factor, overcoming this obstacle meant significant improvements in the production of cereals and horticultural and fruit products, for which there was growing demand. The type of technology used in the expansion of irrigation was still fundamentally based on gravity, with modest infrastructure and little incidence of major hydraulic works. Nor did the technology available to extract water from the subsoil, given its limited operating capacity, lead to destabilising phenomena like those seen later.

The innovations that spread through European agriculture until the end of the 19th century, those to be expected in advanced organic agriculture, brought significant improvements to production and productivity without creating serious tensions with the environment in which they occurred. The degree of *artificialisation* of agrarian ecosystems was relatively modest, and the ways in which the land was worked depended to a large extent on the natural systems that remained the heart of production. With these systems of agriculture retaining their organic base and the consumption of fossil fuels being low, their adverse effects on the environment as a whole were insignificant. Nor was irreparable damage done by fertilisation techniques as these continued to be based on the reuse of organic materials and the consumption of extraneous fertilisers was limited, despite the incipient use of industrial fertilisers. The practice of recycling avoided problems with residue and waste. Integrated land management, with areas devoted to agriculture, stock raising and forestry, and high levels of variety and diversity, continued to guarantee the land's productive potential. The dependency of the farming

[3] For a review of the limits and characteristics of innovation in Mediterranean farming in the 19th century, see Bevilacqua, 1993; Garrabou, 1994; González de Molina, 2001.

sector on the environment and the low intensity of the processes of substitution complicated the widespread use of methods of production that had become the norm in industry. This meant that improvements in production and productivity were lower than those in the economy as a whole.

II. The take-off of a new model of technological change following the end-of-century crisis

Following the agrarian crisis at the end of the 19th century, new lines of technological change acquired growing importance. These developed separately until in mid-20th century they merged into the new technological complex that became widespread during the second half of the century, and which, unlike the technologies of the advanced organic agriculture that had existed until then, have led to major environmental problems. They are to a large extent responsible for the serious ecological crisis currently facing us. How do we explain the change of direction that occurred at this time? Why was there deviation from the lines of innovation that had made it possible to improve agricultural production without harming the physical bases that are vital to guaranteeing the continuity and reproduction of the productive system? The answers are complex, as a wide range of factors was involved. The existence of an institutional climate based on market values which brought pressure to, above all, increase productivity of labour, while at the same time lessening the importance of the functions of the natural world, is undoubtedly one of them. But there were also other causes. For example, the notions of the economy, production and the conceptualisation of the functions of Nature that inspired the new technological model. We should also consider the new principles of agronomic science that were behind the new technologies and the transformations of the secondary sector linked to the second industrial revolution. Finally, we should not forget the situation created by the globalisation of the food system and the problems created in the agrarian sector by intensification of exchange and competition. We will now look at some of these factors.

The development of new technological proposals in the agrarian sector is closely related with the view a given society has of Nature and in particular of the ways in which it conceives its relationship with natural factors. At the end of the 19th century, the answer to both questions is to be found in economic science, constructed from the classical school and its followers, and which had attained great prestige and acceptance, having been recognised as a scientific discipline. What does economic theory have to say about the environment and the economy? J.M. Naredo and H. Immler, who have read the economic theory that dominated at the end of the 19th century from this perspective, raise the follow issues (Naredo, 1996a; Immler, 1993). These authors believe that in conventional economic theory the physical world has in practice been largely left out. The exclusion of Nature from the field of economics was the result of the formulation of a notion of production that became separated from the strictly physical meaning it originally had, and which came to consider labour and capital the only sources of production and wealth. Simultaneously, an economic system was created as a self-sufficient universe of exchange values, where the stabilising elements were of an exclusively financial nature. Production no longer had a material base as consideration was given exclusively to its monetary values. By reducing material wealth to the abstract notion of value and by attributing it to capital and labour, in practice the fundamental contribution to production of the physical world and of natural resources was overlooked. In this way, a conception of wealth that was completely dematerialised was eventually created, with a supposedly unlimited capacity for growth. In the conception of the economy dominant at the end of the 19th century, nature as a single whole and as a vital resource for the reproduction of social life lost its place, with consideration being given only to those fragments of it that had market value.

As a result there was a lack of any concern for managing natural resources according to rules and laws their conservation called for, as they were seen merely as a warehouse of energy and materials to be exploited, as an already built machine, one that was free and which Man could use as he pleased, J.M. Naredo suggests (1996b). Finally, it was all too easy to regard natural resources as being unlimited and inexhaustible, particularly as seen from the neo-classic viewpoint on the principle of replacing factors through technological change, thus opening the way for an exaggerated faith in the possibility of substituting labour and capital for resources. As these ideas on the economy and the relationship between Nature and society took root, there were significant changes in the technologies being proposed. With the physical bases of the production systems not being properly valued, and with the lack of any concern about how they operated, the idea took hold that agricultural production could be increased through an intensification of the 'exploitation' of the environment, which could in fact be ignored, by means of substituting the natural components of systems of agriculture by others from outside the sector.

Another factor involved in the shift in technological change that occurred from the early 20th century onwards was the way agronomic sciences viewed environmental issues. Scientific knowledge of the natural world improved substantially over the course of the 19th century but, as J.M. Naredo has shown (1996b), it was much influenced by narrow mercantile views. Post-Liebig, plants and animals were considered little more than mechanical converters, the food requirements of which it was interesting to ascertain in order to boost their productive capacity, with any deficiencies of the natural environment that acted as a limiting factor or brake on production being made up for. A substantial proportion of agronomic research revolved around the subject of plant and animal nutrition.

With such a focus, plants came to be considered mere converters, the soil a reservoir, with no consideration being given to that fact that both were much more complex and that the interactions established between them could alter their behaviour.

The direction of technological change in the agrarian sector is to some extent determined by the degree of technological development attained by a given society. Until the end of the 19th century, the technologies that had transformed industrial production had barely been seen in the countryside. However, the second wave of technological innovation, known as the second industrial revolution, offered new possibilities, and we should thus regard it as being of major importance in order to understand the new directions taken by technological change from the early years of the 20th century onwards. A few examples will suffice to illustrate this idea. The use of non-organic sources of energy, one of the main lines of innovation in the 20th century, was possible because the internal combustion engine permitted the use of fossil fuels in the form of petrol. The same could be said of the electric engines that opened the way for the use of this new source of energy in agriculture. The substitution of animal and human labour by machines depended to a large extent on the new sources of energy that were available and on the innovations that had been seen in mechanised industry, which permitted its use in agriculture. A similar thing could be said of other lines of change such as that in systems of fertilisation. The introduction of industrial fertilisers was closely linked to the development of a new chemical industry that opened up the possibility of providing increasing amounts of fertiliser at diminishing prices. Innovations in the pharmaceutical industry were also a decisive element in changes in the stock-raising sector, and similar comments could be made regarding the widespread use of pesticides and herbicides, which were unthinkable without the development of a powerful chemical industry. What I wish to show with these examples is that while technological change in agriculture advanced in particular directions, this was only possible because the level of technological development achieved in the second industrial revolution permitted its application in agriculture. To this we might also add the pressure brought by the industrial sectors that were emerging for technological change to occur in particular directions.

The new conditions facing European agriculture following the end-of-century crisis was another of the factors involved in the new wave of technological changes that occurred from the beginning of the 20th century onwards. The creation of world markets for agricultural products intensified competition and led to a drop in farm income, creating deep unease in the European countryside. In such circumstances, technological innovation became a necessary instrument for recovering prosperity. The availability of new techniques that cut labour costs and/or raised income was a necessary condition for righting the situation. The growing social demand for technological change led the state to create research centres, experimental farms and schools that were decisive in the generation and diffusion of technological change.

III. Implementation of the new technological model and environmental conflicts

As we have just seen, from the early 20th century onwards a series of socio-economic, cultural and scientific factors led to a new wave of technological changes that had a twin objective: to boost to the maximum possible extent the physical production of the different crops and, at the same time, to increase the productivity of labour. The two lines of innovation advanced separately until they fully converged in mid-20th century in a new technological complex structured round the green revolution and that of mechanisation which became widespread in the second half of the century. The potential of the new complex is beyond doubt. Productivity of land and labour has seen spectacular increases, well above those achieved during the previous phase. However, in breaking the equilibrium maintained by advanced organic agricultures, the new technologies have caused serious environmental problems. What has caused these conflicts and how have they manifested themselves? I should like to highlight as a major explanatory factor the ideas underlying the notions of economic systems and the relationship between society and nature that I referred to earlier. According to the principle that it was possible to replace production factors with scientific-technical knowledge, the new cycle of innovations in agriculture was much influenced by the idea that improvements in production and in productivity could be obtained by substituting natural inputs with those that could be brought in from outside, with natural ecosystems taking second place. It was a question of developing technologies that, in theory, would bring in production elements that were considered scarce or insufficient or that limited returns. This same principle was then extended to other production factors, with the use of means of production provided by natural ecosystems thus being reduced. In this fashion, the way was opened for substitution to become widespread, with Nature being as far as possible dispensed with. By thus reducing as far as possible the importance the natural environment, the agrarian sector would succeed in increasing production and would operate on lines similar to those of industry, which it had used as its model. The green revolution and other technologies that have become widespread have meant a qualitative leap in forms of human intervention in natural systems and have led to a marked turnaround in the degree of *artificialisation* of farming systems. But why did that process of *artificialisation* lead to environmental problems? The problems arose in the first place when the energy flows of essentially organic origin, with which agriculture had operated until then, were replaced by the use of fossil fuel. This meant growing dependence on non-renewable energy, while agriculture was responsible for a variety of serious pollution phenomena. But above all, problems came about because of the lack of awareness that farming continued to depend on natural ecosystems and that these operated according to their own laws. The natural environment, the soil and animals came to be conceived as being subsidiary to human activity, with no consideration being made of the complex biochemical processes involved. As J.M. Naredo has argued (1996b), the soil was seen as a reservoir into which it sufficed to pour the necessary inputs in order to obtain the desired results. In a similar fashion, animals came to be seen merely as food converters, the fact that they were elements in complex natural ecosystems being overlooked. Along with this growing *artificialisation* often came the under-use of the means of production that Nature provided free. We will now look at two examples of these processes of *artificialisation* that were instrumental in increasing production and, at the same time, the source of serious ecological problems.

IV. New fertilisation techniques and stock production

The question of plant nutrition was one of the major themes of agronomic research. Following the discoveries by Liebig, via the technique of balancing nutrients and his law of minimum, it was established that a series of nutrients had a fundamental role in plant growth and their absence or insufficiency affected levels of return. It was thus believed that increasing the applied doses was one of the most efficient means of increasing farm production. The issue of fertilisation is a good example of the narrow mercantile view that eventually imposed itself in agronomic sciences. A variable in the agrarian system was isolated, in this case plant nutrition, some of its fundamental components were identified, inputs and outputs were quantified and thus the recipe with which to improve yields was obtained. This way of seeing things gave no consideration either to the complex biochemical processes occurring in the soil in order to provide plants with the materials necessary for growth or to the complex network of relationships established with other components of the ecosystem. In fact, the soil was seen as an inert receptacle into which it was sufficient to pour increasing doses of fertilising materials in order to obtain the desired results. Traditionally, agriculture had used a wide range of procedures to replace these, and in the systems of advanced organic agriculture referred to earlier it had been possible to increase the flow of nutrients, in particular with the introduction of leguminous forage plants and the expansion of stock raising. From the mid-19th century onwards, use of guano, some mineral and the first industrial fertilisers had begun. However, the penetration of these new fertilisation techniques was relatively limited and was complementary in nature, with traditional techniques continuing to dominate. The high cost of such fertilisers limited their consumption. It was in the context of the development of the technological complex of the second industrial revolution that new technologies permitted mass production of industrial fertilisers at low cost. In 1910, Fritz Haber and Carl Bosch discovered how to synthesise atmospheric nitrogen and succeeded after the Second World War in synthesising nitrogen by cracking petroleum and natural gas, permitting the widespread use of synthetic fertilisers, which eventually replaced organic fertilisation techniques. The consumption of industrial fertilisers had seen a certain amount of development in such countries as Germany before the Second World War, but it was after the war that the doses took off spectacularly, reaching 200 kg/ha in some countries. The so-called green revolution, with the spread of new varieties of hybrids, was another of the elements involved in this line of technological change, as their ability to respond to increasing doses of synthetic fertilisers was one of the characteristics of new seeds.

The new systems of fertilisation was one form of the new agro-technological complex's 'denaturalisation' referred to earlier, which became widespread from the mid-20th century on, as it eventually implanted a system of fertilisation in which organic consumables were replaced by others of extraneous fossil and mineral origin bought on the market, leading both to high consumption of non-renewable materials and to damage to the natural base sustaining agricultural production.

Increasingly widespread use of new fertilisation techniques that were industrial in origin was one source of the environmental conflicts that have been observed in the last few decades. Growing doses of fossil fertilisers, coupled with a loss in efficiency have lead to very serious problems of pollution. Some of the nitrogen introduced, according to some authors as much as half, cannot be assimilated by plants and is lost by run-off and filtration into the subsoil and eventually contaminates subterranean as well as river and sea water, and brings serious problems for human and animal health. Another of the consequences of the application of such fertilisation techniques is that they eventually cause the mineralisation of the soil, which loses basic properties important to the maintenance of fertility due to the disappearance of organic matter. Traditional systems of fertilisation guaranteed continuous replacement that does not occur with the new technology. The scarcity or non-existence of organic material in soil degrades its structure and texture, it retains less moisture and a significant proportion of the biological life that is a necessary condition for the maintenance of fertility is lost. The progressive mineralisation of the soil, together with an intensification of farming with new machinery, has led to the fragmentation of soil particles and with this the aggravation of erosion phenomena. It has been seen that, in soils with high levels of synthetic fertilisers and a lack of humus, the activity of micro-organisms that fix atmospheric nitrogen is inhibited, with the loss of the possibility of being able to use this means of fertilisation that Nature had freely provided. The use of industrial fertilisers has led to greater water consumption with the result that agriculture has become the top water consumer, while at the same time causing a worsening of its quality. Finally, the widespread use of industrial fertilisers has subjected agriculture to the buying in of and, above all, to the consumption of non-renewable materials requiring large amounts of non-renewable energy for their production and transport.

One area of the food system where there have been radical changes is undoubtedly the stock-rearing sector. The new technologies that appeared, have permitted spectacular increases in the production of meat and dairy produce, albeit at the cost of intensifying pressure on natural ecosystems, with the rational limits of manipulation and *artificialisation* having been surpassed. This has created new sources of environmental conflicts, with vital land use having been destabilised and concerns about food safety having been caused.

Stock raising, from its origins, in that it involved the domestication of animals, implied a certain degree of human intervention in the natural environment. Although under advanced organic farming, with increasing stabling of stock, the practices of selective breeding and the use of artificial fodder, the degree of *artificialisation* intensified and the importance of natural elements was reduced, there was still an awareness of the need to maintain the balance between the advantages of stabling and the need to maintain the natural basis that sustained the sector. However, from the early 20th century onwards, the panorama began to change and by mid-century a new model of stock raising, one increasingly far-removed from Nature, had emerged. Bevilacqua (2002) has made a careful analysis of the content and ideas in the new technologies that transformed the stock-raising sector. In a climate much dominated by productivist ideas, the pressure to increase stock production grew, and in a fashion similar to what was happening in the agrarian sector, the idea took hold that the most efficient means of achieving this was bringing farm production into line with industrial production. This implied increasingly less recourse to Nature and the substitution of natural factors. In this way, a system of stock raising was conceived in which the animals were merely *machines* converting feed. The key to increasing production lay in improving their efficiency at transforming the matter consumed. At this time agricultural research and experimentation focused on establishing what was considered rational or scientific stock feeding. The intention was to determine what form of and how much fodder was best in order to obtain the desired products as well as to ensure the maximum efficiency in transforming the fodder into the meat, dairy produce and eggs. Feeding ceased to be a practice the animals did naturally and became part of a controlled, co-ordinated productive process. While stock raising was increasingly being reduced to the use of herds as converters, the growth of production depended on the availability and quality of fodder for the animals to process. As a result, experiments began with new products that were proposed as being considered more likely to attain the objective of maximising production. Oleaginous seedcakes, the leftovers of distilled products such as the residue from grapes or sugar beet, became fairly commonplace in the early decades of the 20th century. Experiments were also made with dried and powdered meats from Latin America and with fish flours. Although some of these experiments did not always produce the desired results, it is interesting to note that an animal-feed industry emerged that was able to provide diets. These were elaborated according to supposedly scientific principles that had proved to be very efficient. A number of formulae were tried, and to the basic ingredients were added a series of chemical and pharmaceutical products with the objective of accelerating production on a minimum amount of feed. In this fashion, a new model of stock raising emerged that essentially depended on the animal feed industry, and increasingly less on the farm's own production. New technologies in animal feeding meant another form of substituting nature and they were presented as a means of overcoming its imperfections and rigidities. The new stock raising broke ties with systems of agriculture and was able to operate in a fashion similar to industrial production.

Simultaneously with the changes in feeding systems, a new model of stock farming emerged, one that was highly specialised and designed to organise itself in large productive units, taking advantage of the economies of scale that had proved so successful in industry. In this manner, the way was opened for new forms of *artificialisation*, with natural factors increasing being left out of the equation. Stabilisation had already meant a first step towards 'denaturalisation', in limiting the animals' freedom to feed and move about and the free benefits of such things as sunlight. As the size of the new stock farms increased, these trends were accentuated. In aviculture the attempt was made to replace hen runs by large installations housing thousands of animals, which gave rise to *totalitarian* stabling, to quote Bevilacqua. The new poultry farms were organised into batteries inside which the animals were practically immobilised, with the objective of reducing the energy losses produced by the animals moving about to a minimum. In this way, all the fodder eaten could be used for growth of the animal. However, industrial aviculture of this nature led to problems of the animals' health. They were affected by a variety of pathologies, on occasions caused by a lack of exposure to sunlight, deficiencies of diet of different sorts (such as the lack of vitamin E which caused the outbreak of chicken disease in 1937–1938) or problems of force-feeding. The response to these problems, rather than a return to 'natural' space, was to accentuate *artificialisation* even further. New chemical-pharmaceutical knowledge provided the necessary instruments with which to counterbalance the effects of brutal *artificialisation*. The medicalisation of herds, with extensive use of vitamins, antibiotics and drugs of all kinds, became a necessity that had to be regularly applied in order to maintain a system of stock raising that was increasingly far removed from the laws of nature.

Industrialised stock raising also needed to develop another line of technological change, one that would allow it to have a hand in the field of biological material. Scientific advances and in particular those achieved in the field of genetics meant better knowledge of the mechanics of inheritance and allowed selected breeding methods to be perfected. The widespread use of these techniques together with those of artificial insemination and the incubator opened the way for fresh human intervention in natural processes. With intervention in animal breeding dependency on nature was reduced, its imperfections were supposedly limited and stock raising became an increasingly industrial operation.

Industrialised stock raising of this nature has generated a number of issues of environmental conflict. Large stock farms coexist in the same space as arable farms but they operate in completely separate fashion. The ties that with advanced organic agriculture linked stock raising and arable farming and which had made possible balanced land use, with crop rotations and levels of great diversity, of vital importance for maintaining soil quality and defences against pests and plagues, have been broken. In addition, with the industrialisation of stock raising and the spread of new systems of fertilisation, one of the main, historical functions disappeared, namely the provision of a significant proportion of nutrients and organic matter which had to be put back into the soil in order to maintain fertility. Animal dung and manure, which under the previous system provided for free vital input in the form of fertilisers have now become instruments of widespread pollution, given the difficulty of properly recycling the huge amount of produced waste. The pollution of aquifers, the soil and the atmosphere has been one of the undesired effects of the new forms of management of stock raising.

The new technologies that have spread to the stock-raising sector have made possible spectacular increases in the production of meat and dairy products, though the nutritional quality of the products has often been questioned. Under the principle of rational and scientific feeding of stock, experiments have been made with new diets with the aim of obtaining results, without the necessary prudence of observing the effects these might have on the human organism. In addition the growing use of pharmaceutical products of all kinds, in order to prevent certain pathologies, boost growth or obtain certain properties, has led to uncertainty as to the effects this may have on our health. Finally, the new conditions to be found in stock raising today have meant a yawning gap between this and the animals natural life conditions, and reduced them to a precarious physiological state, a veritable hell never previously known.

The environmental impact of the technological complex of an industrialised farm system is not limited to those deriving from fertilisation techniques and the system of stock raising. The widespread use of mono-cultivation, with the abandonment of crop rotation and diversified land use, led to processes of impoverishment of the soil, and a major incidence of plagues and pests required systematic use of pest- and herbicides. The abuse of such chemical products has eventually caused problems of intoxication affecting farm workers, and may have harmful effects on consumers. The widespread use of genetically modified seeds has led to a loss of bio-diversity with the subsequent impoverishment of our resources of genetic material for the agriculture of the future. Growing mechanisation has made the sector one that, for its energy, is subsidised from outside and depends on the consumption of non-renewable energy, so that the food system has come to be responsible for the pollution caused by the consumption of such materials. In short, while the present-day system of farming has attained spectacular success in growth of productivity of labour and physical returns, the negative impact on the environment has reached such a degree of severity that there is good reason to question the likelihood of continuing in this direction. There is sufficient evidence, as has been demonstrated for several decades now, to question the sustainability of this model, as respect for the requirements and laws of natural ecosystems vital for any farming system has been lost.

Historical analysis of technological change in systems of agriculture cannot ignore their impact on the environment, as has been the case for too long. Fortunately, there has been a reaction in this sense. Improved knowledge of the genesis of the technological complex of industrialised agriculture and its ecological effects, bringing fresh information on the principles behind it and on other lines of technological change, could provide elements to find alternatives to the dead-end in which the food system finds itself at onset of the 21st century.

Bibliography

Bevilacqua, P. (1993) *Breve storia dell'Italia meridionale dall'Ottocento a oggi*, Roma.

Bevilacqua, P. (2002) *La mucca è savia. Ragioni storiche della crisi alimentare europea*, Roma.

Chorley, G.P.H. (1981) 'The agricultural revolution in Northern Europe, 1750–1880: Nitrogen, legumes, and crops productivity', in *Economic History Review*, 34, pp. 71–93.

Garrabou, R. (1994) 'Revolución y revoluciones agrarias en el siglo XIX: su difusión en el mundo mediterráneo', in AA.VV., *Agriculturas mediterráneas y mundo campesino. Cambio histórico y retos actuales*, Almería, pp. 93–110.

González de Molina, M. (2001) 'Condicionamientos ambientales del crecimiento agrario español (siglos XIX y XX)', in J. Pujol [et al.] (eds), *El pozo de todos los males. Sobre el atraso en la agricultura española contemporánea*, Barcelona, pp. 43–94.

Immler, H. (1993) *Economia della natura. Produzione e consumo nell'era ecológica*, Roma.

Mazoyer, M. and Roudart L. (1997) *Histoire des agricultures du monde*, Paris.

Naredo, J.M. (1996a) *La economía en evolución. Historia y perspectivas de las categorías básicas del pensamiento económico*, Madrid.

Naredo, J.M. (1996b) 'Sobre la reposición natural y artificial de agua y de nutrientes en los sistemas agrarios y las dificultades que comporta su medición y seguimiento', in R. Garrabou and J.M. Naredo (eds), *La fertilización en los sistemas agrarios. Una perspectiva histórica*, Madrid, pp. 17–34.

3 Environment conditions and biological innovations in European agrarian growth[1]

Josep Pujol-Andreu, Universitat Autónoma de Barcelona

The driving forces behind Europe's economic growth in the nineteenth century are diverse, and not easily prioritised. In economics and economic history, attention has been focused on different institutional and technological variables, and particularly on the dynamic effects of market's expansion (among others Hobsbawm, 1968; Landes, 1969; Pollard, 1981, Maddison, 1995; Mokyr, 2002). In this sense, the middle classes' capacity to foster technological change and the new possibilities of production provided by capitalist institutions have often been underlined. According to some authors, this happened within the framework of intense conflicts around the control of power structures, productive process and/or rent and wealth distribution. But according to others, this happened in contexts in which consensus would have prevailed, given the long-term improvements in standards of living. Nevertheless, it has also been underlined that the evolution of economic activity could not be understood considering only the new production possibilities offered by market economies. As a result, today it is accepted that those processes cannot be explained without considering two additional circumstances: the energy flows that sustained them, and the changes undergone in their transformation (Debeir, Deleage and Hemery, 1986; Wrigley, 1990; Naredo and Valero, 1999; Diamond, 1997; Sieferle, 2001).

In this context, a question arises of special importance. Which was the influence of the biological change in the economic growth? A part of the flows of energy must be made into food, and this transformation can only happen with the *participation* of plants and animals. As Soddy emphasised in 1921, 'The plant world continues to be the only one that can transform the original flow of inanimate energy into vital energy' (Martinez Alier, 1995). In recent years there has been research in this direction, and we should consider the long tradition of biological innovations in the agricultural sector; their important implications; and the final configuration of an important business sector around these processes (Heiser, 1990; Friedland et al., 1991; Goodman, Sorj and Wilkinson, 1987; Goodman and Redclift, 1991; Busch, 1997; Busch, Lacy, Burkhardt and Lacy, 1991; Perkins, 1997). It is also shown that the orientation of this type of innovations and their institutional organisation have become more complex with market expansion, and that their contribution has played a decisive role in the configuration of contemporary economic growth. From this research, in synthesis, an issue can be raised as a working hypothesis. In the study of economic growth, we should consider three processes: 1) the changes undergone in the organisation of society, 2) the flows of energy and materials used and the technical bases of their transformation, and 3) the biological conditions under which the production of food is carried out.

[1] This article is part of a research project financed by the DGICYT 'Food, mortality and Standard of living in Spain (19th and 20th centuries)', SEJ2004-00799. The author wishes to acknowledge the helpful comments of J. Martínez-Alier, R. Nicolau, C. Sarasúa, P. Scholliers and the other researchers that participated in the meeting of Tensions of Europe in Barcelona (March, 2003).

In this context, some clarification is required with respect to our knowledge about the third issue. On the one hand, we know well the processes undergone after World War II, and this circumstance has propitiated unrealistic perceptions about the true possibilities of agrarian change at different times. On the other hand, the handling of the previous issues has advanced notably since the 1980s, but the studies performed have focused especially on the agriculture of the United States. With respect to this country we have excellent analyses of the importance of biological changes in agrarian growth since the 19th century, about its institutional characteristics, and about its relationships with other aspects of technical change (Kloppenburg, 1988; Dalrymple, 1988; Busch and Lacy, 1983; Danbom, 1986; Huffman and Evenson, 1993; Olmstead and Rhode, 2003). For Europe, these contributions have been less numerous. Various circumstances have been involved in this imbalance, such as the different significance of the biological problems in both areas; the traditional interest of US governments in transforming the biological bases of its agriculture; and the hegemony acquired on an international scale by that country's food and biotechnological industries. In any case, for European agriculture, one must remember the studies that have been performed about the wheat sector and different species of livestock, or, about the relations between biological innovations and agrarian change (Martin, 2000). We also must remember two other issues. In the first place, the non-existence of a general framework in economic history for interpreting biological and economic changes over time. Secondly, the need to dispose of more sectoral studies on a national and regional scale, especially with respect to the impact of those innovations on the levels of productivity.

In this chapter I will develop these directions by analysing the biological changes in both the Atlantic and Mediterranean Europe till the 1930s. The first section places those changes in the general framework of the environmental conditions of production. The second section will indicate some of their main characteristics in the wheat and livestock sectors. The third section puts forward some explanations for their differing evolution in different places. The final section relates these changes to other innovations, and underlines their importance in order to understand them better.

Just one word of caution. In this chapter, the institutional variables will be considered only indirectly. Not because they are not considered important but rather because I prefer to focus on certain aspects that are still relatively unexplored in economic history. When biological variables and environmental conditions are considered, some characteristics of the processes of change undergone by European agriculture up until the middle of the 20th century may be better understood.

I. Agrarian systems and environmental conditions

I understand biological innovation to be all the activities performed consciously for increasing the production capacity of the agrarian sector, whether this be by introducing new varieties of plants or animals, or by altering their constitution through different techniques (selection, crossing, etc.). Therefore, from this perspective biological innovations have been one of the main lines of the participation of human societies in the environmental conditions of production, and, more specifically, one of those most used for increasing agrarian production.

In this sense it is important to underline the development of these innovations from the second half of the eighteenth century, as a result of three circumstances. Firstly, the knowledge accumulated on the physiology of plants and animals, the progressive improvements undergone in selection and crossing techniques and the rediscovery of Mendel's Laws in 1900 (Stubbe, 1972; Corcos, Monaghan and Mendel, 1990). Secondly, the ever-closer contacts fostered by the expansion of trade between areas with different natural resources. Thirdly, especially beginning in the second half of the 19th century, the growing availability of new means of production, both chemical and mechanical, linked to the availability of new biological varieties (Walton, 1999; Grantham, 1984).

Another issue to emphasise is more related to the different orientations and possibilities that these innovations could have. As the biological conditions of production depend on the climatic, hydraulic and edaphic characteristics of each area, these innovations were also conditioned by the degree to which these innovations were complementary to the overall environmental circumstances under which the agrarian systems operated. The importance of these circumstances with respect to the two large areas that we will be dealing with, is well known. While in the agriculture of Central and Northern Europe there were high levels of rainfall, deep soils, and mild climatic conditions in the Spring and Summer, in Mediterranean Europe these conditions could be very different. The rainfall was lower, especially when it was needed the most, temperatures tended to be high from the end of the spring on, and agricultural soil was poorer in organic material. These differences are not very dissimilar nowadays, although technical changes have mitigated them (Papadakis, 1966).

As a result, when demographic pressure, institutional changes, and the intensification of exchanges accentuated the expansion of cultivated areas and the processes of specialisation, these processes tended to take different forms from one part of the continent to the other. The first area tended towards very intensive growing systems and increasing integration between agricultural and livestock activities (Tracy, 1982; Grigg, 1992; Van Bavel and Thoen, 1999). In the second area, the expansion of crops was combined with the maintenance of very extensive systems in the grain-growing regions, and growing specialisation in vineyards, olive groves, and fruit trees. Where climatic conditions allowed, and the irrigated surface area could be increased, other orientations must also be underlined. The expansion of vegetable crops, rice, and fresh fruit trees took place where there was more intensive irrigation, and new grain rotations were used in the more irregularly irrigated areas. Broadly speaking, the most important thing in Mediterranean systems was the articulation of an agrarian sector, characterised by few resources of fodder, livestock, and fertilisers; the presence of fallow land in grain areas; and a high presence of vineyards, olives and tree crops in most parts of the territory. On the other hand, livestock farming continued with grazing, and the development of livestock producing milk and meat took place later and was more limited (Simpson, 1997; Bevilacqua, 1992; Garrabou and Sanz Fernández, 1985).

But the influence of environmental conditions on both areas is not only reflected in the different productive orientations that accompanied agrarian growth. Their impact also stands out when we consider the different evolution undergone in two important sectors (wheat production and livestock). The evolution of these sectors has often been used to evaluate the ability of European agriculture to adapt to the expansion of markets, and consequently, its study has played an important role in agrarian history research. Wheat, meat, and milk were also three basic foods for the population, although their importance in this sense tended to vary over the course of time (Teuteberg, 1992; Collins, 1993; Kiple and Ornelas, 2000).

II. Biological innovations during the nineteenth century and the first third of the twentieth century

The first issue observed when we consider biological innovations in the wheat sector is its different evolution according to place. Various research projects have emphasised growing importance of such innovations in the British wheat sector since the 1770s, and the quick spread to other countries of Western Atlantic Europe, especially from the second half of the 19th century on (Walton, 1999; Lupton, 1987; Zeven, 1990; Doussinault, 1995). In Mediterranean Europe, however, these kinds of innovations were not begun until the 1880s, their development was slower, and had fewer repercussions. As a result, while wheat seeds were transformed relatively quickly in Atlantic Europe, this process was later and more limited in Mediterranean Europe, particularly where the dry land conditions were more extreme (Pujol, 1998a).

At the beginning, these innovations were based on the introduction of new varieties from Eastern Europe, and on the intensification of traditional methods of mass selection. Later in the 19th century, three types of initiatives took on growing importance: the spreading of English and Scottish seeds to the continent, the intensification of biological exchanges inside this area, and the progressive substitution of mass selection with individual, along with the growing use of different types of crossing.[2] Consequently, although the new techniques of improvement were still not very precise and on many occasions were not able to stabilise the desired characteristics in the new seeds, the quick spreading of new types of wheat was also observable in many areas of Atlantic Europe by the middle of the 19th century. Two circumstances favoured this process: the autogamous nature of that grain (which limited spontaneous mutations and hybridisation), and the fact that farmers could continue to obtain the seeds for planting from their own productions, once a new variety was accepted.

The fact that the innovations could not be appropriated also meant that improvement activities tended to be very decentralised. Only in special cases were they performed in a new type of company of some size. Even in these cases, it was common that their activities were very diversified, and companies also included the production of other seeds for vegetable or fodder crops among their activities. Two companies of these characteristics were Vilmorin, and Denaiffe, Colle and Sidorot. This situation changed partially between the 1880s and the 1930s. On the one hand, the intensification of competition and exchanges stimulated the demand for seeds that were more productive and resistant to diseases. On the other hand, improvement techniques became more complex and expensive, and their development tended to be concentrated in a new type of institutions, totally or partially financed by the State.[3] In Tables 3.1 and 3.2 of the appendix some of them are listed.

---

[2] Different examples in Lupton, 1987: 53–69; Percival, 1934; Zeven, 1990: 15–99.

[3] Sala Roca, 1948; Walton, 1999: 36–37; Grantham, 1984: 195–202; Lupton, 1987: 53–69; Kamps, 1989; Maat, 2001.

In this context, nevertheless, various issues should be emphasised. While British economic policies tended to limit these innovations until the 1920s, protectionism and/or direct promotion by the state were a driving force behind them in other countries of the continent (Palladino, 1996). The economic and social structures of each area and their different foreign relations probably influenced these options (Offer, 1989, chpt. 5; Tracy, 1989; Koning, 1994). In any case, while these innovations tended to be delayed in British agriculture, in France, Holland, Belgium or Germany they accelerated; and the spread of new wheat and the biological exchange between these countries increased (Simon, 1999; Bonjean and Angus, 2001; Maat, 2001). Additionally, the sources consulted also show that the processes of innovation tended to spread towards the Mediterranean area. Nonetheless, the effects of such processes in this area did not begin to be evident until well into the 20th century: in Italy, particularly in the northern part, towards the end of the 1920s, in Spain, about 20 or 25 years later (Pujol, 2002a). In Figure 3.1 and Tables 3.3 and 3.4 some characteristics of these processes and some of the new types of wheat that tended to be spread, are indicated.

With respect to cattle, horses, mules, and pigs, the information and studies show three issues. Firstly, the biological exchanges and different activities of selection and crossing existed already from the end of the eighteenth century. Secondly, these innovations were important results in Western Europe in the middle of the nineteenth century. Finally, the spreading of these activities in Mediterranean Europe had greater repercussions than with respect to wheat, but their impact was again very limited and concentrated in few regions. Let us see some examples.

Innovations in cattle are probably best known. Different studies have shown that the owners of Swiss and Dutch livestock had already achieved at the end of the 1800s the consolidation of different milk-giving breeds that were improved (for example, the Friesian and the Hölstein in the case of the Dutch, and the Brown Swiss and the Simmental in the case of the Swiss). Soon, with the purpose of reinforcing the uniformity of the new varieties, and focusing their improvement more precisely, they established the Dutch herdbook in 1873 and the Friesian herdbook in 1879. Somewhat later, the Red and White Spotted Simmental Cattle Association were settled in 1890, and the herdbook for the Brown Swiss in 1911. Other varieties that improved for the production of meat were the Charolais and the Limousin from France, and the Hereford from the United Kingdom, for which their respective herdbooks were also established (for example, the Hereford herdbook, published in 1846, two herdbooks for Charolais livestock, in 1864 and 1882, and another one for Limousin in 1886) (Briggs and Briggs, 1980; Felius, 1985; Porter, 1991; Bieleman, 2002). In reference to pigs, two important events are to be mentioned: the successive improvements undergone in different English varieties since the 1770s, and the foundation in 1884 of the National Pig Breeder's Association. As a result of these activities, new herdbooks for the Large White or Yorkshire (1884), the Large Black (1898), and the Berkshire (1884) were established, and the new pigs spread quickly to the continent to give rise to other ones (Hall and Clutton-Brock, 1989; Briggs, 1983). Regarding horses and mules, the changes are more difficult to discern. Despite this problem, the information available also shows that their constitution tended to improve, gaining in height and strength, and that the Percheron, Ardennes, Belgian and Suffolk varieties got notable prestige. Also, in all these cases biological exchanges were very intense, both to directly exploit the new varieties and to generate other ones with successive selections and crossing (Hendricks, 1995; Mason, 1996).

In clear contrast with these processes, those observed in the Mediterranean areas again show important differences. In fact, leaving out the more northern areas with a greater livestock tradition, the information available again underlines the long survival of traditional varieties. The evaluations and comments of different Spanish agronomists and engineers of the end of the nineteenth century are very illustrative. In the 1880s and the 1890s, these technicians still underline two circumstances: the scarce integration of agricultural activities with livestock farming, and the existence of varieties that were not very productive. With respect to pigs, the hegemony of the varieties with dark skin and long snouts, with scarce aptitude for fattening, and slow growth was remarkable. Cattle were apt for working, but with low productivity for the production of meat and milk. In reference to horse and mule species, these engineers pointed out their short stature and light weight and their limited capacity in the operations of cultivation and transport (Junta Consultiva Agronómica, 1892; Junta Consultiva Agronómica, 1920).

This situation changed partially during the first third of the 20th century with the introduction of improved European varieties. In Catalonia in the 1930s for example, a new livestock population replaced the traditional one almost completely, and new varieties of mixed breeds from different places tended to predominate in their composition. Particularly, the characteristics of Yorkshire and Craon pigs, the Swiss and Dutch breeds in cattle, and Percheron, Norfolk and Norfolk-breton in horses and mules spread. These processes are also observed in other agricultural areas of the northern half and the Mediterranean coast, but not so much in the central and southern parts of the territory (Domínguez, 1996; Pujol, 2002b; Castell, 2002).

III. Biological innovations: economic, institutional and environmental conditions

What circumstances would allow the explanation of these differences? The processes that we have just synthesised cannot be explained without considering economic and institutional changes that occurred on a European scale between the second half of the 18th century and the 1930s. Nonetheless, the geographical differences that we saw in the previous paragraphs cannot be explained solely in terms of that type of variables.

The wheat sector, for example, was not only important as a producer of grain but also of straw, and the varieties of wheat had to be long-stalked for this reason. Straw was necessary for the keeping and caring of livestock, especially where fodder was lacking, and also for the preparation of manure prior to its use as fertiliser. Consequently, although greater fertilisation could increase yield in grain and allow more intensive rotations, also facilitated the appearance of lodging. When this happened, it made harvesting operations more expensive, and it could even make mechanical harvesting impossible. With lodging, a part of the production was also lost, and the attack of various diseases was facilitated. In synthesis, to increase grain production and simultaneously mechanise harvesting, it was necessary to have more productive new varieties, resistant to lodging, so that these characteristics became two of the main objectives of biological innovations.[4]

[4] McNeill 2000, 219–225, Walton 1999, 34–39.

In Nordic countries, increasing the resistance of plants to low temperatures also occupied an important place. In contrast, in Mediterranean countries obtaining of earlier-ripening varieties was necessary (Sala Roca, 1948).

The initial interest of European breeders for British wheat is thus not difficult to understand. With the expansion of mixed farming from the middle of the 18th century, British wheat had evolved towards varieties with low gluten content, but which were very productive of grain and straw, and resistant to lodging. This trend accelerated later with the liberalisation of imports and the change to high farming. But, while the institutional framework discouraged new researches in the British case, in Western Europe it encouraged them, and the wheat varieties of Great Britain were used in a wide range of crossings and selections. Three objectives were pursued: to maintain or improve the protein richness of the wheat varieties planted, to increase their yield per seed or surface area unit, and to make their stalks sturdier. In Table 3.4 of the appendix some of the main hybridisation performed are indicated.

Nevertheless, at the end of the nineteenth century the French breeders still indicated the great difficulties when trying to improve wheat seeds in the southern and eastern parts of France, because of climatic conditions. For Spain, the information is even more explicit. Despite various experimental centres created in the 1890s, and the numerous tests performed with the new wheat seeds spread throughout Europe, the results obtained were very poor. The new wheat varieties degenerated quickly if they came from Atlantic Europe, or they did not surpass the results of indigenous ones if they came from other grain-growing areas with similar environmental conditions. Only at times some success was attained, e.g. at the end of the 19th century, with the Italian Rieti and Richella Blanca wheat from Naples, and, already in the 1920s, with some of the new seeds obtained in Italy by N. Strampelli. In reference to these last varieties, we also have to remember two issues. First, those varieties were obtained from a new type of crossing, in which the Japanese variety Akagomushi was used. Second, their dissemination was concentrated in the central and northern parts of the country. In Spain, on the other hand, the improvement of indigenous wheat began in the 1920s, often using new Italian wheat varieties, but their results did not become relevant until after twenty years. It was not until the 1950s that new varieties such as the Aragón 03 spread further, and only again, in the grain-growing provinces of the northern half of the country (Vilmorin and Meunisier, 1918; Nagore, 1935; Pujol, 2002a). In Table 3.5 are listed the main experimental centres that carried out these activities.

In synthesis, two results arise from these experiences: the use of Atlantic wheat was not viable in Mediterranean Europe, because of different environmental conditions in the two areas, and the improvement of the seeds themselves was more difficult to achieve in the Mediterranean areas than in the Atlantic ones.

The problems faced by biological improvements in the livestock sector were different. In this sector, the processes of selection and crossing were easier to perform and to evaluate, and hence their early results in Atlantic Europe during the 19th century. This does not mean that environmental conditions lacked importance. High temperatures throughout a large part of the year, and scarce water also limited the processes of improvement in cattle and pigs in many areas. Also, while the resources of meadows and pastures in Central and Northern Europe were great, in many areas of Mediterranean Europe it was the opposite. This circumstance was aggravated in a large part of the territory by the scarce orientation towards livestock production in the agrarian sector.

This is once again particularly clear in the case of Spain (Santiago Enríquez, 1922; García Bengoa, 1923; Arán, 1933). As we have indicated, both high levels of specialisation in vineyard, olive and other tree crops, and the impossibility of using the crop rotations that were used in the damper parts of Europe, limited the development of livestock in this country. This situation was also fomented by the need to resort to grazing and the scarce resources obtained with this type of farming. With the change of century, various circumstances made the greater development of that sector possible. The changes in agrarian markets, and the expansion of urbanisation were undoubtedly two of them, as they stimulated the expansion of meat and milk consumption in large cities.[5] But the development of new sectors can not be entirely understood without considering two other variables. First, the new production possibilities provided by mineral and chemical fertilisers from the end of the 19th century. Second, the great expansion of irrigated areas undergone at the same time. As a consequence of these innovations, grain rotations were made more intensive, and the offer of fodder resources was more abundant. In a more thorough analysis, nevertheless, it also stands out that the impact of those processes tended to be concentrated in the Mediterranean coast and in other regions of the north-eastern third of the territory, but much less in the central and southern parts of the country (González de Molina, 2001; Fernández Prieto, 2001).

IV. Biological innovations and agrarian growth

The biological changes that I just have synthesised are not the only ones that could be considered. Others affected Mediterranean agriculture very directly, and their impact, in some cases, was also outstanding. The spreading of new seeds is well documented in the rice sector since the end of the 19th century, often in order to tackle lodging and to make more intensive fertilising possible (Calatayud, 2002). Parallel to this, destruction of vineyards by phylloxera led to the transformation of biological bases in this sector, and the spreading of American vines on which the European varieties of *Vitis Vinifera* were grafted (Pan-Montojo, 1994; Garrier, 1989). With regard to other fruit trees, we also have varied information about the spreading of new varieties of plants with three objectives: to improve the quality of final productions and increase yields; to develop new productions; and to better control harvesting operations (Abad, 1984).

Based on these considerations, there are certain questions that should perhaps be raised more clearly in future research. For example, what specific importance did biological innovations have in the different growth processes that took place during the 19th century and the first third of the 20th? Or, what was their role in the expansion of agrarian yields and levels of productivity? These questions are not easily answered. Firstly, because we cannot quantify the biological changes that we have described, and we must limit ourselves to very indirect estimates of their impact and dissemination. And secondly, because biological innovations tended to advance in many cases complementary to other innovations, and it is not easy to isolate their specific effects. Probably we could advance in solving these problems by analysing more carefully the experiments undertaken in the different research centres that were created during those years. Now,

[5] Simpson, 1997, 249–261; Langreo, 1995.

I would only like to stress that the impact of biological innovations may be greater than we usually consider them to be.

Recent research has estimated that approximately 50 percent of the increase in US wheat grain productivity between 1839 and 1909 was caused by the spreading of new seeds of that grain (Olmstead and Rhode, 2003). We still do not have studies of these characteristics for European agriculture. On the one hand, we do not have statistical information on the evolution of planted areas, such as those existing for the USA. On the other hand, biological innovations advanced along with the use of more intensive fertilisations and the expansion of irrigated surface area, so it is more difficult to isolate its effects on the levels of productivity. Nevertheless, in recent studies it has also been suggested that environmental conditions might exercise a greater influence over the dissemination of new means of production, and that among these conditions we should consider the initial biological bases, and the possibility of altering them.

Various research projects allow us to know a fair amount about the dissemination processes undergone by mineral and chemical fertilisers and harvesters. Three issues stand out: the initial spreading of these means of production in British agriculture, especially in the case of harvesters; the intense spreading of the use of these inputs in Continental Atlantic Europe from about the 1880s; and its later and more limited spreading in Mediterranean Europe. In Table 3.6 some of these aspects with regard to the spreading of new fertilisers are shown. In reference to the spreading of harvesters, let us remember the following issues. At the end of the 19th century, 80 percent of the British wheat areas were harvested with machines. In France, on the other hand, this percentage dropped to just under 15, and in Germany to little more than 5 percent.[6] In the rest of the continent these percentages were even lower, and in the cases of Spain and Italy, they were practically negligible. Soon after, the use of harvesters intensified in Belgium, France and Germany, but in the case of Spain and Italy, they did not begin to be significant until the 1920s (Van Zanden, 1991; Gallego, 1986). Besides, the implementation of harvesters ended up being high in the grain-growing areas of northern Spain, but very little in the central and southern parts, and along the Mediterranean coast. Additionally, the spreading of the new fertilisers ended up being quite remarkable in this last area, and other inner regions in the north. On the contrary, they lacked relevance in the central and southern parts of the country. In fact, the use of those materials in a wide part of this area did not even reach 5 kg/ha in the 1930s, when it was often greater than 30 kg/ha in the coast and in the Ebro basin (see Table 3.7) (Simpson, 1987; Pujol, 1998c; Fernández Prieto, 2001).

How do we explain these processes and differences? The sustained expansion of exchanges and the intensification of the processes of industrialisation tended to favour the spreading of new means of production in two ways. One, by improving the conditions of its offering in terms of price, facility of access, and greater adaptation to local needs. The other, by reinforcing successive salary increases, due to the changes caused in the labour markets by these processes. The sustained reduction in the relative prices of new fertilisers, and the improvements that were introduced into the design of harvesters illustrate the first issue very well (Pezzati, 1994). The tendency of agrarian salaries to rise from the last decades of the 19th century, and especially after World War I, is also

[6] Grigg, 1992: 52–55.

well documented (Scholliers, 1989; Martínez Carrión, 2002). These processes are also well known for the Spanish case, and they are illustrated in Figure 3.2. As a consequence of these changes, the threshold of use of these means of production tended to widen over time, and this circumstance reinforced its spreading in a sustained way. But the previous information also shows significant differences in the rhythms and intensity with which the new techniques of production spread, which can not always be explained by the evolution of their offer or by wage pressures.

Evidently, another variable that we must consider is the institutional framework, due to their great influence on the farmers' demand for new production techniques. Numerous studies have analysed these issues and have dealt with the influence of three groups of variables on those processes: 1) the structure of land owning and its changes over time, 2) the size of the farm and the social systems of production, and 3) the agrarian and tax policies. Thanks to this research, today we can better explain, for example, the early spreading of new production techniques in the British agricultural sector during the 19th century, or its intense spreading in France, Belgium, Holland or Germany between the 1880s and the 1930s. Interesting explanations have also been provided for the decline of British agriculture since the 1880s, and for the different orientation of biological innovations in the wheat sector in Atlantic Western Europe (Koning, 1994; Van Zanden, 1994; Offer, 1989, Chpt. 5; Walton, 1999).

But even if we also consider institutional variables, the processes observed in Mediterranean agriculture are not easy to explain, especially considering the intense regional differences between the middle of the 19th century and the 1930s. For this reason, the need to include environmental factors in the analysis has been mentioned on various occasions, and these proposals have often favoured controversial findings. In the case of Spanish historiography they are still being hotly debated.[7]

Recent research on this country sustains what follows. The environmental conditions defined distinct constellations of available techniques in Mediterranean and Atlantic agriculture, and the demand for new means of production also was for this reason, very different. This consideration does not minimise the importance of other variables. The institutional framework undoubtedly delayed the beginning of agrarian changes and contributed to slowing them down, as it realised late and slowly the transformations that the sector needed (irrigation, experimental stations). At the same time, the late development of a new industrial sector, producing fertilisers and mechanical means of production, was another factor that we should not forget. From our perspective, nonetheless, these circumstances cannot satisfactorily explain two issues: the low use of the new agrarian inputs during the 1930s, and their unequal spreading in different places. Moreover, when those agrarian innovations are analysed more carefully, different relations are perceived that should be investigated more precisely. In Table 3.7 the clearest cases are indicated.

Firstly, the close relationship existing between the spreading of new fertilisers and the availability of water. These relationships are shown, for example, in two situations observed in the 1930s: the high consumption of fertilisers in the irrigated areas of the territory and in various northern provinces; and the negligible consumption of these

---

[7] The debate in *Historia Agraria*, 28, 2002: 179–230. See also, O'Brien and Toniolo, 1991 and Pujol et al. (eds), 2001.

same products in wide areas of the centre and southern parts of the country, where precipitation was very little and irrigation was not significant. Secondly, although the surface areas to harvest could be very large, mechanised harvesting tended to be not very significant where the surface areas of vineyards and olive groves were also great, or where the three types of surface areas were relatively near each other. There are certainly exceptions, but the relationship between the spreading of harvesters and cultivation structures is difficult to question and must not be ignored. One of the reasons that has been suggested to explain this relationship is the discontinuity that could be generated by grape and olive-growing specialisation in grain-growing lands. Other reasons are the different problems that the work processes of those crops could generate in the different grain-growing areas. Let us recall that the harvesting of grains had to be performed during a short period of time, between June and July, also coinciding with the reaping of the alfalfa fields and the like, and that the gathering of grapes and olives was done later and successively (grape gathering in September, and olives from November till February or March). These operations also required a great deal of work and could not be mechanised. Therefore, it is not risky to suggest that the pressures to mechanise the harvesting of grains had to be very different according to the structures of crops, and lower in the Mediterranean coast, and in the central and southern parts of the country. Finally, both with respect to new fertilisers and harvesters, in this analytical framework there is another issue: its lesser spread in many areas was also conditioned by the existing varieties of seeds and the difficulty to improve them (González de Molina, 2001; Fernández Prieto, 2001).

V. Conclusions

In synthesis, the transformation of European agriculture during the 19th century and the first third of the 20th should be explained as a result of two large groups of variables. On the one hand, the successive pressures generated by economic and institutional changes undergone during that period, promoted the development of new types of activities, new means of production, and higher levels of productivity. On the other hand, the environmental and biological environments of the different areas, conditioned the productive orientations that could be developed and the available techniques.

In this chapter I have tried to show that the biological characteristics of plants and animals occupied a strategic place in the development of the processes of production, hence the interest in transforming them. In some cases, to mitigate the impact of certain diseases or accidents; in others, to improve the quality of the final production, but broadly to increase the levels of productivity and improve agrarian incomes.

Analysing the case of wheat and different livestock species, nevertheless, we have also seen another issue. Biological and environmental conditions influenced the spreading of other innovations, such as those related to the fertilisation of the soil and the harvesting of grains, and consequently the different patterns of technical change. Therefore, these circumstances should also be taken into consideration to explain the different courses followed by those sectors in the different areas of the continent.

Finally, based on the previous considerations, two working hypotheses could be maintained. First, that the possibilities of agrarian growth until World War II were always fewer in Mediterranean agriculture than in the Atlantic one, although the new offers of means of production and the expansion of irrigation tended to increase them. Second, that these differences did not begin to decrease significantly until the 1960s, and then as a consequence of two groups of innovations: those related to the use of fossil fuel in cultivation, harvesting and threshing, and those related to the use of new seeds and chemical products for the fertilisation of the land and the treatment of plants. That is, when a whole group of new technical possibilities allowed the mitigation of the impact of environmental variables and increased the dependence of agriculture with respect to the industrial sector.

Bibliography

Abad, V. (1984) *Historia de la naranja, 1871–1939*, Valencia.

Arán, S. (1933) *Explotación e industrialización del cerdo*, Madrid.

Bavel, B.J.P. van and Thoen, E. (eds) (1999) *Land productivity and agro-systems in the North Sea area. Middle Ages-20th century*, Turnhout (CORN Publication Series; 2).

Bevilacqua, P. (1992) *Storia dell'agricoltura italiana in età contemporanea. Spazi e Paesaggi*, Venezia.

Bieleman, J. (2002) 'Dutch cattle breeding and dairy farming in transition, 1850–2000', unpublished paper presented at the *X Congreso de Historia Agraria*, Sitges.

Bonjean, A. and Angus, W.J. (eds) (2001) *The world wheat book: A history of wheat breeding*, Andover–Secaucus.

Briggs, H.M. (1983) *International pig breed encyclopedia*, Greenfield (IN).

Briggs, H.M. and Briggs, D.N. (1980) *Modern breeds of livestock*, New York.

Busch, L. (1997) 'Biotechnology and agricultural productivity: changing the rules of the game?', in A. Bhaduri and R. Skarstein (eds), *Economic development and agricultural productivity,* Chaltenham, pp. 241–254.

Busch, L. and Lacy, W. (1983) *Science, agriculture and the politics of research*, Colorado.

Busch, L., Lacy, W., Burkhardt, J. and Lacy, L. (1991) *Plants, power and profit*, Oxford.

Cartaña, J. (2000) 'Las estaciones agronómicas y las granjas experimentales como factor de innovación en la agricultura española contemporánea (1875–2000)', *Scripta Nova. Revista Electrónica de Geografía y Ciencias Sociales*, 69, 16, pp. 1–14.

Castell, P. (2002) 'La ganadería porcina en España antes de la Guerra Civil. Una aproximación a la evolución del sector', unpublished paper presented at the *X Congreso de Historia Agraria*, Sitges.

Calatayud, S. (2002) 'Tierras inundadas. El cultivo del arroz en la España contemporánea (1800–1936)', *Revista de Historia Económica*, 30, pp. 39–80.

Cipolla, C.M. (1972/76) *The Fontana economic history of Europe*, London, vols 5–6.

Collins, E.J.T. (1993) 'Why wheat? Choice of food grains in Europe in the nineteenth and twentieth centuries', *Journal of European Economic History*, 22, 1, pp. 7–38.

Corcos, A.F., Monaghan, F.V. and Mendel, G. (1990) 'Mendel's work and its rediscovery: A new perspective', *Critical Reviews in Plant Science*, 9, 3, pp. 197–212.

Dalrymple, D. (1988) 'Changes in wheat varieties and yields in the United States, 1919–1984', *Agricultural History*, 62, 4, pp. 20–35.

Danbom, D. (1986) 'The agricultural experiment station and professionalization: Scientists' goals for agriculture', *Agricultural History*, 60, 2, pp. 246–255.

Debeir, J.C., Deleage, J.-P. and Hemery, D. (1986) *Les servitudes de la puissance. Une histoire de l'énergie*, Paris.

Denaiffe & Colle and Sidorot, (c.1920) *Les blés cultivés*, Paris.

Diamond, J. (1997) *Guns, germs and steel: the fates of human societies*, New York–London.

Dominguez, R. (1996) *La vocación ganadera del norte de España: del modelo tradicional a los desafíos del mercado mundial*, Madrid.

Doussinault, G. (1995) 'Cent ans de sélection du blé en France et en Belgique', in J. Dubois (ed.) (Réseau Biotechnologies Végétales. Journées scientifiques 4es, Namur, 1993), *Quel avenir pour l'amélioration des plantes?* Paris, vol. I, pp. 3–8.

Felius, M. (1985) *Genus bos: Cattle breeds of the world*, Rahway.

Fernàndez Prieto, L. (2001) 'Caminos del cambio tecnológico en las agriculturas españolas contemporáneas', in J. Pujol [et al.] (eds), *El pozo de todos los males. Sobre el atraso de la agricultura española contemporánea*, Barcelona, pp. 95–146.

Friedland, W.H. [et al.] (1991) *Towards a new political economy of agriculture*, Oxford.

Gallego, D. (1986) 'Transformaciones técnicas de la agricultura española en el primer tercio del siglo XX', in R. Garrabou and C. Barciela y J.I. Jiménez Blanco (eds), *Historia agraria de la España contemporánea. 3. El fin de la agricultura tradicional (1900–1960)*. Barcelona, pp. 171–229.

Garcia Bengoa, J. (1923) *Producción de carne de cebo*, Madrid.

Garrabou, R., Pujol, J. and Colome, J. (1991) 'Salaris, ús i explotació de la força de treball agrícola (Catalunya, 1818–1936)', *Recerques*, 24, pp. 23–51.

Garrabou, R. and Sanz Fernández, J. (eds) (1985) *Historia agraria de la España contemporánea. 2. Expansión y crisis (1850–1900)*, Barcelona.

Garrier, G. (1989) *Le Phylloxéra. Une guerre de trente ans, 1870–1900*, Paris.

González de Molina, M. (2001) 'Condicionamientos ambientales del crecimiento agrario español', in J. Pujol [et al.] (eds), *El pozo de todos los males. Sobre el atraso de la agricultura española contemporánea*, Barcelona, pp. 43–94.

Goodman, D. and Redclift, M. (1991) *Refashioning nature, food, ecology & culture*, London–New York.

Goodman, D., Sorj, B. and Wilkinson, J. (1987) *From farming to biotechnology. A theory of agro-industrial development*, Oxford.

Grantham, G. (1984) 'The shifting locus of agricultural innovation in nineteenth-century Europe', *Research in Economic History*, 3, pp. 191–214.

Grigg, D. (1992) *The transformation of agriculture in the West*, Oxford.

Hall, S.J.G. and Clurron-Brick, J. (1989) *Two hundred years of British farm livestock*, London.

Heiser, Ch. (1990) *Seed to civilization. The story of food*, Cambridge (Mss).

Hendricks, B.L. (1995) *International encyclopedia of horse breeds*, Norman.

Hobsbawm, E.J. (1968) *Industry and empire. An economic history of Britain since 1750*, London.

Huffman, W.E. and Evenson, R.E. (1993) *Science for agriculture. A long term perspective*, Iowa.

Institut International d'Agriculture (1933) *Les institutions d'experimentation agricole dans les pays tempérés*, Rome.

Junta Consultiva Agronómica (1892), *La ganadería en España*, Madrid.

Junta Consultiva Agronómica (1920), *Estudio de la ganadería en España,* Madrid.

Kamps, M. (1989) 'Plant Breeding and Seed Production of Agricultural Crops in the Netherlands', *Prophyta*, 6, 8, pp. 4–19.

Kiple, K.F. and Ornelas, K.C. (2000) *The Cambridge world history of food*, Cambridge.

Kloppenburg, J.R. (1988) *First the seed. The political economy of plant biotechnology, 1492–2000*, Cambridge.

Koning, N. (1994) *The failure of agrarian capitalism*, London–New York.

Landes, D.S. (1969) *The unbound Prometheus. Technological change and industrial development in Western Europe from 1750 to the Present*, Cambridge.

Langreo, A. (1995) *Historia de la industria láctea española: una aplicación a Asturias*, Madrid.

Lupton, F.G.H. (1987) *Wheat breeding. Its scientific basis*, London–New York.

Maat, H. (2001) *Science cultivating practice. A history of agricultural science in the Netherlands and its colonies 1863–1986*, Dordrecht–Boston–London.

McNeill, J.R. (2000) *Something new under the sun. An enviromental history of the twentieth-century world*, London–New York.

Maddison, A. (1995) *Monitoring the world economy, 1820–1992*, Paris.

Martin, J. (2000) *The development of modern agriculture. British farming since 1931*, London.

Martínez Alier, J. (1995) *Los principios de la economía ecológica. Textos de P. Geddes, S.A. Podolinsky y F. Soddy*, Madrid.

Martínez Carrión, J.M. (ed.) (2002) *El nivel de vida en la España rural, siglos XVIII–XX*, Alicante.

Mason, I.L. (1996) *A World dictionary of livestock breeds, Types and varieties*, Wallingford.

Mokyr, J. (2002) *The lever of riches: technological creativity and economic progress*, Cambridge.

Nagore, D. (1934) *El trigo y su selección*, Barcelona.

Naredo, J.M. and Valero, A. (eds) (1999) *Desarrollo económico y deterioro ecológico*, Madrid.

O'Brien, P. and Toniolo, G. (1991) 'The poverty of Italy and the backwardness of its agriculture before 1914', in B.M.S. Campbell and M. Overton, M. (eds), *Land, labour and livestock. Historical studies in European agricultural productivity*, Manchester, pp. 385–409.

Offer, A. (1989) *The First World War. An agrarian interpretation*, Oxford.

Olmstead, A.L. and Rhode, P.W. (2003) 'The red queen and the hard reds: Productivity growth in American wheat, 1800–1940', *Cambridge: NBER Working Paper,* No. 8863.

Palladino, P. (1996) 'Science, technology and the economy: plant breeding in Great Britain, 1920–1970', *Economic History Review*, 49, I, pp. 116–136.

Pan-Montojo, J. (1994) *La bodega del mundo. La vid y el vino en España (1800–1936)*, Madrid.

Papadakis, J. (1966) *Climates of the world and their agricultural potentialities*, Buenos Aires.

Percival, J. (1934) *Wheat in Great Britain*, London.

Perkins, J.H. (1997) *Geopolitics and the green revolution: wheat, genes and the Cold War*, Oxford.

Pezzati, M. (1994) 'Industria e agricoltura: i concimi chimici', in *Annales. Fondazione Giangiacomo Feltrinelli*, Milan, pp. 373–401.

Pimentel, D. and Pimentel, M. (eds) (1996) *Food, energy and society*, Colorado.

Pollard, S. (1981) *The peaceful conquest*, Oxford.

Porter, V. (1991) *Cattle: A handbook to the breeds of the world*, London.

Pujol, J. (1998a) 'Las innovaciones biotecnológicas en la agricultura española antes de 1936: el caso del trigo', *Agricultura y Sociedad*, 86, pp. 163–182.

Pujol, J. (1998b) 'La difusión de los abonos minerales y químicos entre 1890 y 1936: el caso español en el contexto europeo', *Historia Agraria*, 15, pp. 143–182.

Pujol, J. (1998c) 'Los límites ecológicos del crecimiento agrario español entre 1850 y 1935: nuevos elementos para un debate', *Revista de Historia Económica*, 16, 3, pp. 645–675.

Pujol, J. (2002a) 'Agricultura y crecimiento económico: las innovaciones biológicas en la cerealicultura europea: 1820–1940', *Revista de Historia Industrial,* 21, pp. 63–88.

Pujol, J. (2002b) 'Especialización ganadera, industrias agroalimentarias y costes de transacción: Cataluña, 1880–1926', *Historia Agraria*, 27, pp. 191–219.

Pujol, J., González de Molina, M., Fernández Prieto, L., Gallego, D. and Garrabou, R. (eds) (2001) *El pozo de todos los males. Sobre el atraso de la agricultura española contemporánea*, Barcelona.

Sala Roca, E. (1948) *El problema mundial del trigo y el problema del trigo en España*, Barcelona.

Santiago Enríquez, C. (1922) *Las vacas suizas y holandesas en España*, Madrid.

Scholliers, P. (ed.) (1989) *Real wages in 19th and 20th centruy Europe. Historical and comparative perspectives,* New York–Oxford–Munich.

Sieferle, R.P. (2001) *The subterranean forest: Energy systems and the industrial revolution*, Cambridge.

Simon, M. (1999) 'Les variétés de blé tendre cultivées en France au cours du vingtième siècle et leurs origines génétiques', in *C.R. Académie Agricole de France*, 85, 8, pp. 5–26.

Simpson, J. (1987) 'La elección técnica en el cultivo triguero y el atraso de la agricultura española a finales del siglo XIX', *Revista de Historia Económica*, 5, 2, pp. 271–299.

Simpson, J. (1997) *La agricultura española (1765–1965): la larga siesta*, Madrid.

Stubbe, H. (1972) *History of genetics: From prehistoric times to the rediscovery of Mendel's laws*, Cambridge (Mss).

Teuteberg, H.-J. (ed.) (1992) *European food history. A research overview*, Leicester.

Tracy, M. (1982) *Agriculture in Western Europe: Challenge and response, 1889–1980*, London.

Tracy, M. (1989) *Goverment and agriculture in Western Europe, 1880–1988*, London.

Vilmorin-Andrieux [et al.] (1880) *Les Meilleurs blés. Description et culture des principales variétés de froments d'hiver et de printemps*, 2 vols, Paris.

Vilmorin, J. and Meunisier, A. (1918) 'Le blé et sa culture en France', *Revue Générale des Sciences pures et appliquées*, 30–dec., pp. 694–706.

Walton, J.R. (1999) 'Varietal innovation and the competitiveness of the British cereals sector, 1760–1930', *Agricultural History Review*, 47, 1, pp. 29–57.

Wrigley, A. (1990) *Continuity, chance and change. The character of the industrial revolution*. Cambridge.

Zanden, J.L. van (1991) 'The first green revolution: the growth of production and productivity in European agriculture, 1870–1914', *Economic History Review*, 44, pp. 215–239.

Zanden, J.L. van (1994) *The transformation of European agriculture in the 19th Century. The case of Netherlands*, Amsterdam.

Zeven, A.C. (1990) *Landraces and improved cultivars of bread wheat and other wheat types grown in the Netherlands up to 1944*, Wageningen.

Annexe

Table 3.1 European Experimental Centres, members of the International Association of Seed Testing, 1931[1]

	Nº		Nº		Nº
Germany	19	United Kingdom	3	Finland	1
Sweden	8	Spain	2	France	1
Italy	5	Latvia	2	Hungary	1
Poland	5	Switzerland	2	Rumania	1
Czechoslovakia	4	Belgium	1	Netherlands	1
Ukraine	4	Bulgaria	1	Denmark	1
Norway	3	Ireland	1	Estonia	1

[1] All seeds, not only wheat.

Source: Instituto Internacional de Agricultura, *Boletín Mensual de Información Técnica*, 1933: 114.

Table 3.2 Institutions of wheat improvement in Europe, 1880–1938[1]

(1)[2]	(2)	(3)	(4)	(5)[3]
Maison Vilmorin-Andrieux (Verrières)	(II)	FR	1815	M.H. and Ph. Vilmorin
Institut de Recherches Agronomiques	(I)	FR	1921	
C. de Recherches Agronomiques (Versailles)	(I)	FR	1923	
Plant Breeding Station at Gembloux	(I)	BEL	1872	
Station de Selection du Boerenbond (Heverlee)	(II)	BEL	1925	A.G. Dumont
Plant Breeding Institute at Wageningen	(I)	NL	1886	L. Broekema
Station de Recherches Agro (Groningen)	(I)	NL	1889	
Plant Breeding Institut (Munich)	(I)	GER	1872	
Plant Breeding Institut (Breslau)	(I)	GER	1872	
Plant Breeding Institut (Halle)	(I)	GER	1863	
Plant Breeding Institut (Hohenheim)	(I)	GER	1905	
Plant Breeding Institut (Magyarovar)	(I)	HUN	1909	
Plant Breeding Station (Vienna)	(I)	AUS	1903	E.Von Tschermak
Plant Breeding Station (Svalöf)	(III)	SWE	1886	N.H. Nilsson-Ehle
Plant Breeding Station (Landskrona)	(III)	SWE	1904	
Plant Breeding Institut (Cambridge)	(I)	UK	1912	R.H. Biffen, F. Engledow
Ins. di Genetica per la Cerealicoltura (Roma)	(I)	ITA	1919	N. Strampelli
Stazione Sperimentale di Granicoltura (Rieti)	(I)	ITA	1907	N. Strampelli
Ins. di Allevamento per la Cereal (Bologna)	(III)	ITA	1920	F. Todaro

(1) Institution
(2) Type of financing: public (I), private (II), and mixed (III)
(3) Country
(4) Breeder

[1] Institutions and breeders most cited in the source
[2] Other institutions were: Plant Breeding Station at Krizevci (SER), Kaiser Wilhelm Institut of Breeding (GER).
[3] Other breeders were: M. Blondeau (FR), C. Benoist (FR), R. Carsten (GER), C. Krafft (GER), F. Vettel (GER), F. Heine (GER), W. Rimpau (GER), F. Strube (GER) and P.J. Hylkema (NL).

Source: Based on data from Lupton, 1987: 53–61/164–168; Zeven, 1990: 17–99; Institut International d'Agriculture, 1933.

Table 3.3 New varieties of wheat between 1880 and 1938

United Kingdom	Western Continental Europe	Italy
	1880–1914	
Sh. Squarehead, Orice Prilific, Ambrose Standup, Starting II, Little Joss	Lamed, Dattel, Bordier, Strubes, Spijk, Rimpau Früth, Wilhelmina, Japhet, Champlan, Duivendaal, Bon Fermier, Fletum, Hatif Inversable, Briquet Jaune, De Massy, Grosse Tête, Grenadier, Montilleul, Krafft's, Cuiras I, II, Emma, Algebra, Juliana, Concurrent, Jacobs, Géant Rouge, Géant Blanc, Cartens V, Travenant, Milion I, Hylkema, Ceres, Robusta, Kronen	Carlota Strampelli, Undici
	1915–1938	
Yeoman I and II, Holdfast, A1, Premier, Wilma, Steadfast, Quota, Redman, Redman, Warden	Prins Hendrik, Blanka, Des Aliées, Addens, Van Hoek, Extra Kolben II, Van Mansholt, Invicta, Skandia II, Carma, Ideal, Vilmorin 23, 27, 29, Wilobo, Bersée, H. 40, Crown, Jubilée, Mendel, Alba, Astra, Staring, Lovink, Strube 56, Elisabeth, Atle	Senatore Capelli, Ardito, Mentana, Villa Glori, Sestini, Damiano, Fandulla

Sources: Based on data from Vilmorin-Andrieux, 1880; Denaiffe & Colle and Sidorot, c.1920; Percival, 1934, 15-90; Lupton, 1987, 53-61, 164-168; Zeven, 1990, 17-99; Bonjean and Angus,2001, 103-192; Maat, 2001, 126-137.

Table 3.4 Pedigrees of wheat hybrids obtained between 1880 and 1938

	(1)	(2)	(3)	(4)
France				
Chiddam epi rouge	Chiddam			
Chiddam epi blanc	Chiddam			
Gros Tête	Browick			Chiddam Epi Blanc
Massy	Shirreff	Noé		
Bordier	Prince Albert	Noé		
Gros Bleu	Shirreff	Noé		
Bon Fermier	Blé Siegle			Gros Bleu
Trésor	Shirreff			Gros Bleu
Dattel	Prince Albert			Chiddam Epi Rouge
Alliés		Noé		Massy, Gross Tête
H. Inversable	Chiddam			Gros Bleu/Siegle (?)
Vilmorin 23		Noé		Alliés, Persel, Grosse Tête
Vilmorin 27				Dattel, Alliés, H. Inversable, B. Fermier
Belgium				
Jubilée			Iron III (SWE), Vilmorin 23 (FR)	
Netherlands				
Spijk	Squarehead			Zealand White
Wilhelmina	Squarehead			Spijk
Emma	Essex			Wilhelmina
Juliana	Essex			Wilhelmina
Hylkema	Squarehead		Shonen (SWE)	Wilhelmina
Staring			Vilmorin 23 (FR)	Juliana
Germany				
Carstens V	Squarehead			Criewener, Carstens III
Model	Squarehead			Landrace
Rimpau Früh	Squarehead		American Red (USA)	

Braun Rimpau				Model, Rimpau Früh
Strube 56 and 210	Squarehead	Noé		
Sweden				
Grenadier	Squarehead			Schonen
Iron II and III				Grenadier, Kotte
Kronen				Iron, Shonnen
Extra Kolben	Essex		Saumur (FR), Wilhelmina (NL)	
United Kingdom				
Little Joss		Ghirka		Squarehead
Yeoman			Red Fife (CAN)[1]	Browick
Steadfast				Little Joss, Victor, Squarehead
Holdfast			White Fife (CAN)	Yeoman
Italy				
Villa Glori			Akagomughi (JAP), Wilhelmina (NL)	Rieti
Ardito			Akagomughi (JAP), Wilhelmina (NL)	Rieti
Damiano			Akagomughi (JAP), Wilhelmina (NL)	Rieti
Mentana			Akagomughi (JAP), Wilhelmina (NL)	Rieti

(1) From UK (Squarehead also include its selections);
(2) From East Europe (Noé was a selection and include other selections of it);
(3) From other countries;
(4) From the same country.

[1] Originally from Danzig (Poland).

Sources: Based on data from Vilmorin-Andrieux, 1880; Denaiffe & Colle and Sidorot, c.1920; Percival, 1934, 15-90; Lupton, 1987, 53-61, 164-168; Zeven, 1990, 17-99; Bonjean and Angus,2001, 103-192; Maat, 2001, 126-137.

Table 3.5 Wheat improvement activities in Spain, 1880–1935

Public experimentation centers[1]	Date of constitution
Granja Experimental del Jardín del Real de Valencia	1885
Granja Escuela Experimental de Valencia	1888
Granja Experimental de Barcelona	1894
Granja Experimental de Zaragoza	1885
Granja Experimental de La Coruña	1896
Granja Escuela Práctica de Agricultura de Palencia	1908
Campos de Demostración y Experiencias de Segovia	1898
Estación Agronómica del Instituto Agrícola de Alfonso XII	1905
Estación de Ensayo de Semillas de La Moncloa	1908
Escuela Práctica de Agricultura de Jerez de la Frontera	1906
Granja Escuela Práctica de Agricultura de Navarra	1908
Granja Agrícola de Pamplona	1908
Granja Experimental. Badajoz	1906
Granja Experimental. Jaén	1906
Granja Agrícola de Palencia	1909
Estación de Agricultura de Zamora	1919
Granja Regional de Castilla la Vieja	1923
Granja Experimental de Zalla	?
Sección Agronómica de Alava	?
Servei de Terra Campa (Cataluña)	1923/1932

[1] The activities of the Sindicato Agrícola de Guissona beginning in 1932 must also be emphasized.

Source: Based on data from J. Cartañà, 2000; Pujol, 2002a: 77.

Table 3.6 Consumption of N, P_2O_5 and K_2O from mineral and chemical fertilizers between 1880–1936 (Kg/ha)

	1911–1913[1]	1931–1937[1]
Netherlands	163.7	299.2
Belgium	68.4	160.9
Germany	49.9	143.9
UK	28.2	60.1
Denmark	17.9	54.8
France	10.7	40.6
Italy	13.3	26.0
Spain	5.8	16.8
Mediterranean Coast		*32.3*
Northeast		*28.8*
Northwest[2]		*12.9*
Center and South		*9.49*

[1] Different years
[2] Without Cantabrian coast and Galicia.

Source: Based on data from Pezzati, 1994: 398; Gallego, 1986: 195; Pujol, 1998b: 143–182.

Table 3.7 Agronomic conditioning factors and technical change

Mineral and Chemical Fertilizers			Grain Harvesters		
Provinces[1]	(1)	(2)	Provinces[2]	(3)	(4)
Areas of Widespred Use					
Valencia	32.3	75.9	Burgos	8.7	35 to 40
Alicante	24.5	28.2	Palencia	5.6	30 to 35
Almería	23.0	27,7	León	15.9	60 to 70
Lérida	29.2	22.0	Huesca	16.6	30 to 35
Zaragoza	21.1	34.2	Teruel	24.1	30 to 40
Castellón	17.3	33.8	Zaragoza	24.6	20 to 25
Tarragona	15.1	39.5	Gerona	26.4	25 to 30
Areas with Little Use					
Jaén	5.5	1.4	Badajoz	23.6	240 to 260
Ciudad Real	3.8	4.8	Toledo	27.7	>1000
Guadalajara	3.0	7.5	Ciudad Real	37.7	800 to 850
Cáceres	1.6	7.2	Málaga	43.7	200 to 240
Badajoz	0.2	7.2	Córdoba	53.1	350 to 370
Córdoba	0.7	7.1	Barcelona	64.3	620 to 640
Cuenca	0.9	7.9	Tarragona	67.9	565 to 580

(1) Relative importance of irrigated surface areas in 1922.
(2) Kg/ha of mineral and chemical fertilizers around 1933.
(3) Relative importance of surface areas of vineyards and olive groves in the total occupied by these crops, the surface areas sown with grains, and the surface areas of artificial pastures, around 1932.
(4) Hectares sown with grains by harvester, around 1932.

[1] Provinces with little precipitation and high temperatures in spring and summer.
[2] Provinces with different climate conditions.

Source: Based on data from Pujol, 1998b: 160–169; Pujol, 1998c: 657–669.

Figure 3.1 Main flows of wheat seeds between 1830 and 1914

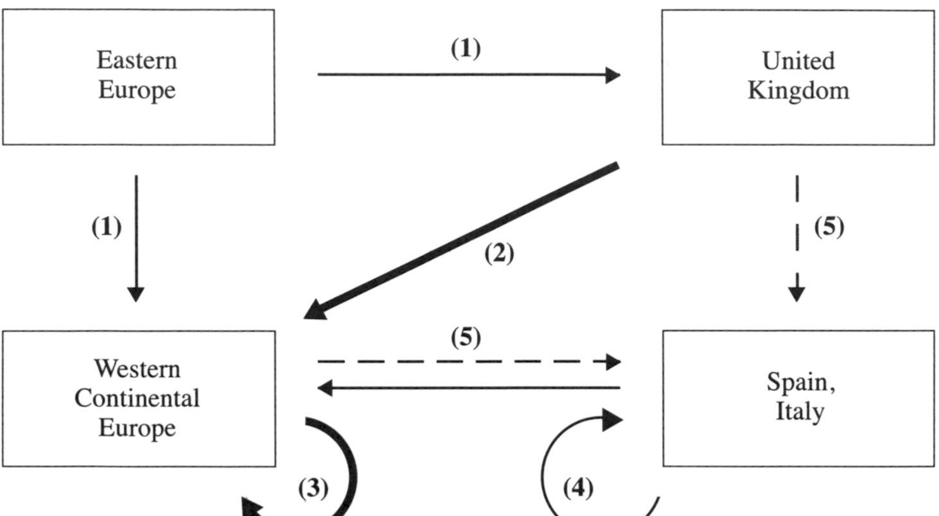

(1) Polish, Odessa, Noé, Chaff Dantzick, Bonte Poolse
(2) Hallet, Hickling, Munsgowell, Goldendrop, Wittington, Hunter, Essex, Chiddam, Prince Albert, Essex, Trump, Spalding, Victoria, Shirreff, Tunsall, Squarehead, Prolific, Standup, Master.
(3) Hatif Rimpau, Perle N. Barbú, Sta. Helène, Gelderse, Japhet, Wilhelmina, Grenadier.
(4) Richelle Bl.Nápoles, Rieti.
(5) For testing.

Sources: Based on data from Vilmorin-Andrieux, 1880; Denaiffe & Sidorot, 1907 (1920); Percival, 1934, 15-90; Lupton, 1987, 53-61, 164-168; Zeven, 1990, 17-99; Bonjean and Angus,2001, 103-192; Maat, 2001, 126-137.

Figure 3.2 Price indexes and salaries in Barcelona (Spain) in constant pesetas, 1865–1935 (1913=100)

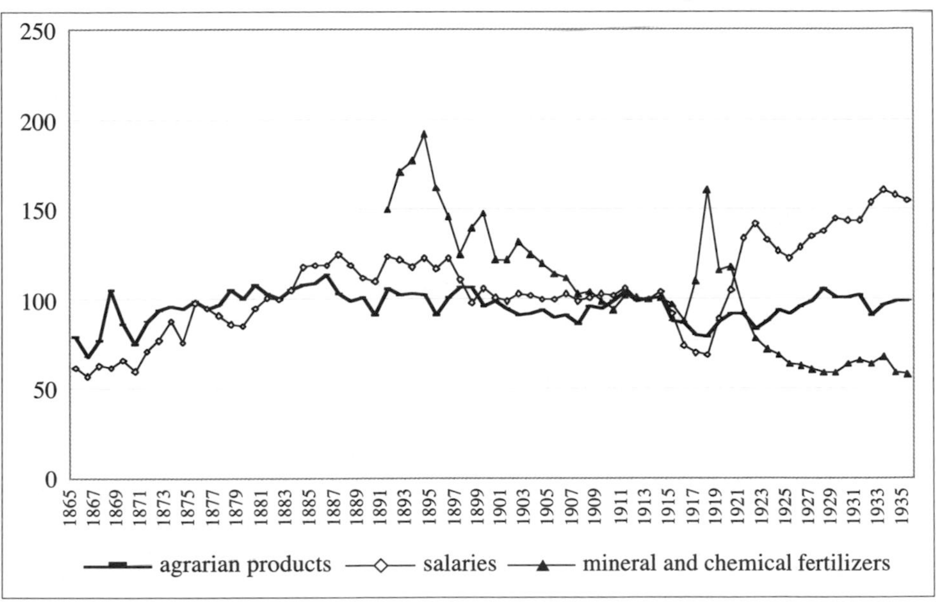

Source: Pujol, 1998c: 180–182; Garrabou, Pujol and Colome, 1991: 23–51.

4 The role of EU policies in technological innovation in agriculture

Konstandinos MATTAS, Aristotle University of Thessaloniki
Efstratios LOIZOU, Technological Educational Institute of Western Macedonia, Florina

I. Introduction

The important role of agriculture is stressed almost everywhere in the literature. According to the World Bank, agriculture is a key sector for economic development, and as shown in history, few countries achieved sustained economic growth without first or simultaneously developing their agricultural sector (World Development Report 1982). Specifically, in most developing countries the agricultural sector is among the most important economic activities that provide income, employment and foreign exchange (Birkhaeuser, Evenson and Feder, 1991). The role of technological innovation in economic or sectoral development is well known. For agriculture, innovations have been since long a major contributor to its progress, and it will continue to influence agricultural inputs, production, processing, distribution and marketing; the innovations contributed to tremendous improvements in the quantity and quality of food and fibre throughout the world (Wagner Weick, 2001).

As reported in Birkhaeuser et al. (1991), rapid technological advances in agriculture have occurred after the Second World War, which induced great changes in agricultural production and highlighted the importance of rapid and efficient diffusion of these advances. On the other hand competitivity is a primary target in contemporary economies and societies, both at micro and macro level for private or public business enterprises or public institutions and regional or national economies. It is well known and generally accepted that competitiveness is a key factor, for maintaining economic development and employment opportunities enhancement. In turn, competitiveness is highly influenced by the ability of an economy, a sector (e.g. agriculture) or a firm to introduce innovative techniques, products or processes in the production process. Technological innovations can result either from a transfer of innovations from outside the country, a region, a sector or a company, or to emerge as the outcome of their own research and technological development.

The agricultural sector, like most sectors in an economy, lies within the above-sketched context. With the assistance of technological innovations in the previous decades, effort was given to the intensification and specialisation of agricultural activities in order to increase productivity, to achieve economies of scale and finally strengthen its competitiveness. This was promoted mainly through the support agricultural policies, specifically the CAP (Common Agricultural Policy – see below), price support and structural measures. Despite the many improvements in the agricultural sector the above context caused significant problems (overproduction, reduction in farm number and jobs, regional disparities, tensions between farming and environment etc), which in turn induced the large reform of 1992 (Van der Ploeg, Renting and Minderhood-Jones, 2000).

Afterwards, a shift of agricultural policies occurred, from the price support scheme to a more integrated agricultural development scheme. New innovative activities are being promoted to help the agricultural sector to become more market oriented, and on the other hand to support alternative innovative activities to the less developed, lagging areas where most agricultural activities are sufficiently competitive. Such innovative activities aiming at integrated development and not development of single sectors, include environmentally friendly initiatives, organic farming (both in crop and livestock production), innovative processing activities to promote traditional local products (PDO, PGI), agro-tourism etc.

The present chapter wishes to present basic issues of technological innovation in relation to EU agricultural policies. More specific, initially some basic types of innovation in the agricultural sector are shown along with the most important sources that induce innovation in agriculture. Next, a brief reference to the issue of technological innovation diffusion and adoption is presented. Finally, EU-agricultural policies and their role in relation to the above-mentioned issues on innovation are recorded, as well as some challenges agriculture will face in the near future are considered.

II. Agriculture and technological innovations

II.1. Basics on innovations theory

According to Joseph Schumpeter (Shumpeter, 1912; Shumpeter, 1942), innovations are one of the three stages in the process of technological change. *Invention* of a new product is the first stage: it is an idea or a model for a new improved product or process. *Innovation* is the transformation of the invention into a product, accomplished after a continual improvement and refinement of the new product or process. In an economic sense, an invention becomes an innovation when the improved product or process is first introduced to the market. Third is the *diffusion,* which is the process of widely spreading of the new product, and the gradual use by other firms or individuals over time.

Technological change is usually induced by public and private actions invested in research and development (R&D). The result of R&D, which leads to innovations, usually called *knowledge capital,* contributes, with other inputs in the production process, to the maximisation of the final outcome of the inventing organisation. As mentioned by Griliches (1979), new innovations sooner or later are adopted and used by competing organisations, which result in positive externalities.

Following the theory of Schumpeter for innovations, Ruttan identifies three major approaches related to the sources of technical change (Ruttan, 1997):

• The induced innovation approach (neo-classical), which suggests that the rate and direction of innovations respond to changes in demand and relative factor prices. This approach stresses the importance of changes in market demand on the supply of knowledge and technology;

• The evolutionary approach, which is inspired by Schumpeterian insights into the process of economic development. Firms that use routines, trivialities in production process, determination of product mix, R&D are aiming at satisfying rather than optimising economic behaviour. Local searches for technical innovations and imitation of other firms' practices are the activities that lead to technical change according to the evolutionary approach;

- The dependence approach; products or systems subject to increasing returns to scale are path-dependent, that is, a sequence of micro level historical events are influencing future possibilities.[1]

Though the term innovation in economic theory seems initially to be well conceivable, it becomes rather complicated if one tries to exactly define a single meaning of innovation. Many definitions can be found in the literature. Using a generally accepted definition, 'an innovation is an idea, practice or object that is perceived as new by an individual or other unit of adoption' (Rogers, 1995). Many definitions in the literature focus on the idea that innovation is a phenomenon that is considered something new. As referred to by Bamberger (1991), using such broad definitions of innovation results, researchers implicitly identify units of analysis that they consider being innovations, and then determine criteria by which their self-identified innovations are defined.

The ambiguity of the definition of the term 'innovation' is also stressed by the Commission of the European Communities (CEC, 1995), following the OECD definition, 'innovation involves the transformation of an idea into a marketable product or service, a new or improved manufacturing or distribution process, or a new method of social service'. The ambiguity arises because, according to the definition, the term *innovation* refers to both the innovation process and the new or improved product or service. Hence, when the dissemination of innovation is examined the above-mentioned question arises: does it mean the dissemination of the process (the method, way that makes the innovation possible) or the dissemination of the result (the new product)?

In the *Oslo Manual* (1997) the OECD distinguishes in more detail innovation in process, product and organisation:

- *Process* innovation occurs when given output (good, service) is produced with less input.
- *Product* innovations require improvement to existing goods or the development of new goods.
- *Organisational* innovations include new forms of management, like total quality management.

The importance of innovations at micro (firm) and macro (country, society) level is noted. Innovations show the firms as a driving force towards ambitious, long-term objectives. Moreover, they lead to the renewal of industrial structures and induce the development of new sectors in the economy. The firms that forward innovations have specific characteristics that can be grouped into two major categories of skills, strategic skills and organisational skills. In the first category, among the basic characteristics are their long-term view, the ability to identify and anticipate the market trends, and the willingness and ability to collect, process and assimilate technological and economic information. The second category includes characteristics such as the undertaking and mastery of risk, the internal co-operation between its operational departments, external co-operation with public research, customers and suppliers, involvement of the whole of the firm in the process of change, and investment in human resources.

---

[1] The three approaches are presented in Loschel (2002), where he surveys the literature on the issue of whether technological change is used as a non-economic exogenous variable or an endogenous one in economic models.

As for technological innovation in the agricultural sector, there are two technologies for the production of an agricultural good: the traditional and the contemporary technology. The basic difference between the two is that technological change is considered as exogenous with the contemporary technology. The latter (exogenous technology) can be the case for the poor, less developed countries, given that they do not contribute to the creation and development of new ideas. On the other hand, this cannot be stated for developed countries (USDA, 2000).

II.2. Types of innovations in agriculture

The importance of technological innovations in agriculture is of growing importance since the end of World War II (Birkhaeuser et al., 1991). Wagner Weick (2001) observes that the role of innovations will become more and more significant, as long as the challenge of satisfying the world's increasing needs for food, feed and fibre is faced (considering the limited natural resources, the concerns for pollution, food safety and the changes in the agricultural and agribusiness sectors),

In recent decades, innovations are much more rapid and multidimensional, i.e. they are not only focused in strengthening competitiveness through productivity reinforcement. R&D investments of private and public (EU funds) organisations are directed to new alternative innovative activities, such as organic farming, food safety technologies, extensive agricultural, environmentally friendly initiatives et cetera. Some of the most important new technological innovations developed, adopted and diffused in agriculture can be divided between the following six broad categories: biotechnology, mechanical and electrical equipment, bio-energy and fossil fuels, environmentally friendly innovations, animal production innovations, and innovations in agro-industries (although many other can be assigned).

A. Innovations in biotechnology and genetics

– New varieties resistant in drought, pesticides and insects are developed through genes transfer from other organisation. Examples are specific varieties of cotton, corn or soya (Pingali and Traxel, 2002).

– New varieties developed using conventional reproduction methods (indirectly related to biotechnology) that give high yield, independently from environmental limitations, most times. This is done by the development of hybrids that can grow in different types of soil, climate and environmental conditions. Many other specialised innovative products and techniques are recently applied, since biotechnology is one of the most innovative, high-tech sectors in agriculture.

B. Mechanical and electrical equipment innovations

This category includes some of the most early and significant innovations in agriculture. Through investments in R&D new effective innovation are developed and applied in agriculture, such as:

– New innovative irrigation systems, more efficient and water sparing.

– Machinery such as tractors with accessories for complicated activities.

– Seeds storage places, greenhouses with energy saving equipment.

– Advanced electronic systems such as geographical information systems, or GIS, telecommunications, computers, internet etc.

C. Innovations in bio-energy or fossil fuels sectors
– Bio-energy systems are used to clean water from salinity and become suitable for irrigation. However, due to the high cost these systems are not widely developed (Chiaramonti, Grimm, El Bassam and Centagorta, 2000).
– Systems producing renewable energy from biomass coming from plants; innovative equipment, which utilises sustainable sources, such as natural gas.

D. Environmentally friendly innovations in production
– More effective and less environmentally harmful pesticides.
– Organic and integrated farming methods.
– Chemical fertilisers highly absorbed by plants, and many other advances that respect the environment and promote sustainable agriculture.

E. Technological innovations in animal production
– Improved veterinary medicines (antibiotics, hormones, vaccines) that augment productivity and output.
– Production of hybrid improved species and advanced technology equipment for livestock activities.

F. Environmental innovations in industries and technological systems in agro-industries
– Such innovations are bio-filters and end-of-pipe filters that control pollution. Reduction of solid and liquid wastes, arising from the production process.
– Recycling methods, reusable materials environmentally friendly. Industrial systems that use more efficiently the production factors, the natural resources etc.

The above-mentioned innovations are some of the most important and recent, although one can find many others.

III. Sources of innovation in agriculture

Well-organised systems support the R&D projects and technological innovations, either public or private. In every country multidisciplinary groups of agents (many of them outside the agricultural sector and often competitive to each other) are involved in the development of agricultural innovations. Whether technological progress and innovations are considered exogenous or endogenous, these important institutions do interfere in the development process. The role of these agents is a crucial one since long.

Possas et al. (1996) noted that the technological regime of modern agriculture involves not only directly related industries, such as chemicals, pesticides, seeds, foods, machinery and mechanical equipment, but also public research and educational institutions, producer organisations, as well as private and public research foundations. In these spheres interfere the EU R&D policies that contribute directly or indirectly, through agricultural and alternative initiatives along with the country members, to the development of technological innovations and to the reinforcement of the sector's productivity and competitiveness.

As mentioned above, innovations in agriculture are not simply a matter of activities concerned with agriculture. Sources of innovations are not singular in agriculture but are rather induced by highly diversified agents. Agriculture cannot be seen as an isolated sector and, thus, cannot be studied as such. This notion is strengthened by Pavitt (1984), in classifying agriculture as a *supplier-dominated* sector. Innovations and technical change in agriculture are almost entirely due to supplier industries, equipment manufacturers and input suppliers (fertilisers, seeds and pesticides) (Possas et al., 1996). This is because such *supplier-dominated* sectors exhibit a very low degree of market concentration, while there is absence of oligopolistic structure. Also, product homogeneity and high level of price competition can be found. And moreover, there are low rates of technical change and very limited innovative activities by its own means, due to insignificant R&D expenditures.

Notwithstanding, agronomic research by institutions is very important, particularly when this research is conducted by public institutions (national ones, but often with the direct and indirect aid of EU) that provide research funds and carry out research activities with innovative results. Hence, someone studying innovations and searching for agents inducing innovations in agriculture should not be limited only to directly related sources. The most important sources of innovation in agriculture, as presented by Possas et al. (1996), are as follows (though many other organisations can provide sources of innovations, especially when particularities for each country are considered).

III.1. Private enterprises and industrial organisations

These organisations are mostly specialised in producing and selling intermediate inputs and machinery equipment to agricultural markets. Their products concern both crop and animal production. More specific, the first category (crop) concerns pesticides, chemicals and fertilisers industries; machinery and other equipment such as tractors, irrigation systems; seeds, vegetables and other industries of such kind. Some of these industries may be found in livestock, veterinary products, animal foodstuffs, hybrid races and equipment for farm construction.

III.2. Public institutions

Public institutions are a very significant source for agricultural innovation. Often, R&D activities are not directly related to profit maximisation, as it is the case for the private institutions, but they are rather focused on the reinforcement of social knowledge. Universities, research institutions and public research enterprises are among those organisations, which are directly involved in or support (finance) indirectly research activities. They are usually conducting basic agricultural research, deal with the development and transfer of technology, and also test and certify the innovative products of private industries. According to Possas et al. (1996), among their basic concerns is to extent the scientific knowledge in agriculture and its related sectors, to improve and develop species in crop and livestock, and to advance new and more efficient innovative agricultural practices.

III.3. Private sources related to agro-industries

These are agricultural products processing industries that are also related with relevant raw materials production; the innovations that they are developing assist mainly the processing stages. Forestry firms, which are making their own plant genetic improvement, provide an example.

III.4. Private non-profit oriented collective organisations

This group consists mainly of co-operatives and associations. Among their operations is the development and providing of new improved seed varieties and various agricultural practices such as new planting methods, chemicals dosage, pest control, animal breeding, irrigation, storage, etc. Although they are not functioning as private, profit-maximising firms (they are non-profit organisations) and do not depend on product sales, they often sell their innovative products or techniques.

III.5. Private sources providing services

Organisations in this group, although being innovative, primarily just disseminate new technology. They are selling services such as support, planning, management and services related to production, crop, storage and animal breeding. The organisations in this group may be divided into two categories according to the type of services they sell: those organisations that provide assistance to agricultural planning, and those that provide specialised services (soil systemisation, insemination, etc).

III.6. Farm production units

These organisations often are the channel through which innovations and new knowledge pass in the learning process. Even if these farmers' organisations are not directly responsible for the development of contemporary technological innovations (as genetic or bio-engineering improvements), they indirectly contribute to the creation of innovations in agriculture via their long experience and knowledge. Their experience and skills assimilated through mainly *learning by doing* along with the feedback with other innovative groups, represent an important factor leading to innovations in agriculture.

III.7. EU policies and organisations

Last but certainly not least, is the EU and its policies and organisations that support and promote R&D and technological innovations, either direct or indirect. Even if most EU organisations may be part to the above-mentioned categories and particularly of the second, it is important to consider the EU as a separate one. R&D activities are since long – actually, from the foundation of EEC – supported by EU. In recent years, due mainly to technological disparities among EU members and the lag of EU compared to the USA and Japan, innovative activities are strongly supported.

EU is both a direct and indirect source of technological creation, adoption and diffusion in agriculture. Through the financing of projects, new knowledge and innovative techniques and applications are developed. Technological knowledge is diffused through the agricultural (and other) policies, either as obligatory or voluntary, to all EU farmers and related organisations. Innovative contributions concern all directions in agriculture: crop and animal production techniques, high-tech evolutions, environmentally friendly farming, supporting and consulting services, etc.

As Possas et al. (1996) mention, the technological structure of agriculture (as any other sector) is very complex; it is very difficult to identify which of the above sources are more important and contribute more to the creation and diffusion of innovations. This process is rather circular and interdependent, and that is why it is very important that all the above-mentioned different groups of organisations are communicating and functioning accurately. Nonetheless, the first two groups (and the EU) can be considered as more important. The first (private ones), which are also called *upstream industries*, and the public ones[2] (to which EU can be added) make up the two principal components that generate technological innovation in agriculture in past and present.

IV. Innovation adoption and diffusion

IV.1. Innovation adoption

The technological innovations process in agriculture is characterised by a top to bottom approach, i.e. researchers develop, intermediaries promote the use, and farmers adopt or reject or accept initially and then reject (Viatte, 2002). Despite the view of linearity in the adoption process, adoption of innovations is a complicated process. Rogers (1995) maintains that the diffusion of new economically superior technologies is not an instantaneous process. It rather follows an S-shaped curve that measures the rate of diffusion of innovation over time. Moreover, according to Rogers, there are five stages in the process of decision and approving a technological innovation: knowledge, persuasion, decision, application and verification. In addition, five groups of users can be specified: innovators, initial adopters, initial majority, late majority and the laggards. Counterparts are the five stages of adoption referred by Leagans and Loomis (1971): information, interest, usage evaluation, testing and finally adoption.

Between the invention and adoption of an innovation a time lag is observed. This lag period, according to Linder et al. (1982), can be separated into three stages:

- The time lag during the invention of the innovation; this is the time period from the availability of the innovation until the information for its existence by the potential adopters of the innovation.
- The time lag during the appraisal period; this is the time period from the information until the experimental use of the innovation.
- The time lag during the testing period; this is the period from testing until the acceptance of the innovation and its wide use.

There are many factors that influence the possible adoption of a technological innovation in agriculture. These factors can be classified according to those that are related with the characteristics of the innovation, and to those that are directly or indirectly related with the characteristics of the farmers. In the first case the characteristics of the innovations are evaluated by the farmers, and influence their decision to adopt or reject the innovation. In the second case, the farmers' characteristics (rich or poor, educated or not, young or old) influence their decision (Altieri, 2002). Those characteristics refer mainly to poor farmers, though the majority of farmers may be interested, even those

[2] A brief discussion on the role of public agricultural research system (the National Agricultural Research System [NARS] of Italy) along with a dynamic empirical model is examined by Esposti, 2000.

with high incomes. Characteristics of innovations that influence their adoption are the possibility of reducing inputs and cost, reduction of risk, expansion of innovations to marginal land and improvement of nutrition, health and environmental problems. Also, the innovation should be economically viable, accessible to local resources, and it must enhance agricultural productivity. Moreover, according to Viatte (2001), such characteristics can also be the simplicity of the innovation and the transparency of its results, the usefulness of the innovation, the satisfaction of an existing need and the low primary investment.

It is crucial that technological innovations in the agricultural sector should be in accordance with existing farming systems. The successful diagnosis of the problems and the ensuring of the solutions alone, do not guaranty the adoption of the innovation. Often, limitations appearing in transferring the technology to the farmers are not due to the inability of the technology to develop innovations, and give solutions to the problems, but to inability to induce farmers to adopt the innovation.

Except from the characteristics of the innovations that influence their adoption, characteristics of the farmers are also very important. Among these are the age of the farmer, his farming experience, his experience with other innovations, the conduct with other farmers that adopt innovations, his willingness to innovate, the education level, etc. (Linder et al., 1996; Ghadim and Pannell, 1999). Accordingly, Pannell (1999) suggests that there are three essential factors for a farmer to adopt an innovative system:

- to understand the innovation,
- to see that a trial test of the innovation is feasible, and
- the innovation must promote the objectives of the farmers, such as the innovative system to be profitable for the farmer; to give solutions to the problem, the time period between the investment and the returns (earnings), and to represent significant social and institutional issues.

In the case of agricultural innovations related to environment issues and organic farming, characteristics such as the availability of information and the education level of the farmers are of specific importance. Non-economic factors are more influential in adopting innovations related to the above issues (Rigby, Young and Burton, 2001).

IV.2. Innovation diffusion

Diffusion is the stage that follows the first introduction of the innovation into the market. Following Rogers's definition, diffusion is the process by which an innovation is communicated through certain channels over time among members of the social system. A basic difference between adopting and diffusing an innovation is that the latter takes place among persons or other units of an area, while the adoption is a matter of a person or a unit (Feder and Slade, 1984; Feder, Just and Zilberman, 1985).

The diffusion process plays a very important role in the success of a technological innovation. It ensures the widespread and use of the laboratory results, and provides the advantages of the innovation to the final users. Contrary to the view that considers the diffusion process as *linear*, Kemp and Rotmans (2001: 27–29) support that the evolution of the diffusion is rather a gradual process, where the nature and the speed of change may be divided into four stages:

- At the phase before the adoption there is limited change but adequate experimentation.

- At the initial adoption phase, the change process progresses and the balance of the system starts shifting.
- At the rapid diffusion innovation phase, many changes for improvements are taking place; moreover at this phase there are many collective processes aiming at teaching/ learning and incorporating the innovation.
- At the last phase of stabilisation, the change process is limited and a new dynamic equilibrium is formed.

Many factors influence the diffusion of an innovation, analogous to those presented above for the adoption. Factors that are related to the nature of the innovation but also agents that are promoting the innovation, influence its successful diffusion. An agent with great significance in the diffusion of innovations is the agricultural extension services. As Birkhaeuser et al. (1991) support, 'agricultural extension services are one of the most common forms of public sector support of knowledge diffusion. Effective extension can bridge the gap between discoveries in the laboratory and in the individual farmer's field'.

Also, agricultural extensions through the diffusion of technological innovations can reinforce productivity, yields and in turn agricultural incomes. Moreover, the role of extensions is twofold, not only to assist the transfer of innovations from labs to farmers but also the reverse. Problems related to agriculture, concerns and valuable information for improving innovations, from farmers to researchers and institutions (see Figure 4.1).[3]

Figure 4.1 Simplified form of the development, transfer and use of technology in agriculture

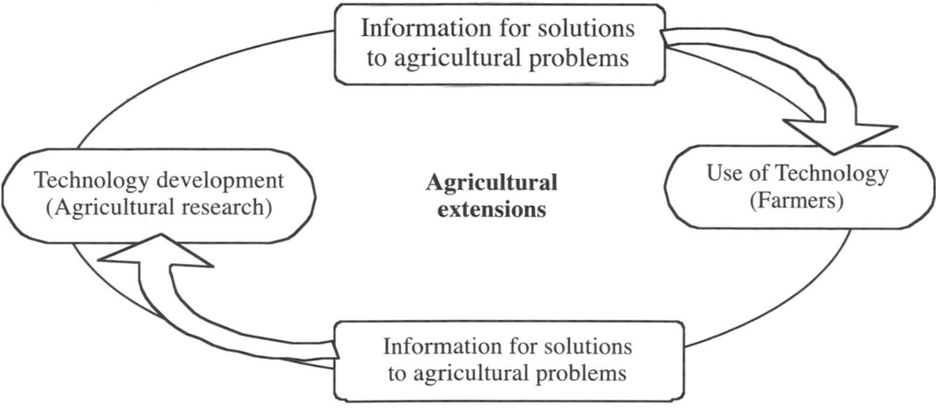

Source: Swanson et al., 1984.

[3] This simplified model of technology transfer was extensively used in agriculture until the beginning of the 1980s, though it is still applied in many developing countries. Since then, more complicated models were developed and are applied, see among others Swanson, Roling and Jiggins, 1984; Blum, 1991; Siardos, 1992.

The so-called *problem areas* (Possas et al., 1996) and production intensification for enhancing agricultural productivity are influencing the way innovations in agriculture develop. Environmental problems arising from agriculture nowadays are characterised as a *new* problem area that guides policies to develop innovative solutions. Identifying these problem areas is strongly needed, and hence the role of agricultural extensions is again very important.

Finally, it can be said that a technological innovation can be characterised as successful if it is diffused sufficiently, i.e. adopted and used by many units above other available or previously existing innovations. Yet, in practice it is very difficult to make clear which is the successful one among competitive innovations, because many other issues are involved.

V. EU agricultural policies and innovations

V.1. Policy role

There are many reasons that induce the regulation of innovations in agriculture by specific public organisations and institutions, either national or EU-based. Increasingly, contemporary socio-economic systems aim at exploiting knowledge intensively (Van Dijk, 1991). On the other hand the knowledge from innovations is often intrinsically characterised by non-separability, unsuitability and uncertainty, which are conditions that restrain competition to allocate sources for innovations efficiently. Also, innovations' success and effectiveness require availability of information.

Consequently, the role of policy decisions and guidance is essential. This role is specially essential in EU-agricultural sector, due to the many particularities of the EU countries: very different physical conditions (climate, soil, specific local crops, natural resources, environmental problems), disparities in technology levels and development, cultural and social differences, etc. These particularities make the role of EU agricultural policies on the issue of innovation, adoption and diffusion very crucial, and indeed a difficult one. Along with national initiatives of innovation, the EU supports and promotes directly and indirectly innovative activities in many ways.

To guarantee success of such policy, it should aim at maximising social welfare through the support of innovations. In achieving this, two policy measures are among the most crucial (Karshenas and Stoneman, 1995): the right on general information level (since the diffusion of an innovation depends on learning and information processes), and subsidies. In the first case, the measures are related to the factors inducing the diffusion of an innovation (easy access to innovations, training for using new technology, effective organisation and management of innovations, etc). In the second case, subsidies can have the form of supporting R&D investments, financing the use of innovative production factors, or products of innovations, etc. In this case the continuing of subsidies is very important.

Regulation of the competition and the market of innovations, property rights and tax motives are some other measures that influence technological innovations. However, there is an alternative view which underlines that the state should not intervene in the market (e.g. with subsidies), even in the case of R&D (Commission of the European Communities, 2001). Intervention with indirect measures (taxes) should be better, specifically in environmental innovations. More diverse views support that any policy manipulations aiming at directing technological innovations have very modest results; because no one can force farmers to adopt for example a new cultivation system. Innovation adoption is a very complex, non- linear and interdependent process, and, moreover, policies are sometimes not suitable and give confusing signals (Pannell, 1999).

Despite the different views about policy measures and the results in adopting an innovation, the role of policy in new technological adoption and diffusion is generally supportive, regulative and directive according to the diffusion stage. One goal is to familiarise with innovations, and guide those that cannot realise initially the present situation and the future benefits (Kemp and Rotmans, 2001). Some specific policies at farm level can be (Phillips, 1985; Lu, 1985):

- To make technology suitable and safe,
- To promote the suitable conditions for transferring the technology from the lab level to the farmer level (especially small and medium farmers),
- To recompense farmers that adopt environmentally friendly technologies,
- To support farmers in cases there are significant changes in agriculture (e.g. organic farming),
- To control farmers in cases of environmental damages.

V.2. EU agricultural policies

As mentioned above, significant technological innovation in agriculture intensified after World War II, with the early appearance of mechanical and chemical means. Since its establishment (then, EEC) with the Treaty of Rome, the EU set up its common policies. The most known among them is still the *Common Agricultural Policy* (CAP). Since then, agricultural policies support innovative activities in the European agricultural sector. This became very clear by the objectives of the CAP as set up by the article 39 of the Treaty of Rome:

- To increase agricultural productivity, through the development of technological progress, rational development of agricultural production and efficient use of production factors, especially labour force,
- To ensure a fair standard of living for the agricultural community, through enhancement of agricultural incomes,
- To stabilise markets,
- To guarantee security supply,
- To ensure that supplies reach consumers with reasonable prices.

Several mechanisms-policies were launched to achieve the above objectives. Through these policies, in order to satisfy the basic objectives, innovative activities were strongly supported either direct or indirect. Especially the first task was very clear, i.e. to increase productivity in the agricultural sector through the development of technology. Until the reform of 1992, the major part of the support to the agricultural sector was connected to production. Hence, the price-support system induced farmers to adopt new innovations (mechanical or chemical) to intensify production, in order to increase productivity, production and, finally, their earnings.

The character of the new technologies was also a factor that induced faster adoption of new technologies. For example, research in genetics led to new, improved varieties, which needed specific production factors (such as fertilisers) to give maximum returns.

The same happened with mechanical innovations: their adoption reduced the employment requirements and, hence, farmers increasingly became innovators in order to reduce further employment requirements. Through all the above, the EU agricultural sector kept a relative high and constant trend in adopting and diffusing new innovations. New improved varieties of plants and animals were introduced, next to fertilisers and pesticides, better veterinary services and healthier animals, modernisation of production with machinery, etc. Until this stage, CAP was primarily successful in implementing its first task, the modernisation and technical progress of the agricultural sector (Roberts, 1985).

However, a totally opposite view exists, advocating that the CAP measures functioned *against* the adoption of innovations. Specifically, innovations are happening without supporting prices. High prices (particularly in cereals) reduce the motivation for investments in new technologies and alternative production systems. It is also stated that most CAP measures in adopting innovations were aiming primarily to increase production rather than reducing inputs. This prevented the farmers to deal with competition and led them to become inefficient, which contributed to the overproduction problem (Tzouvelekas et al., 1999). Another problem was the fact that, after the initial adoption of the innovation, a considerable amount of money was needed to finance the innovation. Small farmers could not afford this, which induced the diffusion of innovations to the large farms (e.g. cattle).

EU supports R&D and innovations through many ways, but the key issue are the five-year Framework Programmes (FP). Until now, five of them were completed and the sixth is in progress. As mentioned by the Commission of the European Communities (2001), despite the fact that the EU has many assets in producing high quality research (long tradition of scientific excellence, a solid fabric of public and private research centres, and academic schools), it is still behind the USA and Japan. EU research suffers from four major weaknesses (CEC, 2001a; CEC, 1995):

- the low share of GDP for research (in 1999 this reached 1,92 percent of GDP in the EU, while in the USA and Japan this was 2.64 and 3.04 percent respectively,
- the comparatively limited capacity of EU to convert technological and scientific achievements into commercial and industrial success,
- research policies in EU are spread into member countries (80 percent of the work is done by national systems),
- EU universities and research centres have to compete with US centres.

Due to such deficiencies, a European Research Area was proposed (the sixth FP is the key for it). This aims at a closer co-ordination of innovation policies in the EU, at more consistent implementation of regional, national and European research programmes, at more efficient use of research facilities, at greater mobility of EU researchers, and at taking into account the relation between science and society.

Innovations were regarded to follow a linear process[4]; i.e. basic research leads to applied research and technology development. In that way public policies were often concentrated on the supply-side, especially in infrastructure with large-scale investments in research centres. Nowadays, this approach is under criticism and especially with regard to less developed regions of the EU. The EU policies that support research and

---

[4] A view that was criticized (Rosenberg, 1988), and the idea that innovations are rather a very complex process is supported.

innovations, and aim at promoting an 'innovation culture', are focusing on the creation of networks to stimulate innovations and maximise their spill-overs.

V.3. New directions and future trends in agricultural policies related to innovations: some conclusions

In recent years, EU agricultural policies related to technological innovations, were redirected following current problems. Technological evolutions in agriculture are influenced by the so-called *problem areas* (Possas et al., 1996). These are problems that agriculture generally faces (for example, in agricultural production such as fertilisers, pesticides, machinery equipment etc). These problem areas have been present in past technological innovations in agriculture[5], but today their role is much more influential.

In the past, attention in agriculture (through CAP) was directed by efforts to intensify production and increase productivity and competitiveness.[6] This was the traditional path of innovations in agriculture. In recent years, global changes in agriculture and generally in economy and society influence to a large degree the orientation of R&D and technological innovations in agriculture. Changes in consumer demands (healthy food, for example), globalisation and competitive international trade, world population increase and enhancement of disparities and needs between developed and less-developed countries, climate changes, environmental problems related to agriculture (turn to environmentally friendly and sustainable agriculture) are some of the changes that have occurred. These influence the direction of agricultural policies (CAP) and, directly and indirectly, technological innovations regime in agriculture.

Most of these evolutions influencing agriculture lead to the creation of a *new problems area*. This involves the needs for more safe food, more healthy and protected environment, conservation of natural resources and quality instead of quantity, issues related to biotechnology, genetically modified products, consist determinants that direct technological innovations in agriculture. The unilateral sectoral (and not integrated) pattern of increasing productivity in agriculture is no longer dominant, though it is still a target specifically in less developed countries. Particularly in mountainous areas where intensified agriculture could not increase productivity, agricultural policies were redirected. Within the new problems framework, agricultural policies now support new alternative integrated directions, such as organic farming, specialisation in traditional products[7], they support innovation in services as agro-tourism (through Leader initiative) that enhance sustainable rural development.

---

[5] Chemical, biological and mechanical is the traditional classification of innovations in agriculture that is used (in neo-classical models of induced innovations) to explain the way innovations in agriculture evolved.

[6] Many studies in the agricultural economic literature are dealing with the examination of the role of technological innovations in the increase of agricultural productivity (see among others, Johnson and Evenson, 1997 and 1999; Schimmelpfenning and Thirtle, 1999).

[7] Such as Protected Designation of Origin (PDO), which are food products that are produced, processed and prepared in a given geographical area, or Protected Geographical Indication (PGI), for these products the geographical issue must occur in at least one of the above-mentioned stages.

The important role of technological innovation in agriculture was shown above. A new challenge is to predict major future evolutions farmers will have to face, and, moreover, the technological innovations that are expected to play a significant role. Wagner Weick performed such an analysis, and some of her results are presented below. The analysis concerns California, but the challenges farmers will face in the near future and the most important technologies expected to dominate agriculture are similar for the European farmers. Such issues are food supply adequacy, food security, land and labour resources, pest management, water availability and management, and generally environmental problems, natural resources exploitation, integrated and sustainable development, etc. All these challenges should direct research and technology in agriculture. New technologies that are expected to induce advancements in agriculture in the future, in order to face many of the above-mentioned problems, are mentioned here (Wagner Weick, 2001):

- global positioning systems (computers related systems), which can help mainly crop production in maintaining efficiency in inputs utilisation and help cost saving,
- Geographical Information Systems, which can help crop production and livestock, as well as research activities related to agriculture; this can help in reducing environmental degradation from chemical pollution, for more efficient use of water resources, and for monitor animals to gather useful information etc.
- Biotechnology advancements, which can be applied in many areas of crop and livestock production (e.g. improvements such as more efficient pesticides, increased genetic diversity, animals with increased resistance to diseases, more efficient vaccines and many other),
- Internet and telecommunications, which can improve marketing processes, enhance education and information related to agricultural activities and assist agricultural research.

These are considered as the technologies with the most important role in the near future for agriculture to confront new challenges. Those technologies concern crop production, animal production, research services, marketing and management and the environment and natural resources. Within this framework of new challenges and new advancements in agriculture the role of policy in controlling, supporting and promoting innovative technologies becomes truly indispensable.

Bibliography

Altieri, M.A. (2002) 'Agroecology: The science of natural resource management for poor farmers in marginal environment', *Agriculture, Ecosystems and Environment*, 93, pp. 1–24.

Bamberger, P. (1991) 'Reinventing innovation theory: Critical issues in the conceptualization, measurement and analysis of technological innovation', *Research in the Sociology of Organization*, 9, pp. 265–294.

Birkhaeuser, D., Evenson, R.E. and Feder, G. (1991) 'The Economic Impact of agricultural Extension', *Economic Development and Cultural Change*, 39, 3, pp. 607–650.

Blum, A. (1991) *The agricultural knowledge transformation cycle*, Jerusalem.

Chiaramonti, D., Grimm, H. P., El Bassam, N. and Centagorta, M. (2000) 'Energy crops and bioenergy for rescuing deserting coastal area by desalination: Feasibility study', *Bioresource Technology*, 72, pp. 131–146.

CEC (Commission of the European Communities) (2001) 'Corporation tax and innovation: Issues at stake and review of EU experiences in the nineties', *Innovation Papers,* No. 19.

CEC (Commission of the European Communities) (2001a) *Research and technological development activities of the European Union 2001.* Annual Report, COM 756, final.

CEC (Commission of the European Communities) (1995) *Green paper on innovation*, Brussels.

Dijk, J.W. van (1991) 'Foresight studies: A new approach in anticipatory policy making in the Netherlands', *Technological Forecasting and Social Change*, 40, pp. 223–234.

Esposti, R. (2002) 'Public agricultural R&D design and technological spill-ins: A dynamic model', *Research Policy*, 31, pp. 693–717.

Feder, G., Just, R.E. and Zilberman, A. (1985) 'Adoption of agricultural innovations in developing countries: A survey', *Economic Development and Cultural Change*, 33, pp. 254–297.

Feder, G. and Slade, R. (1984) 'The acquisition of information and the adoption of technology', *American Journal of Agricultural Economics*, 66, pp. 312–320.

Ghadim, A.K. and Pannell, D.J. (1999) 'A conceptual framework of adoption of an agricultural innovation', *Agricultural Economics*, 21, pp. 145–154.

Griliches, Z. (1979) 'Issues in assessing the contribution of research and development to productivity growth', *Bell Journal of Economics*, 94, pp. 92–116.

Johnson, D.K. and Evenson, R.E. (1997) 'Innovation and invention in Canada', *Economic Systems Research,* 9, pp. 177–192.

Johnson, D.K. and Evenson, R.E. (1999) 'R&D spillovers to agriculture: Measurement and application', *Contemporary Economic Policy*, 17, pp. 432–456.

Karshenas, M. and Stoneman, P. (1995) 'Technology diffusion', in P. Stoneman (ed.), *Handbook of the economics of innovation and technical change*, Oxford, pp. 265–297.

Kemp, R. and Rotmans, J. (2001) 'The measurement of the co-evolution of technical, environmental and social systems', Paper at the International Conference "*Towards Environmental Innovation Systems*", September, 27–29, Garmisch-Partenkirchen.

Leagans, J. P. and Loomis, C.P. (1971) *Behavioral change in agriculture: Concepts and strategies of influencing transition*. Ithaca.

Linder, R.K., Pardey, P.G. and Jarret, F.G. (1982) 'Distance to innovation sorce and time lag to early adoption of trace element fertilizer', *Australian Journal of Agricultural Economics*, 26, pp. 98–113.

Loschel, A. (2002) 'Technological change in economic models of environmental policy: A survey', *Ecological Economics*, 43, pp. 105–126.

Lu, Y.C. (1985) 'Impacts of technology and structural change on agricultural economy, rural communities and the environment', *American Journal of Agricultural Economics*, 67, pp. 1158–1163.

OECD (1997) *OECD Proposed guidelines for collecting and interpreting technological innovation data. Oslo manual*, Paris.

Pannell, D.J. (1999) 'Social and economic challenges in the development of complex farming systems', *Agroforestry Systems*, 45, pp. 393–409.

Pavitt, K. (1984) 'Sectoral patterns of technical change: Towards a taxonomy and theory', *Research Policy*, 13, 6, pp. 343–373.

Phillips, M.J. (1985), Microeconomic impacts of energing technologies', *American Journal of Agricultural Economics*, 67, pp. 1164–1169.

Pingali, P.L. and Traxel, G. (2002) 'Changing locus of agricultural research: Will the poor benefit from biotechnology and privatization trends', *Food Policy*, 27, pp. 1164–1169.

Ploeg, J.D. van der, Renting, H. and Minderhood-Jones, M. (2000) 'The socio-economic impact of rural development: Realities and potentials', *Sociologia Ruralis*, 40, 4, pp. 391–408.

Possas, M.L, Salles-Fino, S. and da Silva, J.M. (1996) 'An evolutionary approach to technological innovation in agriculture: Some preliminary remarks', *Research Policy*, 25, 6, pp. 933–945.

Rigby, D., Young, T. and Burton, M. (2001) 'The development and prospects of organic farming in the UK', *Food Policy*, 26, pp. 599–613.

Roberts, I. (1985) *Agricultural policies in the European Community: Their origins, nature and effects on production and trade*, Canberra.

Rogers, E.M. (1995) *Diffusion of innovations*, New York.

Rosenberg, N. (1988) *Exploring the Black Box*, Cambridge.

Ruttan, V.W. (1997) 'Induced innovation evolutionary theory and path dependence: Sources of technical change', *The Economic Journal* 107, pp. 1520–1529.

Schimmelpfenning, D. and Thirtle, C. (1999) 'The internationalization of agricultural technology: Patents, R&D spillovers and their effects on productivity in the EU and US', *Contemporary Economic Policy*, 17, pp. 457–468.

Shumpeter, J. (1912) *The theory of economic development*, Cambridge.

Shumpeter, J. (1942) *Capitalism, socialism and democracy*, New York.

Siardos, G. (1992) *Agricultural extensions*, Thessaloniki.

Swanson, B.E., Roling, N. and Jiggins, J. (1984) *Extension strategies for technology utilization*, in B.E. Swanson (ed.), *Agricultural extension: A reference manual*, Rome, pp. 89–107.

Tzouvelekas, V., Giannakas, K., Midmore, P. and Mattas, K. (1999) 'Decomposition of olive oil production growth into productivity and size effects: A frontier production function approach', *Cahiers d' Economie et Sociologie Rurales*, 51.

USDA (United States Department of Agriculture) (2000) *Changing climate and changing agriculture*, Washington.

Viatte, G. (2001) *Adoption of technologies for sustainable farming systems: An OECD perspective*, Paris.

Wagner Weick, C. (2001) 'Agribusiness technology in 2010: Directions and challenges', *Technology in Society*, 23, pp. 59–72.

World Bank (1982) *World development report*, Washington DC.

5 Recent innovations in the horticultural production system in the Southeast of Spain

Carmen NAVARRO DEL AGUILA, José LÓPEZ-GÁLVEZ and José SALAZAR MATO,
Universidad de Almería

I. Introduction

An indicator of the maturity of a country, in any given sector, is its capacity to produce adequate technology and manage to suit its needs. The problems posed by any territory are usually quite unique and do not lend themselves to technical solutions imported from other places, but should be resolved by appropriate means within its surroundings. Research in this case should be aimed in the first place at resolving the deficiencies that are typical of this area, with the creation, importation, copy and adaptation of knowledge and techniques. However, this does not usually happen in countries with a low degree of technological development, since their objective tends to be to enforce modernisation directing national technology towards certain internationally famous fields, whilst neglecting others which are closer to the specific realities of the area, its resources or its inhabitants.

The lack of useful inventions or their insufficient dissemination contributes to the decline of agrarian production systems. When we speak about technology, we are referring to an accumulation of techniques, procedures or ways in which economic activity is arranged. The competitiveness and, therefore, the modernisation of the agrarian sector bears relevant innovation contributions, and the technological change in this sector takes place with the addition of new production techniques to those already existing. In the horticultural sector, the introduction of new techniques can improve the efficiency of the use of water and soil, two of the most constraining factors in the Southeast of Spain.

This chapter analyses technical, economic and environmental aspects that propel technological innovations in the fruit and horticultural sector of Southeastern Spain. To this end, the evolution of the greenhouse cultivation system is described, contributing economic data about this activity and evaluating the environmental impacts caused by the cultivation systems employed in the area. Likewise, we analyse technological innovations that aim to reduce environmental impact and to improve production yields in greenhouses by the addition of diverse techniques.

II. Aspects underlying technological innovations

This section describes the need for change in the greenhouse system of the Spanish Southeast. The evolution of the production system and the knowledge about techno-economic and environmental aspects are crucial for the implementation of innovations. Fifty years ago the Spanish Southeastern coast was a desert area. Agricultural practices were rendered impossible or were severely hindered by extremely hostile edaphoclimatic conditions. The intense sun heat, high temperatures, strong winds, lack or irregularity of rainfall and lack of surface water, as well as the poor quality and permeability of the soil, did not allow the expansion of many agricultural practices in this area.

The development of agriculture in these areas became possible when the old limitations turned into new advantages, by means of a series of *appropriate* innovation techniques (Naredo, 1988). These innovations were aimed at mitigating the main constraining factors (soil and water) and taking advantage of the more abundant factors. Thus, the intense sun heat and the lack of frost allow a certain development of the photosynthetic function during the winter, thus avoiding additional energy costs in artificial heating or lighting of the crops. It is true that the parameters mentioned above are far from the best for cultivation. This situation leads to a decrease in growth intensity during the winter months and consequently in fructification rate. The permeability of the soil has generally allowed savings in drainage works. Even the strong winds partly aid ventilation in greenhouses without the need for mechanical means or energy expenditure for this purpose.

The key for this applied technology lies in relating water control (by irrigation), appropriately to soil control (by sand culture), and environmental control (by using the Almería vineyard-type greenhouse). Thus, the system can be defined as irrigation agriculture, developed on sand soil in greenhouses covered with plastic film and without heating.

II.1. Irrigation

In the 1950s, the National Institute of Colonisation (INC) initiated the launch of irrigation changes in Almería, and the area under irrigation still continues to increase. Irrigation systems have undergone a remarkable modernisation since then, and the process is still ongoing at the present time. The success of vegetables grown in this area (tomatoes, peppers, cucumbers, melons and watermelons, mainly) explains the vitality of its agriculture. This is partly due to the developments in plastic technology in general, and in particular to the arrival of dripping irrigation systems.

The flexibility of the irrigation calendars in the traditional systems improved with the construction of regulation reservoirs by the users themselves and this, with the systematisation of irrigation by dripping, has contributed to a remarkable improvement in the efficiency of horticultural production, and particularly in the use of irrigation water. However, the latter with similar water and soil conditions varies greatly from some users to others, which agrees with the obvious (and nonetheless, insufficiently acknowledged) fact that the results of irrigation depend not only on systematisation characteristics, but also on the criteria held by users about water application. Thus, it is logical to estimate that the achievements attained are still well below the potential that could be reached.

The data available on the use of water are only estimated, but they allow us to state that the hydric resources are mainly subterranean. Thus, out of 25 hm^3 of water annually supplied by the irrigation community Sol y Arena (SYA), relatively the most important one in the area, only an average of 6 hm^3 comes from the surface waters of the Benínar reservoir. This fact must be related to the contribution that technological development has made to drillings and pump installations to increase irrigation. In fact, the increasingly intensive exploitation of bore holes drilled in the past by the INC or by the Institute of Agricultural Reform and Development (IRYDA) corresponds to the bore holes ever growing in number and depth which often substitute the old ones. However, it should be observed that this intensification has reached limits that have shown a distinctive degree of over-exploitation of the waters, to the point of endangering the sustained irrigation that had been attained. Thus, there is an evident risk of a vegetative growth in irrigation, which will be incapable of maintaining its rational development.

The difficulties that the State administration encounters to ensure a permanent exploitation of the hydric resources available for irrigation show a tendency towards an excessively liberal exploitation. The tendency to delegate responsibility on the part of the State is already obvious, since it keeps delegating on irrigation communities that are excessively interconnected or have scarce technical assistance. In fact, irrigation communities are merely regarded as water supplying firms by their users, and the simple routine and indolence of the procedures for managing resources discourage active participation of the users in the management. The few members who integrate the management board suffice for this, so it hardly works except by delegation.

Irrigation water is distributed in Almería by communities and other users, which are over several hundreds, just in Campo de Dalías there are over 100, adopting several forms of constitution.[1] A community can have the use of one or various bore holes and surface waters. The supply to users (with some exceptions, such as Sol-Poniente (S-P) and some SYA sectors, is distributed by turns, of which the frequency can exceed one week, manned by dike-keepers (pump workers and time keepers) and charged by the hour. The turn is imposed according to the operative requisites of each community. The water divided in the above mentioned way was applied on foot to crop fields, but in most of them irrigation by dripping has substituted the previous method, becoming the most commonly used one in the area. To this effect, it became necessary to increase the flexibility of irrigation timetables. The user has solved this problem with regulation reservoirs. The presence of these reservoirs has become a characteristic element of many of these agricultural exploitations, but it tends to disappear when supply works on demand with enough pressure.

In order to regulate the discretionary application of irrigation water, there are two basic types of reservoir. One which is manufactured or made of concrete, with vertical armoured walls and a capacity of 300 to 500 m^3/ha, or even higher (according to the way in which the water supply to the farms is organised); or one with inclined slopes which is lined with butyl rubber or some other plastic material. Generally, these are only used to store vast volumes of water.

The economic effort of building reservoirs and the tertiary networks that receive water from them, undertaken by the community members themselves, is worthwhile in view of how their management of the water stored is facilitated. This procedure, starting from an irrigation infrastructure and from a distribution methodology thought out and planned for irrigation by flooding and by furrows, made the adoption of new irrigation techniques by dripping possible: the water is still supplied to the farmer by the system described above, distributed in turns; but this takes place at the tap points, instead of being released directly into the cultivation fields, it is stored in the above-mentioned

[1] The casuistry collected in various notaries' offices about titularity of irrigation and sales regime, which conditions exploitation, is very varied. It should be pointed out that only S-P has the characteristics of a public right corporation appointed to the CHS basin organism, according to articles 74.1 and 199.1 of the Law of Waters and the RD 849/1986, respectively.

reservoirs, which allow its regulation, to be used at a later stage, at the farmer's discretion, after pressing the pressure networks. The distribution of irrigation water can be practised in a discretionary way or on demand, and this is the case of S-P. In this community, the water from the Benínar reservoir passes into the pressure breaking chests and from there into the regulation reservoirs or storage units, where the water from the community wells is also taken in order to be guided via fibrous cement pipes which are systematically ramified to cover the whole area belonging to the community, and to be supplied to the farmers. At the exit of the regulation storage units, the community had meters installed, placed between the main pipes which form the distribution networks, at the intersection and derivation points of which the control and distribution of water is made possible by butterfly faucets or floodgates.

Above the main ramifications and pipes, there are 430 outlets, which are distributed throughout the community area and each has the capacity to feed 3 ha. There is a meter for a maximum use of 20,000 L/h. The outlets for the users' taps are situated in them and the set is located inside a chest. This is where the community ownership ends.

The private irrigation installations of each user start from the outlets, where farmers usually have 75 to 90 mm diameter PVC pipes connected, which join systems with their own meters situated inside the greenhouses. These, together with the elements of a standard irrigation dripping exploitation, have a meter or a volumetric key which regulates pressure and limits use. This meter is supplied by the community and belongs to the farmer.

II.2. Soil

The limitations of the agrarian soil, as it happens in most of the areas with greenhouse agriculture, have been mitigated by the use of artificial soils. This technique has extended to different areas of the Spanish geography, thus we can find sand cultures in the Southeast of the peninsula, sandy shores in the province of Cádiz and 'sorribas' and charcoal in the Canary Islands. The creation of sand soil has been shown as an appropriate way to grow crops in Almería, being considered outstanding because of its fertility and its effect on the efficiency of the use of water, as loss by evaporation decreases.

Cultivation techniques on substrate go further using the soil only as a support on which the containers are placed. The substrates used are quite varied, both in their physical and chemical characteristics and in their volume. Substrate crop systems can be classified in two categories, with lost drainage, and with re-circulated drainage. Practically all the installations in Almería come under the first category. The volume of the drained dissolution (fertilisers plus irrigation water ions) is the function of irrigation management, quality of the water used, substrate characteristics and salinity tolerance of the crop. The most widely used substrates are perlite, rockwool and coconut fibre, other substrates are turf, pine bark, sand, etc.

II.3. Greenhouses

The appearance of plastic materials and their application for greenhouse cover allowed a different conception of the classic structural frame used until then: glass. The expansion of this protection system for crops was slow initially, due to the fact that plastic presented the teething problems that usually occur in any new industrial application. The main problems were the lack of uniformity in the thickness of the film, fast deterioration, scarce mechanical resistance, and excessive transparency for long wave infrared radiation.

The oil crisis in 1973 led to an improvement in agricultural production processes, with the aim of decreasing the amount of energy used per unit produced. The new situation encouraged the petrochemical industry to work on the improvement of plastic for its application to agriculture and cattle raising. With regard to cover materials for greenhouses, all the above mentioned flaws were corrected, and nowadays there is such a wide range in the market that it supplies the appropriate material for each crop, agroclimatic location and sanitary problem.

The new materials together with the use of low cost structural frames facilitated the expansion of this production technique. The craft greenhouse consists of an agricultural construction, with a structural frame made out of bamboo, wood, iron or other materials, covered with a more or less transparent material and with sufficient height, especially if crops of indefinite size are grown in it, to allow people to work comfortably inside it. The greenhouse designed in this way, acts passively against certain negative climate effects. The components of this structure are diverse, such as wood, wire, bamboo cane, iron, concrete, and recycled plastic. The cover material is flexible plastic. The main function of these greenhouses is to act as a cover which maintains a more or less enclosed environment, to generate a partly controlled microclimate where evapotranspiration is reduced, the harmful effects of certain climatic variables, such as rain, hail, wind are limited and the thermal regime is slightly improved increasing or decreasing the temperature. The shape, orientation and cover material determine to a great extent, the microclimate generated inside it.

The typology, as one can see is very varied, in fact it is very hard to find two greenhouses which are the same in terms of size, dimensions, number of porches, orientation and other aspects. If we consider the construction procedure employed we find two types of structure, which can be defined as classic pillar-support and lintel, and tension-structure. The metallic structure models, known as industrial, prefabricated in a workshop and delivered to be assembled on location, belong to the first type and we can highlight two varieties: the Dutch Venlo type with glass cover and the multichapel variety with plastic film cover.

The Venlo greenhouse is designed for cold areas. The multichapel greenhouse is very popular for production in warm climates, with acceptable yields in most crops. There are some with pointed arches or ogives, seeking a greater slope so that anti-dripping films can fulfil their mission better, or with arches which has the disadvantage of accumulating condensation in the upper part of the summit and in this case the shape can be symmetrical or asymmetrical.

The craft greenhouse has had a great development in many countries, and especially in the Mediterranean basin. Its structural frame corresponds to the other two types cited above. A common characteristic of most of them is that the structure is not calculated, the resistance of the materials and the load distribution are not taken into account, thus they are built empirically, by trial and error, improving the characteristics with time. One of the most widely used in Almería is the vineyard type, of which over 30,000 ha exist all over the world. It is given this name because its structural frame is taken from the one used to support vines, a crop that used to be very important in Almería, which consists of 50 x 50 cm wire squares, situated at a height of 2 m, supported by wooden posts in the perimeter and inside the plot. Vine shoots were guided and tied over the wire web, while grapes hang in the shade of the foliage, without the risk of being scratched. The great vineyard construction tradition has contributed to the vast development of this type of greenhouse in Almería. The transformation of the vineyard structure into that of the greenhouse was achieved by placing plastic film over two wire webs.

The classic structural frame of a vineyard greenhouse is closer to a tension structure than the classic structure. Its perimeter is formed by a network of wooden prop supports, on concrete blocks joined by wires and anchored to the ground by tension wires which give stability to the construction and, inside by wooden supports which also stand on concrete blocks. Most of these greenhouses were built with flat roofs. Because of this, the film cover had to be perforated in order to let the rainwater out, which had a negative effect on plants. In the beginning they were low, not reaching over 2 m in height. The introduction of indefinite growth plants led to greenhouses being built with a greater height, which allows for longer props and for a certain volume of air between the plant cover and that of the greenhouse.

Later evolution of this type of green house, with regard to structural materials, has substituted wood with metal, mainly with galvanised tube. This change has increased investment costs remarkably, due to the fact that hardly any of the components are prefabricated, thus, the joints have to be welded in situ, which makes assembling the structure more expensive. The shape of the cover has also changed from flat to pitched roof, both symmetrical and asymmetrical, in its different orientations. One variety of pitched roof, which is interesting because of its low cost and easy construction, is the one known as ridge and valley. In it only the ridge pole is supported by stanchions, while the intermediate points coincide with the two adjacent slopes, it is anchored to the ground by plaited wire, steel cable or iron tree-nail which works as a tension wire and supports a small canal which collects part of the rain water.

The problems confronted by this structure are its lack of water storage and difficulty for automation. The use of double mesh to hold the film cover with stitching, to stop it from tearing by friction, even in pitched roofs, allows rainwater to come in resulting in sanitary problems for the crops. As it is hardly air-tight, there is heat loss by convection, which makes the use of adequately efficient heating systems difficult, whilst on the other hand they put constraints on certain cultivation techniques, such as the integrated control of pests and diseases. The use of thermal screens or of rain channels poses problems in terms of their location. Ventilation systems, with windows situated mainly on the lateral facings are usually rudimentary and manually operated. There is a tendency to place manually operated roof windows, in the various types. Scarce ventilation and lack of air-tightness make this type of greenhouse suitable only for areas with low rainfall and frequent winds.

II.4. Plants

The different kinds of edible vegetables came traditionally in the 70s from local varieties selected by the farmers themselves. Thus, vegetable material was produced which was adapted to local conditions, and, conversely yielded small crops and heterogeneous fruit. However, with intensification growing techniques, crop yields improved as sanitary problems increased.

In recent years, genetic improvement has produced higher yield seeds, which are resistant to certain pathogens. Long-lived varieties have signified a remarkable improvement. The production technique of plantules has also had a noticeable evolution. Traditionally, a farmer would produce his plants in rustic installations called nurseries or hotbeds and when it was time to transplant them, they would be pulled out to be planted bare-rooted in permanent settling grounds. The next step was the use of pressed turf cubes or pots for seeding first and later growing the plant, transplanting it with the substrate in which it had grown.

The arrival of firms dedicated to the production of plantules was a very important technical leap. These incorporated new seeding techniques, pregermination, and introduced automatic equipment for irrigation control, fertilisation, heating and sanitation. Thus, it became possible to produce plants with guaranteed homogeneity, strength and health and with a pre-set delivery date. Moreover, the grafting technique was also made possible.

II.5. Health control

The evolution of health control methods has run parallel to that of the other production techniques. Health problems are minimised by the combination of different control methods, using phytosanitary products, physical means (anti-insect mesh, quilted soil, lagging of water regulating tanks and others), biological means and cultural measures, resulting in healthier crops for consumers and a lower incidence on environmental problems, both from the point of view of the workers and of the natural environment.

Soil or substrate disinfecting has encountered new application methods, such as: the incorporation of phytosanitary products with irrigation water, solarisation and the application of methyl bromide and chloric picrine.[2] The ways in which phytosanitary products are used in crops vary, spraying being predominant, while dusting, nebulisation, application of phytosanitary mixtures to the neck of plants, without pressure, and the incorporation of phytosanitary products through the irrigation system are also used. Integrated control is based on the simultaneous use of biological and chemical means, besides cultural techniques to improve the health of plants. The evolution of this practice has been relevant in the last few years due to commercial and institutional pressure on its development.

II.6. Social-economic aspects

The rapid evolution of this agrarian system has probably been aided by the great importance of its home market. Figure 5.1 shows the evolution of production and exportation in Almería, and in it one can see its great effectiveness and how at present 50 percent of production is destined to home markets.

Next, data on evolution are presented for the seasons 90/91[3], 93/94 and 98/99; investment costs (Table 5.1), cultivation and production alternatives (Table 5.2), and internal yield taxes (Table 5.3). The investment required the setting in motion of the crop system described above, compared with the 90/91 season, was almost 37 percent higher in the 93/94 season and over 200 percent in the 98/99 season. The price of land has increased

[2] This procedure is used less and less due to the environmental problems it causes.

[3] Data for the 1990/91 season come from Naredo and López-Gálvez, 1992.

Figure 5.1 Evolution of production, home market and export in the Almería agrarian sector

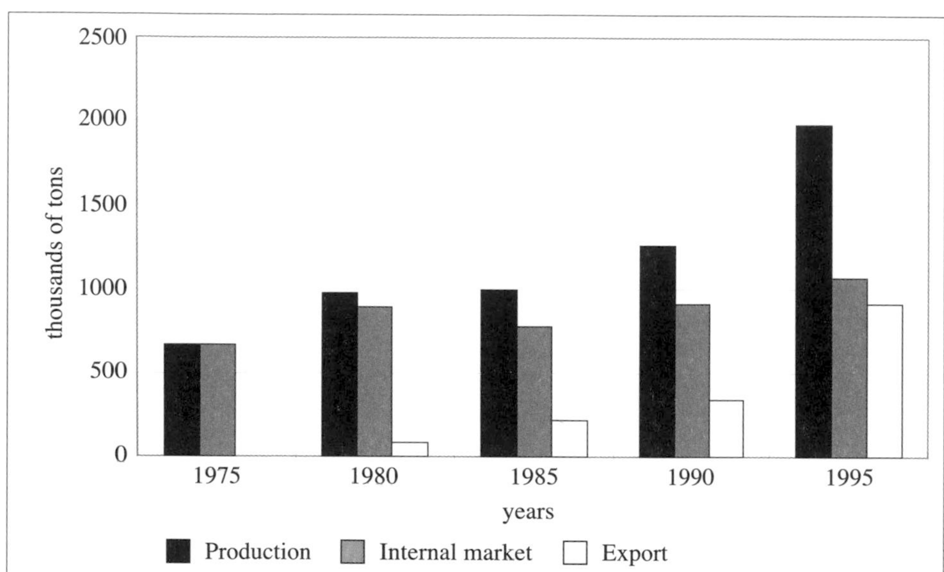

Source: Chamber of Commerce of Almería.

fourfold, which means that whilst it was 25 percent of the total investment, in the first season referred to, in the last one it has gone up to 45 percent. The other concepts have either maintained their proportion or gone down with regard to the 90/91 season.

Table 5.1 Cost and destination of the investment (Euros/m²)

	90/91 season	93/94 season	98/99 season
Price of land	2.70	4.21	10.82
Preparation of the soil and sanding	4.43	3.52	4.21
Greenhouse construction	2.85	4.81	6.01
Installation of irrigation, storage and others	1.92	2.40	3.00
Total	10.91	14.94	24.04

Source: López-Gálvez and Salazar, ongoing research.

Table 5.2 shows the area of values where crop yields, prices and income move and the most common alternatives in the region. The first three rows correspond to alternatives with two-cycle crops in the season and the following two to the one-crop option adopted by some exploitations. Yields have improved in the period analysed: tomatoes by 20 percent, peppers and courgettes by 40 percent, while the rest remain stable. With regard to income, there is an average improvement, slightly over 10 percent.

Table 5.2 Production (kg/m²) prices (Euros/kg) and income (Euros/m²) from the main cultivation alternatives (1990/91, 1993/94 and 1998/99 seasons)

		90/91 season			93/94 season			98/99 season		
	Crops	Production	Prices	Income	Production	Prices	Income	Production	Prices	Income
1	Cucumber	8.5	0.32	2.70	8.5	0.40	3.37	8.5	0.44	3.78
	Green bean	3.5	0.56	1.95	3.5	0.54	1.89	3.5	0.53	1.85
				4,65			5,26			5,63
2	Courgette	3.5	0.56	1.95	6.0	0.42	2.52	6.0	0.43	2.60
	Melon	5.0	0.30	1.50	4.0	0.40	1.61	4.0	0.31	1.25
				3.45			4.13			3.85
3	Pepper	3.5	0.69	2.40	4.4	0.71	3.12	5.0	0.63	3.13
	Water- melon	6.5	0.21	1.36	5.0	0.26	1.32	6.0	0.17	1.00
				3.76			4.44			4.13
4	Aubergine	6.0	0.52	3.14	6.0	0.59	3.53	6.0	0.48	2.88
5	Tomato	7.0	0.45	3.16	7.2	0.42	3.03	8.8	0.42	3.7
Mean				3.64			4.08			4.04

Source: López-Gálvez and Salazar, ongoing research.

Table 5.3 summarises investment costs, ordinary costs and income. This shows how the financial profitability of the system has collapsed, due to an important increase in investment costs as a consequence of the high price of land and the scarce increase in income.

Table 5.3 Evolution of internal yield taxes (Euros/ha)

	90/91 season	93/94 season	98/99 season
Investment	109,083.70	149,351.51	240,404.84
Ordinary costs	17,549.55	20,494.51	25,543.01
Income	36,361.23	42,070.85	40,388.01
Internal Yield Taxes	16.5 per cent	3.3 per cent	2.1 per cent

Note: In the 1990/91 season the investment return period is 15 years, and in the 1998/99 season the investment return period is longer than the useful life of the investment reaching negative values for the actual net value (ANV).

Source: López-Gálvez and Salazar, ongoing research.

The following rows present aspects of the greenhouse system which show production evolution and labour force needs in Almería. The relevance of the labour force in this crop system is due to its great demand. Labour force costs represent over 40 percent of the total expenses in terms of the demands of the different crops. Another aspect to be considered is the difference in productivity between family labour and hired labour. Table 5.4 shows the high productivity attained by family labour (40 kg cucumbers/hour of work) as opposed to that attained by hired labour (17 kg of cucumbers/hour of work). The main conclusion we can draw from this is that profitability is very sensitive to work productivity and that this sensitivity decreases as wages go down.

Table 5.4 Work required by cucumber crops in greenhouses in two exploitations, one with family labour and the other with hired labour

Exploitation	h/m^2	Cucumber kg / m^2	Cucumber kg/h
Farm 1	0.21	7.96	37.5
Farm 2	0.19	8.06	43.4
Mean	0.20	8.01	40.1
Farm 3	0.53	8.98	16.9

Note: concerns 1 and 2 are worked by a family and concern 3 by hired hands.

Source: López-Gálvez et al., 1993.

From Figure 5.2, which shows the evolution of fruit and horticultural production in tons and months, one can see that the need for labour in the July to September period is minimal. During these months cultural work is limited to punctual operations, most of them carried out by service firms, such as soil disinfection, greenhouse repairs, etc. In the months of October and November the need for a labour force gradually increases up to December and January when it reaches its highest point in the Fall-Winter crops. In April, May and June the need for a labour force peaks.

Figure 5.2 Evolution of fruit and horticultural production in Almería, season 1996/97

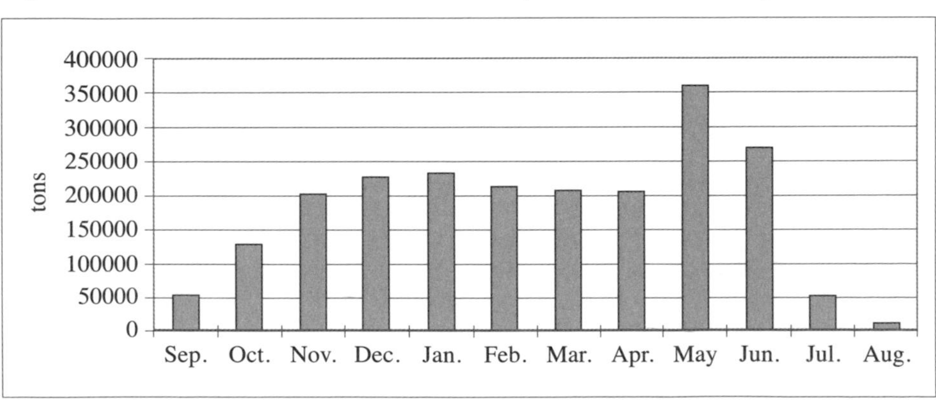

Source: López-Gálvez and Salazar, ongoing research.

Figure 5.3 shows the distribution of working hours in the production of cucumber, grown in the September-January period, and green bean, from February to July, the data come from an experimental study.[4] This study highlights the difficulties in terms of labour force availability, which would arise if the crop preferences of great collectives were to coincide. This situation would mean that if all the greenhouse exploitations in Almería (30,000 ha) adopted this cycle, the amount of daily wages would be 370,000 for the month of May, which would make this crop cycle impossible.

Figure 5.3 Distribution of working hours in the cucumber and green bean season 1987/88

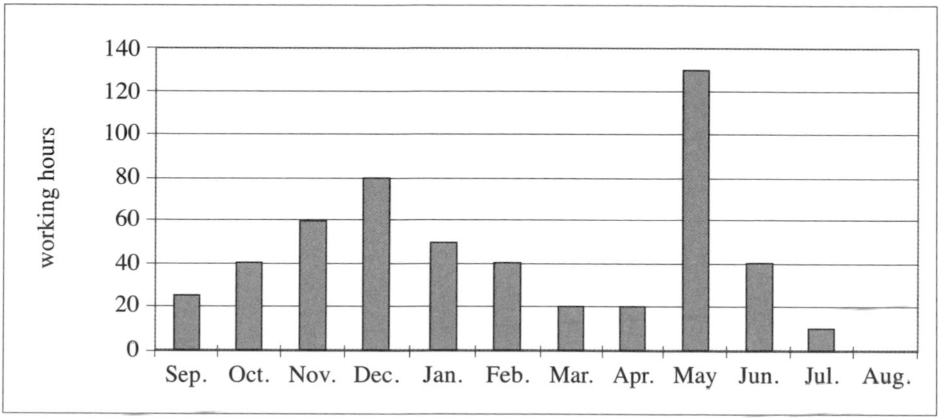

Source: J. López-Gálvez, J. Sánchez-Carreño, J.M. Naredo and N. Castilla (1993) *Technical economic analysis of alternative structures of the flat-roofed Almería-vineyard green-house. Agrarian Research and Vegetable Protection.*

In Figure 5.4 we can see an experiment developed in the north of the Balanegra locality in the municipal jurisdiction of Berja (Almería) during the 1998/99 season.[5] The area under study was one ha, transplantation was carried out on 15/09/98 and the uprooting at the end of May 1999. The tomato variety grown left 8 pieces of fruit per inflorescence for its commercialisation in bunches. The problems that arose are the following:

1) Difficulties in labour force management due to having to dismiss workers for some months only to hire them in the following months. If we bear in mind that the existing labour force is hardly qualified and that these workers are hired in other areas when they are dismissed, so that it will be difficult for them to go back to their previous job, we can understand the decreasing efficiency of these workers, since the manager has to let them go once they have acquired a certain level of skills.

2) With the intention of giving the labour force some stability, which is necessary in the months when they are most in demand, exploitations are increasing productivity in the months with less activity adopting the use of technology. The consequence of these

[4] López-Gálvez, Sánchez-Carreño and Castilla, 1993.

[5] Unpublished data of ongoing research by J. López-Gálvez and J. Salazar.

initiatives have increased the difficulties in labour force availability, rather than solving them, since these become more acute due to the demand created.

Figure 5.4 Estimation of labour force needs in one ha of tomato crop grown in a greenhouse without heating, season 1998/99

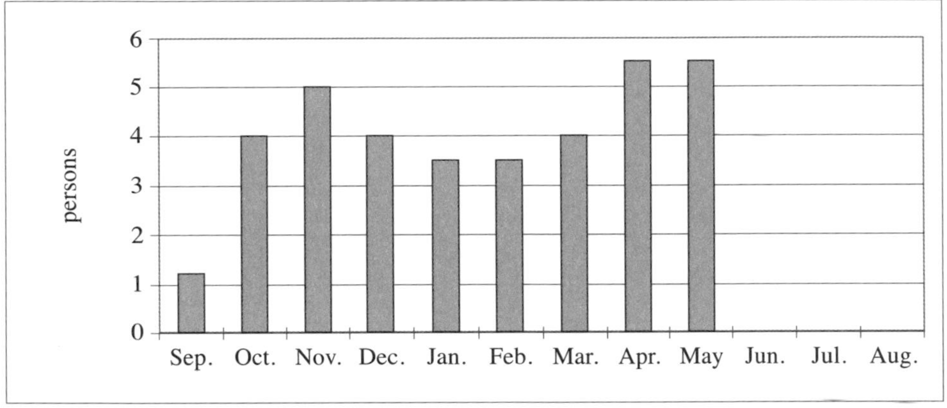

Source: López-Gálvez and Salazar, ongoing research.

III. Environmental aspects of the agrarian system

Traditionally, in most countries, farming activities were not regarded as environmentally harmful. However, the intensification of crop systems and landless livestock rearing, concentrated in certain spots of the territory, have been causing over-exploitation problems and contamination of surface and underground waters. The consequences of this practice, with regard to agriculture, are mineralisation, loss of soil structure, high salinity, water contamination by leaching, increasingly difficult to solve pest and disease problems and depredation of the territory with the resulting habitability problem. This situation usually carries with it a progressive decrease in yields, compromising not only the agricultural future of many areas (significant examples are the South-east of Spain, Morocco, etc.), but also having an impact on fresh water sources, which of old have quenched the thirst of our ancestors, endangering population maintenance in the territory. All this takes place a century after the birth of Liebig's 'minimum law' which tried to identify those elements in the soil which limited the development of plants with their scarcity and to make farmers aware of the advantages of generously replacing the plants that have been taken out. At present, a greater awareness of soil and water deterioration is promoting a concern for the improvement of the efficiency of cultivation systems. The aim of this is to define maximum recommended irrigation and fertilisation limits capable of making high yields and reduced contamination compatible. The European Union nitrate directive is a good example of the above comments.

Crop systems, which use fertirrigation techniques, undergo water and fertiliser losses through leaching or drainage of up to 20 to 50 percent of the total nutrient solution supplied (Schroder, Schwarz and Kuchenbuch, 1995). The resulting contamination, especially in the case of surface waters, will cause the use of closed systems or of those which re-circulate the nutrient solution in 100 percent of the cultivation area to be enforced by law. Modern greenhouse techniques use substrates for plant support, such as rockwool, perlite, coconut fibre, etc. This practice is becoming, generalised with loss of the filtering produced, which causes environmental problems due both to this filtering and to substrate waste. The following data obtained from Belgium and Holland can be seen as an example, where the volume of leached solution is nearly 2.000 m^3/ha (with 20 percent drainage) for 1 ha of tomato crop and 5 t/ha of fertiliser loss (Benoit and Ceustermans, 1995). To the figures given above we need to add the environmental risk derived from the waste mass of 60 m^3 of rockwool slab, 12 m^3 of wool cubes and 0.5 t of plastic per ha per year.

The vast concentration of greenhouses in certain geographical areas, as in the case of some zones in the Almería province, is turning the soil into a constraining factor, which together with the lack of territorial order, has caused its depredation. This situation has led to a truly remarkable concentration of greenhouses, which clashes with the most basic aesthetic and health rules. The fact is that over 30,000 ha said to be dedicated to greenhouse crops in Almería, mostly on sanded soil, generates up to 60 t (ha/year) of vegetable residue (crop remains), plastic material up to 1 t (ha/year) and a leaching volume (water and fertilisers) of 300 m^3 (ha/year), with nearly 0.6 t/ha of fertilisers (López-Gálvez et al., 1993). Likewise, we must bear in mind the rapid growth of substrate cultivation areas in Almería, Alicante and Murcia, which generate 40 m^3 (ha/year) of rockwool, or in those cases where perlite is used, 70 m^3 (ha/year). To these amounts we need to add 2,200 m3 (ha/year) of leached residues, with almost 7.5 t (ha/year) of fertilisers (López-Gálvez et al., 1993). To the previous figures we need to add the plastic used as substrate wrapping. Supposing the totality of the greenhouse area in the Spanish Southeast grew crops on substrate, these figures would have to by multiplied by over 35,000. These data prove that the generalisation of open system crops on substrate is incompatible with the conservation of water resources in the Southeast, an area with scarce rainfall. In Almería, over-exploitation and contamination problems caused by the technology of substrate cultivation increase when compared to the usual system of cultivation on sand, which requires less water and produces less contamination. The only way of reducing the negative impact on the environment of the substrate cultivation system is to re-circulate the leaching, in order to use its fertilising potential and mitigate its contaminating effect, associated to the recycling of substrates.

The difficulties presented by this type of cultivation with regard to solution recirculation and substrates recycling have awakened an interest in techniques such as NFT (Nutrient Film Technique), which besides promoting the re-circulation of leaching, will stop using substrates. Table 5.5 shows the amount of non-vegetable residues generated in an experimental tomato crop grown in Almería in which NFT is used. The residues produced by a tomato crop grown on rockwool substrate are also shown for comparison. In these figures, those referring to greenhouse plastic covers have not been taken into consideration, since it is a residue, which is common to all the systems used. Due to poor management of the NFT system, there was a 20 percent loss of the total water and fertilisers supplied (in Table 5.6 we show the efficiency of the use of water and fertilisers).

Table 5.5 Residues generated yearly by tomato crops grown on the NFT system and on rockwool substrate

Residues	NFT	Rockwool[a]
Leachings		
Water (m^3/ha)	1,085.5	2,307.0
Fertilisers:		
– In U.F.[b] (t/ha)	0.9	4.1
– In commercial product (t/ha)	2.5	9.1
Solids		
Rockwool cubes (m^3/ha)	11.3[c]	6.3[c]
Rockwool slabs (m^3/ha)	–	35.0[d]
Cultivation channel plastic (t/ha)	1.2	–
Substrate wrapping plastic (t/ha)	–	0.25[d]

[a] Source: López-Gálvez et al. (1996).
[b] U.F. = Fertiliser units: sum of pure elements (N, P, K, Ca and Mg).
[c] Density in NFT is of 2 plants/ cube, the cube measuring 10x10x7.5 cm, and on rockwool 1 plant/ cube, the cube measuring 6.5x6.5x7.5 cm.
[d] The useful life considered is of 2 years.

Table 5.6 Efficiency of the use of water and fertilisers, and contamination relations between tomato crops grown on the NFT system and on rockwool substrate

	NFT	Rockwool[a]
Efficiency		
Kg crop per m^3 of water	35.7	22.0
G crop per g of fertiliser:		
– In U.F.[b]	41.8	25.8
– In commercial product	15.8	11.6
Contamination		
L of leached water per crop kg	5.6	12.3
g of fertiliser per crop kg:		
– In F.U.[b]	4.8	21.5
– In commercial product	12.7	48.4
g of fertiliser (residue) per m^3 of leached water:		
– In F.U.[b]	854.1	1755.5
– In commercial product	2264.6	3944.5
L of substrate (rockwool slabs and/or cubes) per crop kg	0.06	0.22
kg plastic (cultivation channels or substrate wrapping) per crop kg	6.2	1.3

[a] Source: López-Gálvez et al. (1996).
[b] U.F. = Fertiliser units: sum of pure elements (N, P, K, Ca and Mg).

IV. Technological innovations

The new production techniques seek to improve crop efficiency both financially and environmentally. Some of the technological innovations described in this section are being developed in research programs implemented in the Spanish Southeast.

IV.1. Nursery techniques

The present nursery techniques provide farmers with healthy plants. The grafting technique is used to minimise the problems caused by the biological contamination of the soils (tomato) and/or to improve productivity (watermelon) (López-Gálvez, 1995). Generally, plants leave the hotbed (nursery) to be transplanted to a permanent settling ground when they are scarcely developed.

Modern nursery techniques try to supply a sufficiently developed plant in order to shorten the period between transplantation and the beginning of crop production. This new approach requires a change in production infrastructure, and also a type of management that allows the obtention of plants with a balanced development between plant and root.

The hotbed where these plants are produced, consists of an industrial greenhouse with automatised windows, shading mesh with a 30 percent light passage and radiating floor heating. There are growing tables where the tomato or pepper seeds are placed within polyexpan trays, remaining there until the first real leaf appears. The irrigation method is by aspersion. When direct sowing is done, the cucumber seeds pre-germinate in a hothouse, except green bean seeds, both of them being placed on rockwool cubes on the nursery floor. The concrete floor is shaped as trays which slope towards the centre of each nave. This allows them to be filled and emptied, and irrigation is done by flooding, with re-circulation of the nutrient solution (Figure 5.6). The slope at the bottom of the trays must be perfectly accurate so that the distribution of the nutrient solution remains even. The aim pursued is to avoid the formation of puddles or dry patches in some areas, and to guarantee the appropriate height of the water film. Maximum solution heights of 2 cm with a minimum of 0.5 cm allow plants to grow without root problems. This situation advises that the slopes on the floor should be around 4 percent. The efficiency of the nursery improves if small fishbone-shaped channels are built in the floor, converging towards the drain (Figure 5.5).

In all the seasons, and especially in summer, the flooding of trays must take place very often, to avoid stress problems, facilitate root development and improve soil temperature. The use of a nebulisation system, situated near the floor, will help to lower the high temperatures of the summer months. The decrease in filling and emptying time is fundamentally conditioned by the volume of water supplied. An automatic system connected to a computer has controlled the filling and emptying of the trays, besides the pH and electric conductivity (EC) of the nutrient solution.

The operating system is direct sowing, with pre-germinated seed or otherwise, and replanting, depending on: species, seed value and adoption or not of the grafting technique. Direct sowing consists of placing the seed in the hole in the middle of the rockwool cube, covering it with vermiculite. When replanting is practised, the sowing takes place in polyexpan trays with turf substrate. When the first real leaf appears, the plantules are sown in rockwool cubes, filling the hole (or holes, since sometimes more than one plant is sown per cube) where the plants have been placed with vermiculite. If the grafting technique is implemented, the stages at which it is done and at which the cube is re-sown are generally more advanced and depend on the species being used.

Figure 5.5 Sketch representation of the nursery fertigation system

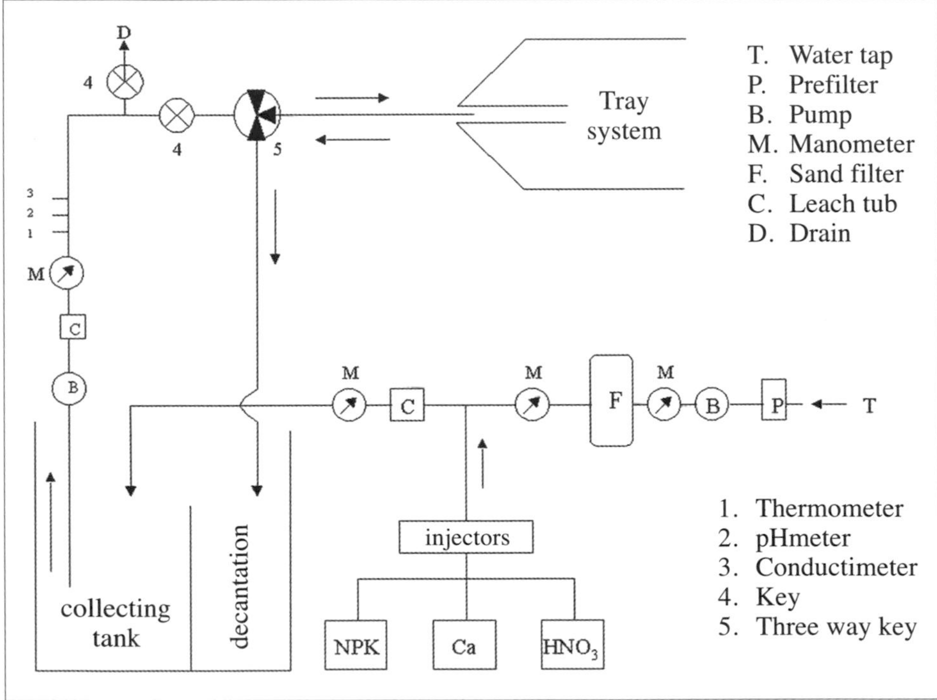

Figure 5.6 Plant of a seeding tray with irrigation by flooding

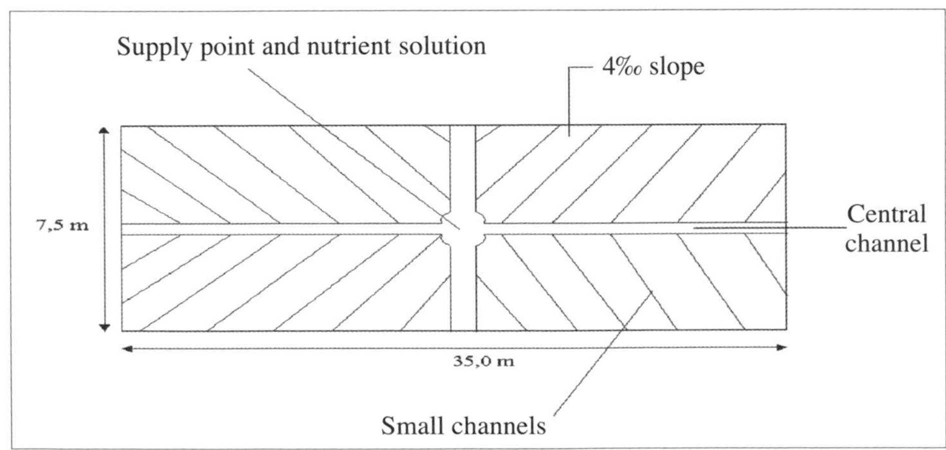

Plant density within the nursery must be adapted at all times to species, time of year (light availability) and degree of development required at the time of transplantation. As an example, the interception of PAR radiation (photosynthetically active radiation) in tomato plants with 6 real leaves, with a density of 25 and 12.5 plants/m^2, in the autumn was 84 percent and 53 percent of the one that incides respectively. Following the objectives of shortening the period between transplantation and beginning of the harvest, these plants should stay in the nursery until they have 12 and 14 real leaves. The previously mentioned data must be borne in mind in order to decide the most suitable density, to avoid competition for light, and consequently, to prevent the annihilation of plantules inside the nursery. Table 5.7 shows data on plant density within the nursery, the relation nursery area/greenhouse, for different crops, in different seasons and crop conditions.

Table 5.7 Plant densities within the nursery and in the greenhouse, relation between nursery surface and greenhouse surface for different crop conditions

Crop	N° plants/ rockwool cube	Dimensions of the cube (cm)	Number of leaves when transplanted	Density in the nursery (plants/m^2)	Density in the greenhouse (plants/m^2)	Relation nursery greenhouse (m^2/m^2)
Tomato	1	10 x 10 x 7.5	12–14	11.1	2.0	1 : 5.0
(Winter)	2	10 x 10 x 7.5	12–14	12.5	2.0	1 : 5.7
Tomato	1	10 x 10 x 7.5	12–14	16.6	3.0	1 : 5.0
(Summer)	2	10 x 10 x 7.5	12–14	22.2	3.0	1 : 6.7
Green bean *(Winter)*	1	7.5 x 7.5 x 7.5	3–4	100.0	1.3	1 : 70.0
Cucumber *(Winter)*	1	7.5 x 7.5 x 7.5	3–4	100.0	1.5	1 : 60.0
Cucumber/ Green bean *(Summer)*	1	7.5 x 7.5 x 7.5	3–4	100.0	2.0	1 : 45.5

Modern nursery techniques, whereby tomato or pepper plants remain in the nursery for a long period of time, until they reach the degree of development required to be transplanted, need to have lateral shoots pruned and flowers nipped at this stage, if plants are to have a balanced development in terms of vegetative and reproductive growth, which will have repercussions on the quality of the plants produced.

IV.2. Environmental improvements in the greenhouse

A greenhouse is conceived to protect crops from negative environmental effects and it generates a microclimate which is different from the outside one. Temperature and air humidity control becomes a problem that needs solving. Its solution is complex, as too many variables are involved. Besides, temperature and air humidity are related. The humidity content of the soil, the species, the phenological state of the crop have an influence on the content of the water steam through the evapotranspiration process. Cultivation techniques condition the microclimate inside the greenhouse, thus high plant density or an increase in the number of shoots per plant, will increase the air humidity content and will modify the temperature. The following paragraphs will attempt to describe techniques to improve the microclimate in greenhouses, bearing in mind that high or low temperature is the factor to be controlled.

Plants require maximum temperatures that must not go over a certain level if production is not to be affected. A reduction in air temperature can be one of the main challenges for greenhouses located in areas with high temperatures. High temperatures can be corrected in the following ways: by ventilation, refrigeration and shade.

Temperatures below a certain threshold not only make the achievement of production objectives impossible, but can also be lethal for them. We should aim to correct this situation so that we can approach the optimum point for each species in greenhouse cultivation conditions. The thermal day-night regime should be appropriate for each species, there are some with low demands, such as lettuce, spinach, strawberry, for which the night temperature threshold is between 5 and 10° C, whilst for the majority of fruit vegetables this threshold is between 16 and 18 ° C.

The improvement of environmental conditions for the development of crops is obtained with the change from the vineyard type structure to industrial greenhouses of the multi-tunnel type, with automatic climate control. The use of heating and thermal reflective shading mesh contribute to a more effective control of the greenhouse microclimate, if compared with the scarce ventilation in the traditional vineyard structures in Almería, where this is presented as the only means of intervention in environmental conditions.

Studies using the new production technology were carried out in industrial multi-tunnel greenhouses, with pitched symmetrical cover, with five naves 7.5 m wide by 40 m long, with a surface of 1,500 m^2 each and north-south longitudinal axe. The height up to the guttering is 4.0 m and to the ridgepole 5.8 m. The cover material is thermal polyethylene, with a 200 µm thickness.

Greenhouses have roof ventilation by means of zip windows located on its eastern side and lateral windows that can be rolled up in the south and north sides. To improve their microclimate they have shading mesh with a 50 percent light passage which is also partially thermal isolating, and a water heating system which circulates through iron pipes placed on supports situated 10 cm from the ground and which work as rails on which carriages run to facilitate work on the crops. The ground is covered with two-layer white-black plastic. A computer controls the opening and closing of windows, the folding and unfolding of shading meshes, the on and off switch of the heating system, according to pre-established parameters for the different crops and the different climatic conditions.

IV.3. Soilless crops: the nutrients film technique (NFT)

The cultivation of plants on a re-circulating nutrient solution film was first developed for research. In its origins, NFT is a return to the principle of actual hydroponics, in which plants are grown solely in a fluid medium. The principle of this hydroponic system consists of the constant circulation of a very thin nutrient solution film, which passes through the roots of the crop. It takes place in channels, without loss or leaking of the nutrient solution to the outside, which constitutes a closed system. The high relation surface/depth of the solution guarantees good aeration and the fine film reduces the volume of roots totally immersed in the solution. Likewise, the re-circulation of the solution increases aeration and ensures that the roots receive an appropriate supply of oxygen and nutrients (Cooper, 1979).

The flexibility of the NFT system has made possible its adaptation to a wide range of crops. These are always characterised by high yield and quality production, but improvements have been made over the years. The easy control of the root environment with heating practices, variations in electric conductivity and programmed flows of the nutrient solution allows an effective crop control. The relatively small amounts of water and nutrients employed have made this a highly desirable technique in areas with scarce water resources. A reduction in the use of materials and the high automation level attained, allow a high economic yield and a rapid investment return.

The NFT system examined in this study (Figure 5.7) is formed by the following parts:

a) a series of flexible polyethylene white-black parallel channels, where the plants are grown. The channels have a 2 percent slope, for the nutrient solution to flow constantly in the system at 3 L/min with a 5 mm water film height;

b) a collecting tank (aljibe), to store the solution. The tank has two impulsion pumps to carry the nutrient solution up to the distribution network which is located in the cultivation channels at the highest level;

c) a collecting pipe which gathers the nutrient solution and carries it back to the tank.

d) an automation system connected to a computer, which controls the temperature of the solution, the pH and the electric conductivity (EC).

The objective of the collecting tank is to act as a thermal regulator for the nutrient solution and, also, to improve the oxygenation conditions of the solution due to the leap that occurs on its way back. The introduction of the rockwool cube in the system is a consequence of the fact that when scarcely developed plants are transplanted, the roots cannot reach the solution circulating at the base of the channel. Likewise, when the plants are well developed, the roots are not attached firmly enough in the NFT system. These reasons led to the development of an adaptation, which consists of growing plants in rockwool cubes to be placed in the cultivation channels. None the less, the use of rockwool cubes violates the basic NFT principle introducing a solid rooting material in the system. Therefore, the cube must be as small as possible.

The solutions used must be adapted to the different crops, phenological phase, time of year. Table 5.8 shows examples of nutrient solutions used for tomato, green bean and long-type cucumber in Almería, where the pH varied between 5.5 and 6.3.

In high temperature periods, the nutrient solution poses oxygen deficiency problems, due to higher respiratory demands. When the solution temperature is 18°C, the oxygen dissolved, at sea level, is 9.5 ppm: at 25°C it is 8.4 ppm; at 30°C it is 7.7 ppm; at 35°C it is 7.1 ppm and at 40°C it is 6.5 ppm. High temperatures also stimulate the respiration of plants. The respiratory demand for oxygen of the roots doubles every time the temperature rises by 10°C, up to 30°C. During August 1998, in the trials carried out, the temperatures of the re-circulating solution were measured and they varied throughout the day between 29°C and 36°C.

Figure 5.7 The NFT system

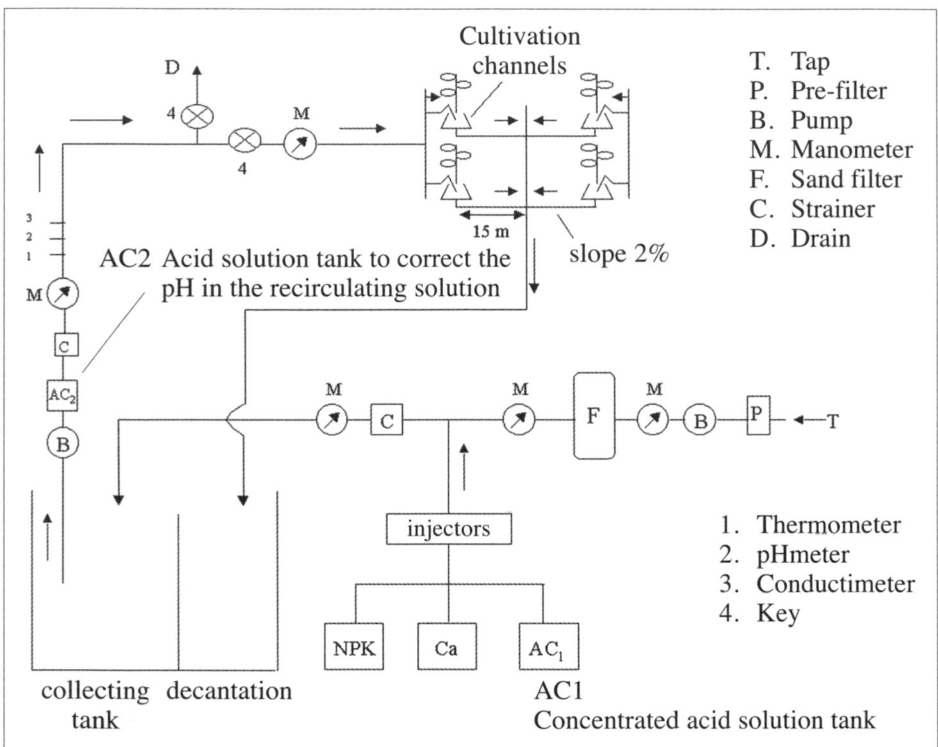

Table 5.8 Composition and characteristics of the nutrient solutions used for the different crops grown on NFT

	Nutrient (mmol/ l)							Relation	CE
Crop	NO_3^-	NH_4^+	$H_2PO_4^-$	K^+	Ca^{+2}	Mg^{+2}	SO_4^{-2}	N/ K	(dS/m)
Tomato (spring/ initial solution[a])	12.75	0.50	1.50	9.50	5.63	2.06	4.25	1/2.00	2.5
Tomato (spring/ fructification[a])	13.75	0.50	1.50	12.75	6.08	2.23	6.00	1/2.50	3.0
Green bean (winter)	15.00	0.50	1.25	8.00	4.50	2.00	1.10	1/1.44	2.0
Cucumber (winter)	15.50	1.25	1.25	8.00	4.45	1.75	1.38	1/1.33	2.0
Cucumber (spring and summer-fall)	13.75	0.50	1.50	8.50	4.25	2.00	1.75	1/1.70	2.1

[a] The initial nutrient solution is used in the first two weeks after transplantation. From the third week the fructification nutrient solution is used.

IV.4. Crop management

A technological innovation such as the one represented by greenhouse crops with climatic control and a structural frame different from those traditionally used in the Southeast of Spain, together with the adoption of the NFT technique, require the adaptation of these practices to the plants, which will ultimately determine the efficiency of the system.

Transplantation must be carried out with special care to avoid plantule losses, since their replacement means a lack of uniformity in the crop, and additional expenses with regard to seed, labour and others. Care is of paramount importance to ensure strong crops and high yields, especially when fairly well developed plants are taken to their permanent settling ground, as in the case of tomato crops. None the less, even for cucumber and green bean crops, which are transplanted when their development is still at a very early stage, great care should be taken. Plants must be strong, with a thick stem and not weakened. They must not be allowed to bend down and touch the channel plastic after transplantation, in order to avoid the appearance of burns on the leaves, therefore, transplantation and totally vertical propping with tense string must take place simultaneously. Likewise, plants should come from the nursery with well-developed roots, which gives them a greater capability to absorb the nutrient solution. This is fundamental at the post-transplantation phase if adequate physiological functions and buoyant plants are to be maintained.

Previous to transplantation, the plastic channels must be taut and situated in the centre of the slope, unevenness and creases at their base must be eliminated, so that the nutrient solution can reach the roots of all the plants. At the time of transplantation, the cubes must be placed exactly in the centre of the cultivation channel. This measure also contributes to avoid creases and maintain an even distribution within the channel. This action should be performed at the end of the afternoon and the nutrient solution should be supplied as soon as it is finished, especially in hot-weather seasons.

In the plantation management the arrangement (frame) which will be the function of the crop cycle and the degree of mechanisation. Crops grown in the NFT system should optimise: a) radiation intercepted by crop (fall-winter-spring-summer cycles) by adapting the frame, initial density and the number of shoots per plant (seeking grater production and defense from high temperatures); b) the correct balance between vegetative and reproductive growth, by means of an appropriate planting density, flower nipping and pruning; c) the duration of the crop cycle, which can be very long, so that one may have to introduce stem letting down techniques (tomato and cucumber).

For all the year round tomato and cucumber crops, and for green bean spring and summer crops, the arrangement of the channels is in double-paired lines. Figure 5.8 shows how these are situated in the cultivation ground. However, green bean crops with low planting densities present excellent results in fall-winter, when the arrangement is in single equidistant lines, situated at 1.88 m from each other, due to a better distribution of light within the canopy. Maximum fruit vegetable production is attained when there is an appropriate balance between demand and the contribution of assimilates and an optimal proportion of vegetative growth during the cycle to sustain the photosynthetic capacity of the crop.

In principle, it is possible to maintain the size of fruit within the range favoured by the market throughout the crop cycle. The number of fruit per shoot should vary in order to coincide with the changes in the assimilates contribution which brings about changes in

Figure 5.8. Arrangement of the double-paired lines in the cultivation channels

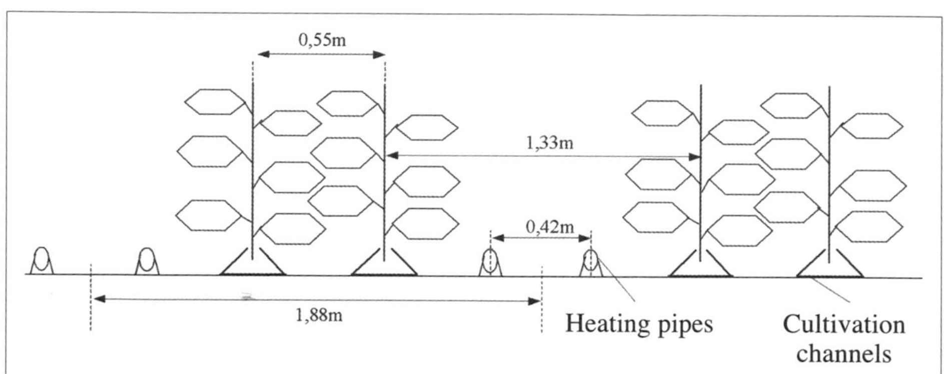

the incident solar radiation. This can be achieved by an increase in the density of the shoots, increasing number of the fruit per area unit at the times when solar radiation is high, or by limiting the number of fruit per plant when radiation is low. An adaptation of the number of shoots to the radiation conditions will increase the shading and transpiration of the plant canopy, decreasing indoor temperature in the greenhouse and facilitating an improvement in the crop conditions in hot periods, resulting in an impact on crop quantity and quality.

Thus, the addition of lateral shoots within a range of 2.4 to 3.5 shoots/m^2, in a spring cycle tomato crop, increases early production and gives fruit a uniform size, not affecting its average weight or total production. However, comparing these with plants grown in high density from the beginning of the cycle (3.0 plants/m^2) the addition of lateral shoots means a 20 percent reduction in total yield, but a 21 percent saving in the number of plants required for the same crop area in NFT (Marins-Peil and López-Gálvez 1999).

Short cycle cucumber plants (3–4 months), can be guided by one shoot throughout the whole cycle, buds being nipped when they reach a 2.8 m height, or it can grow as one shoot and then be nipped at leaf number 20 (around 1.8 m high) and from then on two shoots are left which are nipped when they reach 2.8 m. This second option is used in low density planting conditions, with stronger plants in spring, summer and autumn cycles in the northern hemisphere (high solar radiation). For long cycle one year crops, planting is done with a density of 1.8 plants/m^2, being nipped at the sixteenth leaf, leaving 2 shoots per plant, with which the density reached will be 3.6 shoots/m^2 (Van Os/IMAG-DLO; Jansen/Research Station for Floriculture and Glasshouse Vegetables; ongoing research in Holland).

It is necessary to regulate the number of fruit per plant, that is to say, to regulate competition for assimilates, by nipping the inflorescence (in the case of tomato crops) or the flowers (in the case of cucumber crops) according to the time of year (radiation and temperature conditions) and the degree of development of the plants in order to obtain uniform fruit and of the appropriate calibre for commercial demands. This task has to be carried out in the blossom phase, before fruit begins to take shape, thus avoiding allotting assimilates to the fruit that is going to be discarded.

Trials carried out in Daniela, Atletico and Habana vine tomato cultivars, indicate that during the winter period the number of flowers has to be limited to four per inflorescence; during the spring, summer and fall periods, the number of six flowers per inflorescence is the most appropriate (López-Gálvez et al., ongoing research in Almería).

For the long type, short cycle cucumber crop fruit is left between the sixth and the tenth leaf, depending on the length between knots, in a relation of one fruit for each leaf, removing the rest of the flowers which appear on the knot, thus a correct balance between vegetative and reproductive growth is guaranteed. In situations with scarce root and/or vegetative development (low solar radiation and temperatures which are too high or too low), a relation of two pieces of fruit for every three leaves is recommended, to improve the formation and quality of the fruit harvested. In long crop cycles, one fruit is left for every two leaves through the year.

With regard to pruning, lateral shoots have to be removed systematically before their degree of development compromises the distribution of assimilates. This also facilitates the manual completion of this task with which the transmission of diseases, such as botrytis, can be avoided. This dwells in the dead vegetable tissue remaining in the shoot, due to poor pruning. This problem is particularly important in long cycle tomato crops, with let down stems, but it also occurs fairly frequently in cucumber and green bean crops with shorter cycles. In the same way, leaf nipping has to be practised in tomato and cucumber crops, targeting those leaves situated under the last inflorescence or harvested fruit. In long cycle tomato and cucumber crops all the leaves situated under the bunch or harvested fruit need to be removed systematically, in order to facilitate better ventilation at the plant base, which leads to a better phytosanitary control. In the case of the short cycle cucumber, the intensity of the stripping of leaves is lower and the leaves removed are those which are completely shaded by those next to them and are, therefore, starting to become yellow.

Indeterminate growth tomato productive cycles usually extend over a long period of over 8 months in NFT crops. This situation requires that shoots are let down about once a week. The shoots surround the double crop line, reaching a length of around 15 m. In the same way, in long cycle cucumber crop, this practice is carried out when the shots reach a height of 2.5 m, so that the developing flowers and fruit receive enough light. In this case, plants can reach a length of over 22 m, thus requiring new rootings of the stem every 6 to 12 m. This technique, which for cucumber is implemented in rockwool, has as yet no record of having been developed for this crop in NFT.

IV.5. Yields obtained with the new production technology

Table 5.9 reflects a summary of the main results obtained until now with tomato, green bean and long type cucumber crops grown on NFT, in the Almería conditions. We need to emphasise the resolution of the problems caused by the high temperatures of the nutrient solution in the summer months and the lack of oxygen arising from this situation. During August 1998 temperatures over 35°C were measured in re-circulating solutions, which remarkably affected the behaviour of the cucumber crop, with a decrease in commercial crop yield, reflecting worse yields of the use of water and fertiliser in this cycle than in spring.

Another important aspect to consider is the use of water with reference to the production of dry matter in the crop with the new production technology. The spring cycle biomass yield of the long type cucumber crop in NFT, 4.06 g of dry matter per liter of water supplied, is higher than that of the same crop grown on sand in Almería, which is of 2.82 g (Díaz-Alvarez, López-Gálvez and Tallón-Calíz, 1997).

V. Final considerations

The physical geography of the region conditions the availability and use of its water resources. It poses difficulties for the exploitation of irrigation, but also has positive aspects that have contributed to its development. The result is a high technology and productivity intensive agriculture, about which there is no sustainable criterion knowledge with regard to available resources.

The technical media at the farmer's disposal offer sophisticated irrigation equipment with the capacity for an efficient use of water resources, and the irrigation and production yields are very high. However, the relatively easy access to these water resources, with potent drills and pumping equipment, capable of extracting them from the Earth is faced with a lack of means or policies to protect them effectively. Up until now, there is only one legislation destined to give permanence to irrigation, which does not provide the expected results because the ways to enforce it have not been found: the Hydrographic Confederation of the South (CHS) behaves as if it did not have the means that correspond to the scientific and technological development at its disposal for the control of the evolution of water resources.

Table 5.9 Main results obtained with tomato, green bean and cucumber crops grown on NFT

Crop[a]	Duration of the cycle[b]	Commercial yield (kg/m²)	Use of water yield (kg/m²)	Use of nutrients yield (g/g)[d]
Tomato (spring)	9/3–2/7/1998	19.4	35.7	41.8
Green bean (fall-winter)	27/10/1998–8/3/1999	6.0	12.1	15.0
Cucumber (winter)	4/12/1997–16/3/1998	12.9	32.5	45.4
Cucumber (spring)	2/4–24/6/1998	16.8	54.0	63.7
Cucumber (summer-fall)	13/8–15/10/1998	8.9	31.6	44.0

[a] These data are the mean of the results obtained with the following crops: *Tomato*: Radja; *Green bean*: Mantra, Donna, Pursan, Helda y Festival; *Cucumber (winter)*: Majestic, Miura not grafted and grafted on courgette and on hybrid 6001, Palmera not grafted and grafted on pumpkin; *Cucumber (spring)*: Bellísima y Atlanta; *Cucumber (summer-fall)*: Atlanta.
[b] From transplant to last harvest.
[c] Crop kg per m^3 of water supplied.
[d] Crop per fertiliser g unit (sum of the pure elements N, P, K, Ca and Mg).

The outcome is that farmers respond to the productivity of their businesses with new drillings and with the substitution of their salinised explorations, encouraged by the economic results of an enterprise that ignores its effects on third parties. However, someone will have to pay for these 'externalities', probably the present irrigation users or their descendants. This is not the heritage they received from previous farmers.

If the above mentioned tendencies continue, the future is disturbing. Concerns remain about the irregularity of procedures whereby users grant each other concessions of underground resources and manage them themselves, without public administration exercising its responsibilities.

The change from a family-run agrarian system to an entrepreneurial one calls for innovations to improve exploitation management. Greenhouse crops, even those which are very mechanised, require a fairly large, somewhat specialised labour force. This becomes one of the factors that need good management, reducing casual labour force to a minimum, especially if their availability is scarce, as it happens at present in the Spanish Southeast. The introduction of heating systems seeks to increase the productive intensity of the crops, and consequently, to stabilise labour.

The NFT technique poses technical and diffusion problems that need to be resolved with regard to innovation. The greatest technical problem is the high temperature of the solution during the summer months. The transference of the new production technique can encounter scarce commercial motivation which will hinder its dissemination. However, as it is an innovation which generates less residue than the ones currently in use, the institutional framework should encourage its application.

Bibliography

Benoit, C. and Ceustermans, N. (1995) 'Growing cucumber on ecologically sound substrates', *Acta Hortic*, 396, pp. 55–66.

Cooper, A.J. (1979) *The ABC of NFT*, London.

Díaz-Alvarez, J.R., López-Gálvez, J. and Tallon-Caliz, M. (1997) 'Economy of physical resources: evaluation of the importance of irrigation water management in a greenhouse crop', *Horticulture Acts*, 17, pp. 412–418.

López-Gálvez, J. (1995) 'Technical influence on crops', in J.R. Díaz-Alvarez (ed.), *Post-crop and commercialization of horticultural products: Distribution techniques and channels*, Almería, pp. 31–41.

López-Gálvez, J. [et al.] (1996) 'Comparative experimental study of crop systems on sanded soil and on substrate', in J. López-Gálvez and Naredo (eds), *Production systems and environmental impact of crops on sanded soil and on substrates*, Fundación Argentaria, pp. 154–271.

López-Gálvez, J., Sánchez-Carreño, J. and Castilla, N. (1993) 'Technical economic analysis of alternative structures to the flat roofed Almería-vineyard greenhouse', in *Agrarian research. Production and Protection of Vegetables*, pp. 50–71.

Marins-Peil, R. and López-Gálvez, J. (1999) 'Tomato crop with the nutrients film (NFT) in the Southeast of Spain', in *Spanish American Conference of Plastic Materials Application used in agriculture*, Panama City, pp. 61–84.

Naredo, J.M. (1988) 'About own, adopted and future technologies', *BIT*, 52, pp. 66–70.

Naredo, J.M. and López-Gálvez, J. (1992) 'Crops grown under plastic cover in the Almería area', in *XII Acts, International Conference on Plastics in Agriculture*, Granada, pp. 130–154.

Schroder, F.G., Schwarz, D. and Kuchenbuch, R. (1995) 'Comparison of biomass production of tomatoes grown in two circulating systems', *Gartenbaumwissenschaft*, 60, 6, pp. 294–297.

6 Nitrogen in modern European agriculture

Vaclav Smil, University of Manitoba

There is no life without nitrogen. The element is an irreplaceable constituent of amino acids that are required for the assembly of proteins. These, in turn, provide building blocks of all metabolising tissues and of all enzymes that control the chemistry of life. Nitrogen is also present in the nucleotides of nucleic acids (DNA and RNA) that store and process all genetic information. Adults cannot synthesise eight amino acids found in proteins, and children need two more for their tissue growth. Consequently, these ten essential amino acids must be ingested in food – and hence agriculture, as Justus von Liebig (1840) so aptly noted more than 160 years ago, is all about producing digestible nitrogen.

Natural photosynthesis is a prolific fixer of carbon, with the global net primary productivity nearly 60 billion tonnes of it on land, and close to 50 billion tonnes in the ocean (Smil, 2002). But most of the terrestrial carbon is stored in woody phytomass as cellulose, hemicellulose and lignin: none of these abundant polymers contains any nitrogen, and except for the ruminants they are not digestible by mammals. Among uncultivated species only tree nuts and seeds of cereal and leguminous plants contain relatively large amounts of nitrogen because of their unusually high protein content. Eventual domestication of cereals and legumes became the nutritional foundation of all settled societies. Harvests of these crops were restricted by a variety of environmental (water, nutrients, pests) and management (tillage, weeding, crop rotations) factors, but in most traditional agro-ecosystems one limitation was nearly always the most common reason for low yields: inadequate supply of nitrogen.

I. Pre-1900 realities

During the early modern era (1500–1800) two regions at the opposite ends of the Eurasian continent – namely East Asia (coastal China, Korea and Japan) and Western Europe – experienced most acutely the combination of relatively high population densities and limited availability of additional good-quality farmland. Hence, they had the greatest stake in increasing crop yields. Traditional farmers had three ways in which to supply additional nitrogen to their crops: recycling of organic wastes (including crop residues, mainly straw, and animal and human wastes); practising crop rotations including nitrogen-fixing leguminous grains (beans, peas, lentils, soybeans); and planting of leguminous cover crops that were plowed under as green manures (Smil, 2001). These cover crops included most commonly two clover genera (*Trifolium* and *Melilotus*) and several vetches (*Vicia*), but field peas (*Pisum*), chickpeas (*Cicer*) and lupines (*Lupinus*) were also used in some European regions (Pieters, 1927).

Traditional extent of crop residue and manure recycling was very similar in parts of East Asia and Western Europe. Typical French rates were around 20 t/ha, and the Netherlands were the early paragon of intensive manuring with as much as 30 t of manure applied annually per hectare of cropland. However, nitrogen content of this organic matter was low, as manures were often recycled only after many months, or even several years of storage, losing a large share of the nutrient through volatilisation and leaching. Post-applications losses were particularly high when the manures were spread during fall and winter and on bare, sloping land. These losses typically reduced the original nitrogen content by at least half, often by two-thirds and even by more than four-fifths of the initially voided nitrogen. As a result, nitrogen from recycled manures actually available to crops amounted often to just a few kilograms per hectare, or no more than the nutrient in atmospheric deposition.

Rotation of grain legumes (peas, beans, lentils) also left behind only small amounts of nitrogen (fixed by the symbiotic *Rhizobium* bacteria) for the subsequent non-leguminous crop. In contrast, cover legume crops, whose inclusion into often complex rotation schemes became more common only after 1750, could supply large amounts of the nutrient for the following cereal, tuber and oil crops. Rapid adoption of new cultivation schemes (the best known form example was the English Norfolk cycle of rotation that included wheat-turnips-barley-clover, but there were other similar sequences in other countries) led to higher nitrogen supply (Chorley, 1981). By the middle of the 19th century cover legumes (with clover dominant) commonly accounted for as much as a quarter of all planted area, or double the share before 1750, a change that had at least tripled the rate of symbiotic nitrogen fixation.

Consequently, Chorley (1981) suggested that this neglected innovation was comparatively as important as was steam power in Europe's economic development in the period of early industrialisation. A detailed nitrogen balance reconstructed for a large grain-and-dairy Dutch farm of the early 19th century shows the importance of symbiotic fixation: it was the largest nitrogen input (almost 50 kg N/ha), followed by the recycling of manure and animal droppings on the pasture (total of nearly 40 kg N/ha), and by the return of crop residues (about 20 kg N/ha) (Frissel, 1978). In total, about 3/4 of all nitrogen supply on this farm (about 100 kg N/ha) came from managed inputs, compared to no more than a third in a typical pre-1750 operation.

Two new options of nitrogen delivery emerged just before the middle of the 19th century: imports of guano (droppings of tropical birds that accumulated on arid tropical islands and contained up to 14 percent N) and Chilean nitrate (Smil, 2001). While the deposits of high-quality guanos were largely exhausted by the 1870s, mining and processing of Chilean nitrates ($NaNO_3$ has 15 percent N) continued to increase and by the year 1900 it reached 1.3 million tonnes (Mt). Western Europe, led by Germany and the UK, was the largest importer of this inorganic fertiliser, buying 65–70 percent of all exports during the last decades of the 19th century. In addition, by 1900 Germany and the UK were also the world's leading producers of ammonia recovered from coke ovens that were first introduced in Western Europe during the 1860s and 1870s and whose more efficient versions spread rather rapidly during the 1880s.

Western Europe's combination of organic recycling, imported guano and Chilean nitrate and newly available by-product ammonium sulfate was able to support the world's highest staple crop yields. In spite of this achievement the region was also the world's largest importer of wheat. Annual imports were about 5 Mt in the UK and almost 1 Mt/year in Germany, and there was no significant amount of potentially arable land to be converted to cropping. Given this reality it was not surprising that William Crookes (1899) made a much publicised appeal to the continent's chemists for coming up with a radically new method of securing agricultural nitrogen – or else facing a 'deadly peril of not having enough to eat' perhaps as early as the 1930s.

II. *Die gründer Jahre* (1900–1913)

This German subtitle is entirely appropriate as Crookes's fears were rather promptly allayed and the foundations of modern agricultural productivity and prosperity were laid down thanks to the perseverance, dedication and innovation of German chemists. The quest for the synthesis of ammonia, the simplest of all stable inorganic nitrogen compounds, was not an exclusively German affair: there were some notable pre-1900 contributions by French and English scientists, but the later German effort was decisive. In March 1900 Wilhelm Ostwald, at that time already one of the world's most respected chemists, notified the General Director of the Badische Anilin- & Soda-Fabrik (BASF) in Ludwigshafen that he succeeded in synthesizing NH3 from its elements in the presence of an iron catalyst. His claim was soon exposed as a contamination artifact, but by 1904 Fritz Haber, a young Privatdozent at the Technische Hochshule in Karslruhe began his work on the synthesis.

Haber, aided by Robert le Rossignol, experimented with different catalysts, pioneered the use of high pressure in the synthesis, and achieved acceptable reaction yield by re-circulating the reagents (Stoltzenberg, 1994). After some inevitable setbacks he demonstrated his success to the BASF representatives in July 1909. A no less remarkable effort followed as Carl Bosch and his colleagues solved some unprecedented technical challenges and were able to transform the laboratory process into a full-scale commercial operation in just four years (Smil, 2001). The world's first synthetic ammonia plant began operating in Oppau near Ludwigshafen in September 1913, and the compound was used as the feedstock to produce solid fertiliser in the form of ammonium sulfate, $(NH_4)_2SO_4$.

III. A long interlude (1914–1950)

During these years both the Haber-Bosch synthesis of ammonia and applications of nitrogenous fertilised derived from the gas remained an overwhelmingly European affair. The WWI interrupted further expansion of ammonia synthesis for nitrogenous fertilizers but it led to the construction of the world's second ammonia factory at Leuna whose output was used, as was that of the Ludwigshafen plant since 1915, as the feedstock for the synthesis of nitrates needed for explosives. After the WWI several notable modifications of the Haber-Bosch process were developed in France (by Georges Claude), Italy (by Luigi Casale) and Germany (above all by Friedrich Uhde), but the BASF kept its primacy in ammonia synthesis during the 1920s, and Germany dominated its production. By1928 its annual output just surpassed 1 Mt NH_3, accounting for about 90 percent of the world's total, but this ascent was followed by a sharp downturn, stagnation and slow growth that was brought by the global economic crisis: German output was nearly halved by 1931, and it surpassed the 1928 peak only by 1939.

But throughout this period Germany remained the leading user of new fertilisers in absolute terms, although their share in the country's total nitrogen supply remained low: during the 1930s, when manure remained the country's largest source of agricultural nitrogen (followed by biofixation), it amounted to less 20 percent of the total and to no more than 20 kg N/ha in absolute terms (Smil, 2001). Dutch agriculture was by far the most intensive pre-WWII user of nitrogen fertilisers, with applications averaging above

40 kg N/ha during the late 1930s, (Erisman, 2000) and reaching 50–60 kg N/ha in parts of the country. In contrast, the average US rate was less than 3 kg N/ha.

Throughout the 1930s European consumption of nitrogenous fertilisers remained above half of the world's total while the continent's arable land accounted for only 12 percent of all cropland. But the applications were heavily concentrated in Germany, Benelux, England and France: although ammonia-based nitrates were commercially available throughout Europe, their applications outside the Northwestern part of the continent remained marginal. After a deep plunge of production during the WWII their synthesis and applications began to grow impressively only during the early 1950s, but a faster growth elsewhere began to erode Europe's primacy.

IV. Decades of expansion (1951–1989)

Large expansion of new ammonia production capacities during the 1950s was based on only incremental technical improvements. A fundamental cluster of technical innovations came only during the late 1960s when single-train natural gas-based plant equipped with centrifugal compressors (the first one was built by M.W. Kellogg in 1963) began replacing the multi-train coke-based process that relied on reciprocating compressors (Smil, 2001). The two most important consequence of this development were enormous energy savings in ammonia synthesis and rapidly rising average capacities of ammonia plants. During the early 1950s typical energy cost was above 80 GJ/t NH_3, a quarter century later they fell below 35 GJ/t for the best plants, and typical plant capacities rose from less than 200 t NH_3/day in the early 1960s to 1200–1400 t/day during the 1980s, when the largest plants surpassed 1600 t/day (Smil, 2001).

European consumption of nitrogenous fertilisers stood at 1.8 Mt N in 1950, 20 years later it reached 9.6 Mt N, more than quintupling in a generation, and it had expanded by almost 50 percent during the 1970s; only then it slowed considerably, growing by only about 12 percent between 1980 and 1988 when it peaked at 15.98 Mt N. For the EU15 the peak consumption took place three years earlier when the applications of 11.2 Mt N were 12 percent above the 1980 level (FAO, 2002). As a result of this virtually uninterrupted growth (there was only one year during the nearly four decades of expansion when the continent's total nitrogen consumption did not increase) Europe's share of global nitrogen applications remained disproportionately high when measured against the continent's share of arable land.

However, rising demand in North America, and later in Asia, reduced it from about 50 percent in 1950 to about 38 percent in 1960, to just over 30 percent in 1970 and less than 20 percent during the late 1980s when Europe cultivated roughly 10 percent of the world's agricultural land. In relative terms average applications rose nearly tenfold between 1950 (about 12 kg N/ha) and 1988 when they peaked at 118 kg N/ha of arable land and permanent crops. The highest national rate rose twice as high as the continental mean: by the early 1980s the Dutch application averaged just over 250 kg N/ha and they were the highest in the world.

V. Recent contraction (1990–2002)

Two trends combined to bring Europe's total nitrogen applications from their peak of nearly 16 Mt N in 1988/89 to just 11.6 Mt N by the year 2000, roughly a 28 percent contraction. Collapse of the Communist regimes halved the applications in the central and eastern parts of the continent where all agricultural inputs, be they fertilisers or field machinery, were previously used with staggering inefficiency. And a more efficient use of the nutrient and, in some countries, environmentally driven limits on its use lowered the Western Europe's applications by nearly 20 percent. This shift brought the continent's share of the global use of synthetic nitrogenous fertilisers from 18 percent in 1990 to 15 percent in the year 2000, still about twice as high as Europe's share of the world's cropland.

This pullback lowered the average European applications to 99 kg N/ha by the year 2000, a nearly 20 percent from the 1988 peak, and the Dutch applications declined to about 185 kg N/ha by the year 1995. Although the Dutch had retained their global primacy China's nation-wide mean of the late 1990s was only about 5 percent below the Dutch average (China is now both the world's largest producer and user of nitrogenous fertilisers). And several Chinese provinces (above all Jiangsu and Guangdong), whose areas and populations are considerably larger than the Dutch national totals, are now ahead of the Netherlands, applying annually in excess of 250 kg N/ha of arable land (Smil, 2001). Moreover, Chinese and Dutch figures are not directly comparable, as virtually all of the Chinese fertiliser goes to crop fields, but a large share of Dutch nitrogen is spread on pastures.

VI. Consequences for food production

Europe's intensive fertilisation, whose post-1950 costs were increasingly supported by rising agricultural subsidies, was reflected in a steady growth of average crop yields. As already noted, staple grain yields in parts of the continent were the world's highest even before the discovery of ammonia synthesis. By 1900 wheat harvests were around 2 t/ha in the UK, as well as in the Netherlands and Denmark, and close to 2 t/ha in Germany. In contrast, they were only about 1 t/ha in the Austro-Hungarian empire, about 0.8 t/ha in Italy (that was also the US mean), and less than 0.6 t/ha in European Russia, the last rate being almost as low as the typical medieval mean in Western Europe.

Yields continued to increase after the WWI with higher nitrogen applications, but the gains became particularly impressive after the introduction of new, short-stalked cultivars during the 1960s. Again leading the world, Dutch wheat yields rose from 4.3 t/ha during the early 1960s to 7.3 t/ha two decades later, and English and French yields were not far behind. But the most remarkable development concerning the crop productivity began unfolding during the 1980s as the Western Europe's declining applications of nitrogen were not accompanied by declining harvests; instead, higher rates of the nutrient uptake were translated into slowly, but steadily, rising yields. One of the best illustrations of this trend was the British experience.

Europe increased yields naturally brought a higher degree of self-sufficiency in staple food production (EU has actually become a net exporter of wheat) but increasing demand for animal foods has led to rising imports of feedstuffs dominated by corn, soybeans and dried cassava. Continent's net imports of all feedstuffs were about 27 Mt in the year 2000, representing an addition of at least 6 Mt of protein whose metabolism by domestic animals further burdens Europe's environment.

VII. Changing diets

According to FAO's food balance sheets Europe has by far the highest per capita food availability in the world. The entire continent averages nearly 3,300 kcal/day and the EU mean was about 3,500 kcal/ day in the year 2000 (FAO 2002). Some poorer EU countries are actually ahead of the mean, with both Greece and Portugal just above 3,700 kcal/day, well ahead of France and Germany. Given the continent's ageing population (metabolic requirements decline with age) and the increasingly sedentary way of urban life, the actual daily food requirements range mostly between 1,500–2,000 kcal/capita for adult females and 2,000–2,600 kcal/capita for adult males, and weighted means for entire populations are rarely above 2,000 kcal/person. Large gaps between average availability and actual consumption – averaging daily more than 1,000 kcal/capita – can be accounted for by two major trends: overeating, leading to higher rates of obesity (although Europe has nowhere near the epidemic proportions of North American obesity) and increasing food waste (Smil, 2000).

Greatly expanded food production also led to some very important qualitative dietary changes, including a rapid transformation of Mediterranean diets that have been extolled for decades as perhaps the best choice for maximising longevity and minimising the risk of many chronic civilisational diseases including, most notably, cardiovascular illnesses. But by the end of the 20th century this desirable diet had become more of a myth than a reality in the region's most populous nations as typical Spanish and Italian ways of eating have been dramatically shifting in the direction of less salubrious Northern eating. This means (often rapidly) declining intakes of bread, fruit, potatoes and olive oil, and increasing consumption of dairy products and meat. Most notably, Spanish per capita supply of meat is now nearly six times higher than it was two generations ago, and the intake of animal fat is more than three times high.

So far, this shift has had no negative effect on Spanish health, resulting in a paradox of declining cardiovascular mortality even as intakes of meat and dairy (and hence of saturated fats) have been increasing, and consumption of olive oil and foods rich in complex carbohydrates has been dropping (Serra-Majem et al., 1995). The best explanation appears to be expanded access to health care, improved control of hypertension and reduced smoking; increased consumption of fruit and fish may be a contributing factor. However, a generation from now the outcome may be different.

VIII. Environmental impacts

Many traditional agricultures that did not practice intensive recycling and rotations with leguminous species were constantly mining soils by steadily reducing their nitrogen content. Large parts of the sub-Saharan Africa are now the most worrisome example of this degradation that reduces the productive capacity of agricultures. In contrast, as fields in Europe's intensively cultivated regions began to receive higher applications of nitrogenous fertilisers in addition to the recycled organic matter and to rising levels of atmospheric deposition, their nitrogen balance turned slightly positive, and eventually they became very substantial net recipients of the nutrient.

Already a generation ago a study of nitrogen balances that included 13 European agro-ecosystems found gains in soil organic nitrogen (the pool which usually contains more than 90 percent of the nutrient present in soil) in 10 (76 percent) of them, with annual increases adding commonly to 30–70 kg N/ha, and ranging as high as 280 kg N/ha (Frissel, 1978). However, without synthetic fertilisers most of these agro-ecosystems would have had net nutrient losses. Two more recent national agricultural nitrogen balance sheets, for Germany and the Netherlands, found average annual accumulations of, respectively, 47 and 38 kg N/ha (Smil, 2001). Obviously, gains are much higher in many of the most intensively cultivated regions.

Because of inevitable post-applications losses, the fertiliser nitrogen must be applied in amounts exceeding the nutrient requirements of crop. Actual uptakes of the nutrient by crops vary widely depending above all on climate, soils, crops grown and agronomic practices. European studies of nitrogen uptake efficiencies demonstrate the expected difference between the nutrient's utilisation in rainy Northwestern part of the continent and in much more arid Mediterranean climates (Jenkinson and Smith, 1989). High-yielding winter wheat recovered 39–57 percent of nitrogen from the applied urea and 38–70 percent from ammonia in France, 52–65 percent in England, and 52–62 percent in the Netherlands; in contrast, nitrogen uptakes ranged between 27–40 percent in Portugal, and only between 18–37 percent in Greece. Good approximations would thus use rates of 55 percent for the best applications, 45 percent for standard good practices, and 35 percent for substandard uses. Consequently, even a generous, weighted mean for the continent would come up with no less than 40–45 percent of all applied nitrogen lost before it reaches the fertilised crops.

Part of the nutrient that is not taken by crops is immobilised *in situ* by the increased mass of soil micro-organisms, but this fixation is only short lived. However, a significant share of the nutrient may be stored for much longer periods (decades to centuries) in long-lived organic compounds that form the soil humus. While it is highly desirable to boost nitrogen content of soils by proper agronomic management, their capacity to store the nutrient is obviously finite, and once a new equilibrium, determined by a complex interplay of many local environmental and management factors, is established the excess nitrogen will have to leave the agro-ecosystem.

Only one route of nitrogen loss is largely, though not completely, innocuous, as denitrification bacteria convert highly reactive nitrates first to nitrites and then to dinitrogen whose basically inert molecules rejoin the enormous atmospheric pool of the gas. The only unwelcome complication of the process is that the reduction does not always proceed entirely to N_2 but stops at nitrous oxide (N_2O) which is a much more efficient absorbers of the outgoing long-wave radiation: its global warming potentials relative to CO_2 is about 60 times higher during the first 20 years after its release.

Volatilisation of ammoniacal compounds and emissions of nitrogen oxides from soils will carry the nutrient downwind to be deposited on other fields as well as on natural ecosystems. Inevitably, emissions of ammonia correlate highly with the intensity of fertilisation and with the density of domestic animals, and atmospheric deposition of ammonia and nitrates is now taking place on such unprecedented scales that nitrogen inputs to natural ecosystems have become significant even by agricultural standards.

Total mean annual deposition of nitrogen averages close to 10 kg N/ha for the entire continent, twice the global mean (Van Egmond, Bresser and Bouwman, 2002). Fluxes over the Northwestern Europe are between 20–60 kg N/ha a year, and the peaks, in the Netherlands, are over 80 kg N/ha a year. Annual rates around 60 kg N/ha are as high as average fertiliser applications to North American spring wheat, and they are several times higher than the fertilisation means in most sub-Saharan countries.

During the mid-1990s critical nitrogen loads (that is amounts below which there is no known damage to ecosystems) were somewhat lower than during the mid-1980s, but they were still greatly exceeded in nearly all of Germany, Switzerland, northern Italy, northern France, Belgium and, of course, in the Netherlands (Posch et al., 1999). Ecosystem that evolved under the conditions of low nitrogen inputs are particularly affected by these high deposition fluxes. Boreal growth that taken place after the end of the last Ice Age on nitrogen-poor soils, and heathlands are the two ecosystems where even relatively low inputs of atmospheric nitrogen may equal or surpass the quantity of the element made available through net mineralisation of organic matter in the forest floor. As a result, supercritical nitrogen loads will alter species composition by favouring fast growing and nitrofilic plants.

Nitrogen oxides released by soils are eventually converted to nitrates which, together with sulfates, are the major causes of virtually continent-wide acid deposition. Emission of agricultural NO_x are a minor source of the gases compared to the combustion of fossil fuels, but acidification is also intensified by the deposition of ammoniacal compounds once their converted by nitrifying bacteria in soils to nitrates. But by the far the most acute problems with nitrogen leaking from excessively fertilised agro-ecosystems is the leaching of highly soluble nitrates into surface and ground waters.

The best available estimates show that 87 percent of nitrates accumulated in ground waters of the EU countries are leached from agricultural soils; in addition, a substantial share of leaching from non-agricultural lands is also caused indirectly through the nitrification of deposited ammonia which was volatilised from fertilised fields and from the wastes of farm animals. European experience also shows that it may take not merely months, but many years or decades, before leached nitrates reach deep aquifers and before their concentrations, in the absence of de-nitrification, rise to potentially harmful concentrations. Parts of western Europe where heavy nitrogen applications pre-date the WWII provide excellent examples of this gradual process: nitrate concentrations began to rise quickly in both ground waters and surface waters only during the early 1970s.

By the early 1980s nitrates were either near or above the EU's maximum contaminant limit (MCL) of 50 mg NO_3/L in a number of regions, especially in England and the Netherlands. Average concentrations of nitrates in the most affected European rivers – the Thames, Rhine, Meuse, and Elbe – are now two orders of magnitude above the mean of unpolluted streams, and in the early 1990s more than a tenth of western Europe's rivers had NO_3 levels above the MCL. In December 1991 the Council of the European Communities issued its nitrate directive (91/676/EEC) that requires the member states to reduce the nitrate load from the agricultural sector to acceptable levels in all nitrate-sensitive zones and to avoid further pollution from that source. This directive applies not only from nitrogen from inorganic fertilisers but also from manures. The quantities of nutrients applied on individual farms are to be controlled by levies on nutrient surpluses defined by mineral accounting systems and manure disposal contracts (Henkens and van Keulen, 2001).

IX. Summarising the achievement and the challenge

Europe has been at the forefront of the 20th century most important agricultural revolution as the invention of ammonia synthesis from its elements eliminated the most common natural limits on crop productivity. German synthesis of ammonia dominated the pre-WWII global output of nitrogen fertilisers. Europe's post-WWII share of nitrogen fertiliser applications has been falling steadily, but the continent still uses a disproportionate amount of this key nutrient. As a result, average rates of nitrogen applications in several European countries remain among the highest in the world, and they make it possible to harvest the world's highest yields of several staple crops, particularly of wheat.

In spite of higher post-1980 efficiencies of nitrogen uptake, about half of the applied fertiliser still does not get taken up by growing crops, and only a part of this excess can be stored in long-lived soil organic matter. The rest is lost to the environment. The price Europe pays for its obscene surplus of food thus goes beyond the staggering cost of irrationally high agricultural subsidies and the health effects of changing diets. Environmental costs of excess food output are actually more worrisome as many of these impacts would persevere even if the subsidies were to be miraculously cut. Excessive nitrogen in the continent's agro-ecosystems must rank high on any list of such concerns because its impacts range, literally, from high atmosphere (rising N_2O concentrations) to deep wells (nitrate contamination) and include long-term changes of those natural ecosystems that are particularly sensitive to relatively high levels of nitrogen deposition.

X. Agenda for research

Since the mid-1990s human interference in nitrogen cycle has been receiving a great deal of research, as well as public policy, attention, and so in many respects there is now much less uncertainty regarding many basic flows, levels and trends than it was just a decade ago. But, as is always the case with a rapidly advancing research, there is a growing need to integrate this information in an effective and revealing manner. As a part of this effort I would particularly favour preparation of the best possible long-term quantitative summaries of all major trends as well as of all realistic management options that could be used in a variety of environmental and national settings. In addition, attempts to monetise at least the major costs and benefits of the intensive use of nitrogenous fertilisers should be particularly revealing. Such exercises would run into the usual difficulties of quantifying the burdens of environmental and health impacts but they should be of help to policy-makers who need to assess better the cost-benefits ratio of either contemplated actions or of benign neglect.

Bibliography

Chorley, G.P.H. (1981) 'The agricultural revolution in Northern Europe, 1750–1880: Nitrogen, legumes, and crop productivity', *Economic History*, 34, pp. 71–93.

Crookes, W. (1899) *The wheat problem*, London.

Egmond, K. van, Bresser, T. and Bouwman, L. (2002) 'The European nitrogen case', *Ambio*, 31, pp. 72–78.

Erisman, J.W. (2000) *De vliegende geest: ammoniak uit de landbouw en de gevolgen voor de natuur*, Bergen.

FAO (Food and Agriculture Organization) (2002) *FAO Statistical Databases*, Rome (http: //apps.fao.org).

Frissel, M.J. (ed.) (1978) *Cycling of mineral nutrients in agricultural ecosystems*, Amsterdam.

Henkens, P.L.C.M. and Keulen, H. van (2001) 'Mineral policy in the Netherlands and nitrate policy within the European Community', *Netherlands Journal of Agricultural Science*, 49, pp. 117–134.

Jenkinson, D.S. and Smith, K.A. (eds) (1989) *Nitrogen efficiency in agricultural soils*, London.

Liebig, J. (1840) *Chemistry in its application to agriculture and physiology*, London.

Pieters, A.J. (1927) *Green manuring*, New York.

Posch, M. [et al.] (eds) (1999) *Calculations and mapping of critical thresholds in Europe*, Bilthoven (http: //arch.rivm.nl/cce/datasummary.html).

Serra-Majem, L. [et al.] (1995) 'How could changes in diet explain changes in coronary heart disease mortality in Spain? The Spanish paradox', *American Journal of Clinical Nutrition*, 61 (Supplement), pp. 1351S–1359S.

Smil, V. (2000) *Feeding the world: Challenge for the 21st century*, Cambridge, MA.

Smil, V. (2001) *Enriching the earth: Fritz Haber, Carl Bosch and the transformation of world agriculture*, Cambridge, MA.

Smil, V. (2002) *The earth's biosphere: Evolution, dynamics and change*, Cambridge, MA.

Stoltzenberg, D. (1994) *Fritz Haber. Chemiker, Nobelpreisträger, Deutscher, Jude*, Weinheim.

7 Modernisation and the international food system: re-articulation or resistance?

David GOODMAN, University of California, Santa Cruz
Michael REDCLIFT, Kings College, London

This chapter examines the evolution of the agri-food system in Europe. Beginning with an analysis of the political economy of food production in the nineteenth century, it goes on to explore the way in which challenges to this system emerged, and in ways that were largely unanticipated. Increasingly it has been the quality and safety of food, and the sustainability and resilience of ecological systems linked to agriculture, which have been the rallying points for opposition to the large-scale agri-food system. However, while globalisation has conferred a certain legitimacy on some new forms of environmental policy intervention, for example, in areas such as climate change and energy use, in the case of food it has been viewed as a process undermining legitimacy. Food represents a key area in which many of the precepts of the Risk Society have come to be questioned (Beck, 1992). This is particularly, but not uniquely true of genetic modification of crops and animals, and their effects on ecosystems.

In addition, decisions about food preparation are bound up with time and space equations for households, and located within specific social formations. They are often far removed from the rationale, or logic of food production. The chapter examines the partial success of alternative food networks in illuminating some of the costs, and inconsistencies, of the way in which food is produced in Europe. At the same time, it is evident that the mainstream food system has a powerful hold upon consumers, culturally as well as economically. As a result, we argue that consideration for the social forces unleashed by the agri-food system should not exclude the institutionalised forms of behaviour, and cultural associations, which have served to buttress major food conglomerates and trading interests. The international food system is being actively re-articulated as well as resisted.

At the conclusion to this chapter we point to just such a locus of resistance and re-invention in the performance of the body. Concerns with health, diet, ageing, and the risks attached to food, have taken the food system closer to the corporeal sphere, to which it is always linked, although the links are not always acknowledged. Contests surrounding food are part of wider social and cultural constructions, and at the same time often linked to radical changes in genetics and biotechnology, that carry wider political implications for human security than the external environment alone (Redclift, 2001).

I. The historical context

Rural development and the transformations of agri-food systems in twentieth-century Western Europe have been conditioned fundamentally by the region's insertion in global commodity markets. The significance of these interactions, shaped in turn by defensive state policies to protect farming from overseas competition, first became apparent with the consolidation of a new international division of labour in agriculture after 1870. The formation of world grain markets that followed the economic integration of New World agricultural frontiers effectively extended the 'resource base of the western world' (North, 1958: 557), and initiated broad processes of change in rural Europe. The historical dynamics of these processes are inscribed on rural social structures and state-agricultural relations whose past iterations continue to influence current developments. Here we focus briefly on selected 'moments' of these historical processes in order to identify significant inflection points and continuities in change that frame contemporary problematics.

The flood of New World imports after 1870 undermined grain prices in Europe and triggered 'the great agrarian depressions of the 1870s and 1880s' (Hobsbawm, 1979: 70), which, with the exception of Britain, unleashed the 'first wave' of agricultural protectionism (Tracy, 1982). European responses to the collapse of world grain prices from the mid-1870s are neatly summarized by Kindleberger: 'In Britain, agriculture was permitted to be liquidated. In Germany large-scale agriculture sought and obtained protection for itself. In France (...) agriculture (...) successfully defended its position with tariffs. In Italy the response was to emigrate. In Denmark grain production was converted to animal husbandry' (Kindleberger, 1951: 37). The fundamentals of the European 'farm problem' can already be discerned in the structural over-production crisis of the later nineteenth century and, a generation or so later, protectionist trade measures were uniformly adopted as part of efforts to insulate agriculture during the Great Depression of the inter-war years.

Further chapters in the complex political economic history of Western European agricultural protectionism include the price supports and investment incentives of the post-war settlement and the consolidation of the Common Agricultural Policy (CAP). By 1970 'state-agriculture relations presented an intricate web of protection, support systems, rising surpluses and formidably strong farmers' organisations' (Goodman and Redclift, 1991: 122). Somewhat paradoxically, these regulatory and corporatist structures epitomise the protracted shift in the role of national agricultures in Western European social formations. That is, from a position of systemic importance in capitalist accumulation, recognised in the 'agrarian question' problematics of classical political economy, to re-definition as a sectoral problem characterised by lagging incomes, poverty, out-migration, and degraded agri-environments.

Broadly speaking, and despite protective trade barriers, integrated global markets eroded the competitiveness of large-scale capitalist agriculture in Western Europe, setting in train adjustment processes that have accentuated the preponderance of family-labour forms of production in rural social structures. These processes, already discernible to Kautsky at the turn of the twentieth century (Kautsky, 1988[1899]), variously encompassed rural exodus, shifts to livestock production and dairying, self-exploitation of family labour, part-time farming, and the formation of agricultural co-operatives. Kautsky also presciently observed the subsumption of family-labour producers to agro-industrial capitals strategically located upstream in farm input sectors and downstream in off-farm processing. These developments, magnified from the 1930s by agricultural technoscience and agro-industrial concentration, initiated the marginalisation of farming activities in the value chain, a process subsequently accelerated by the centralisation of retail capital and associated practices of contract production, supply chain management and global product sourcing.

I.1. Agricultural modernisation

Structural change in the farm sector is indelibly marked by the institutionalisation of agricultural price support policies after World War II and their later harmonisation under the CAP, which has shown remarkable resistance to reform, from the Mansholt Plan proposals of 1968–1972 to the present-day. The effect of these policies, grosso modo, has been to attenuate production risk and moderate farm income fluctuations, thereby creating powerful incentives to lower production costs and raise output by adopting yield-enhancing innovations. In the inter-war years, efforts to increase crop yields built on the rediscovery of Mendelian genetics, culminating in the innovation of hybridisation, initially with corn, and launched a trajectory that has led to changes in intellectual property rights to facilitate the private appropriation of plant genetic resources. The commodification of the seed and its manipulation as a capitalist force of nature are manifest today in the form of transnational agri-chemical corporations, with their proprietary plant germplasm portfolios, and social struggles to prevent the diffusion of genetically modified plants and food in Western Europe.

Hybridisation mobilised synergies between mechanical, chemical and genetic technologies, resulting in capital-intensive integrated crop management systems or 'technological packages.' In the case of corn, for example, technological convergence led to the development of hybrid varieties, which were high-yielding, fertiliser-responsive and adapted to mechanical harvesting. These advances, in turn, encouraged monocultural production to exploit scale economies, further extending markets for plant protection chemicals and other agro-industrial inputs. Farmers who embarked on this modernisation project were caught on a 'treadmill' of competitive innovation in the effort to capture 'first mover' advantages. Moreover, the capitalisation of subsidies and productivity gains in rising land values stimulated farm concentration, giving further momentum to this treadmill of technological efficiency and intensification. In this state-subsidised model of farm competition and capital- and energy-intensive agro-industrial development, agrarian restructuring proceeded by social differentiation, declining farm numbers, reduced farm employment, and out-migration. Conversely, surviving 'modernisers' responded to these cost-price imperatives by increasing farm size, intensification and specialisation.

However, it is important to recognise disparities in the rate and degree of adoption of the modernisation paradigm in the countries and regions of Western Europe as the result of resource endowments, inherited social structures and uneven capitalist development. These disparities are expressed institutionally in EU/CAP policies to support agriculture and rural activities in less favoured regions. Nevertheless, if the transition to capital-intensive agriculture occurred at different speeds and exposed regional disparities, 'the essentials never changed: the capitalisation of farms and a massive reduction in the agricultural workforce; specialisation according to locality and within the production process (arable crops, breeding, livestock fattening); a technical revolution through the use of industrial inputs (pesticides, fertilisers) and genetically improved seeds; and the intensification of livestock production (poultry, pigs and cattle) on the basis of industrially-processed feed (cereals, soybean)' (Tubiana, 1989: 26).

I.2. 'Cheap' food, mass consumption

By removing price uncertainty and absorbing surpluses, state intervention after World War II effectively institutionalised over-production and created conditions favourable to the sustained growth of output and productivity. Concomitantly, with farm price policy divorced from income support programmes, the state found leeway to pursue cheap food policies, centred on the expansion of the grains-livestock complex and the transition to animal protein-based diets. This transition was accelerated by falling real prices of animal protein, particularly of poultry, reflecting productivity gains, surpluses, and declining prices of feed crops in the wake of agricultural modernisation. This conjuncture stimulated the dramatic reorganisation of livestock production, achieved by the mutually-reinforcing effects of innovations in genetic inbreeding-hybridisation techniques, animal nutrition and disease control, and the introduction of intensive grain-fed systems and high-density confinement housing of modern pig farms and battery poultry units.

Western Europe has followed the lead of the United States in climbing the 'protein ladder,' with the corollary of the increasingly indirect consumption of food and feed grains and oilseeds, one of the hallmarks of modern food systems. As exemplified by the case of animal feed, modernised, capital-intensive farming has been integrated into complex agro-industrial systems as a source of low-cost, standardised inputs for processing sectors, which, in turn, provide intermediate products for the food manufacturing industries and retail chains. Over the past fifty years, the modernisation paradigm inexorably has heightened the importance of cheap, standardised and industrially processed food in mass consumption diets. This transformation has consigned agriculture to a subordinate, indirect role in modern food systems and shifted control to oligopolistic transnational food industry conglomerates and retail multiples.

The costs of these profound structural changes in the face of Western European agriculture and food systems typically are abstracted from the calculus of economic efficiency and find expression, for example, in environmental pollution and the destruction of wildlife habitats, declining norms of animal welfare, food 'scares,' and the health problems, notably heart disease and obesity, associated with cheap and abundant supplies of industrially mass-produced, processed food. The degradation of surface and ground water supplies by nitrate leaching and point- and non-point source pollution has attracted increasing EU regulation over the past twenty years. Moreover, with farm diversification, including agro-tourism, and the influx of ex-urban, higher income residents, productivist agriculture has encountered rising opposition from the new interests of the 'consumption countryside' (Marsden, 1999), exacerbating conflicts over these 'externalities.' In short, the increasingly acute contradictions of the modernisation process and centralised agro-industrial food systems have infused 'green issues' into European debates on farm and food policy. The environmental and public health critiques of these intensive systems have been developed insistently since the 1980s by environmental organisations and a wide constellation of urban food movements and pressure groups.

In summary, the following themes describe synoptic 'moments' in the historical development of Western European agriculture and food systems in the twentieth century:

- Protection from foreign competition.
- Agriculture's loss of systemic importance in capitalist accumulation and the emergence of the European 'farm problem.'
- Preponderance of family-labour production units, state-supported modernisation,

social differentiation, and rural exodus.

- The 'treadmill' of technological innovation and structural change by technological competition.
- Institutionalised surpluses, cheap food policies, and ascent of the 'protein ladder.'
- Marginalisation of agriculture and corporate control of food systems by agro-industrial and retail capitals.
- Broadly articulated environmental, health and social critiques of capital-intensive, centralised agro-food systems have created space for a variety of contemporary alternative agro-food networks (AAFNs).

At the turn of the millennium, these critiques are encapsulated in the notion of food quality; that is, how and where is food produced and by whom. These questions now are finding expression in a variety of institutional innovations and alternative agro-food networks, which for some observers mark the emergence of a new rural development paradigm to challenge the modernisation model. Several competing characterisations of these 'new economic spaces' and their significance are explored in the remaining sections of this chapter.

II. Rural Europe redux? CAP reform, food 'scares' and the quality 'turn'

Two related, mutually-reinforcing contextual influences stand out when mapping the contemporary features and trajectories of change in Western European agriculture and food systems. These two influences help to frame rural futures by problematising the institutional apparatus and practices of capital-intensive agriculture and its role in rural development.

II.1. CAP Reform

The first of these contextualising influences involves the attempts to re-orientate the CAP from a production-centred sectoral programme for agriculture, towards a broader territorial vision of integrated rural development. That is, a more inclusive and multi-dimensional conception of rural development that encompasses *inter alia* new actors, policies of rural re-vitalisation, enhanced local governance, and the promotion of ecologically sustainable activities. Significant steps on the long march to CAP reform include expenditure cuts to reduce mounting food surpluses and policy debates about the targeting of EU Structural Funds in the 1980s, and the 1992 MacSharry reforms in response to the inclusion of agriculture in the Uruguay Round of the GATT international trade negotiations. These reforms reduced commodity support prices towards world market levels and, in partial mitigation for these cuts, introduced direct compensation payments to farmers, arable set-aside, and agri-environmental programmes.

The 1996 Cork Declaration on sustainable rural development and the European Commissions's Agenda 2000 proposals attempted to further consolidate the territorial orientation of the CAP by accelerating the decoupling of farm support payments from commodity production, strengthening agri-environmental measures, and re-focusing EU Structural Funds, especially Objective 5b, more directly on lagging rural regions. Although the Agenda 2000 negotiations in 1997–1999 failed to close 'the divide between liberalizing and protectionist factions' and 'must be judged a missed opportunity to reform the CAP' (Lowe, Buller and Ward, 2002: 4), the adoption of the Rural Development

Regulation (1257/99) placed the question of 'multi-functional,' integrated rural development firmly on the agenda.

Despite the October, 2002 Franco-German agreement to maintain current CAP commodity subsidies until 2006, and permit the overall CAP budget to rise by 1 percent annually in 2007–2013 to accommodate new EU member countries, the onset of the Doha round of world trade negotiations is maintaining the pressure for reform. That is, to switch EU farm spending from production-linked activities, in the form of tariff protection and export subsidies, to programs of income support and rural development considered to be non-trade distorting according to WTO rules. Expansion of such 'green box' payments at the expense of these forms of border protection is the centre-piece of new CAP reform proposals presented in January, 2003 by the EU agricultural commissioner in preparation for the Doha round meetings. These proposals envisage the transfer of 80 percent of EU farm spending into non-trade distorting categories, while leaving the overall level of payments to farmers unchanged. Although these proposals face opposition, notably from France and Ireland, such a massive decoupling of farm support from production would signal a sea change in EU agriculture and rural development policy. In mid-2005 the United Kingdom Government signalled its intention to use the rotating Presidency of the European Union to initiate such a change in policy.

II.2. Food 'scares'

The second contextualising influence is constituted by the combined and cumulative effects of periodic 'food scares,' the supposed exhaustion of the farm modernisation paradigm and its intrinsic logic of social exclusion, and new rural livelihood opportunities afforded by rising demand for local 'quality' foods.

In recent years, Western Europe has been assailed by a series of 'extreme food events' that have created a palpable sense of unease and mistrust in conventional farming and food processing, notably in livestock production. These episodic food 'scares' include the 1986 outbreak of 'mad cow disease' (BSE) in Britain and its pandemic translation to the EU in 2000–2001. There have also been recurrent food contamination events, such as the Belgian dioxin crisis and 1999 reports of untreated sewage, septic tank residues and abattoir effluent contaminating animal feed processing plants in France. Public unease is exacerbated by periodic incidents of food poisoning by E. coli 0157: H7, reports of adulterated meat, and international pressures to condone the use of growth hormones in livestock production and to accept genetically modified foods.

Nor does it appear that this mistrust can be dispelled by introducing appropriate regulatory measures and adopting better risk management techniques. These extreme 'events' have placed industrial food provisioning under much closer public scrutiny and strengthened ethical opposition to the consequences for animal welfare of standardised industrial livestock practices. Food 'scares' also have heightened public awareness of the centralising and homogenising forces of globalised agro-food systems, and the threat these represent to the material and symbolic content of foodways, and thence to culture identity. To modify Jean Brillat-Savarin's well-known aphorism, there is unease about what we eat, how it is produced, and what it means for what we are and what we might become. Food 'scares' powerfully reinforce political, social and cultural struggles around the metabolic relations between nature and society, field and table, and what these signify for the material and discursive metrics of everyday life.

More directly, food 'scares' are one potent factor in the opening up of 'spaces of resistance' to industrial food provisioning, and which provide economic opportunities for alternative forms of food production. Exploitation of these new spaces is represented by the 'turn' to quality, which embraces alternative production practices and institutional innovation in modes of food provisioning.

II.3. The quality 'turn'

The quality 'turn' away from industrial food provisioning broadly comprehends organic production, other 'alternative' agro-food networks and short food supply chains (SFSCs), quality assurance schemes, and the proliferation of territorial strategies to promote local foods. These include the long-standing A.O.C. designation and the more recent EU labelling scheme (2081/92), which confers Protected Designation of Origin (PDO) on food products. The academic literature on the 'turn' to quality in food consumption, which has grown explosively in the past few years, is focused overwhelmingly on the local. Accordingly, it is primarily micro-analytic and ethnographic in its investigation of place-based alternative food practices, and social embeddedness, trust and place are among the key concepts deployed in this research. Somewhat paradoxically, however, this literature is dominated by theoretical frameworks that privilege the production 'moment' in agro-food circuits, with the result that consumption is 'used' mainly to talk about production (Goodman and Dupuis, 2002). Consumers in the quality 'turn' are conceptualised in abstract, economistic terms as 'discerning', 'affluent', and so on, leaving little room to acknowledge a politics of consumption or recognise the importance of producer-consumer alliances in ensuring the economic viability of AAFNs. This focus also is seen in recourse to Salais and Storpor's (1992) 'worlds of production' version of convention theory to conceptualise the quality 'turn' as a contested process of transition (Murdoch and Miele, 1999, 2002). That is, a transition from the 'industrial world', with its heavily standardised quality conventions and logic of mass commodity production, to the 'domestic world', where quality conventions embedded in face-to-face interactions, trust, tradition and place support more differentiated, localised and 'ecological' products and forms of economic organisation.

Such meso-level theoretical perspectives are the exception, however, and this literature is characterised mainly by empirically-grounded analyses of alternative food practices, institutional mechanisms of rural governance and policy, and the potential of AAFNs/SFSCs as catalysts of rural economic growth. European AAFN research intersects here with public debates on food safety and CAP reform, and contested alternative imaginaries of rural economy and society. These debates provide the background to the advocacy of AAFNs as innovative precursors of a paradigm change, a more endogenous, territorialised and ecologically-embedded successor to the allegedly exhausted and crisis-ridden modernisation model of conventional industrial agriculture (Van der Ploeg [et al.], 2000; Marsden [et al.], 1999). The following paragraphs explore these claims of a paradigm shift and note some political economic limitations of territorial value added strategies as the innovative motor of the 'new' rural social economy.

III. Quality food networks and the value chain

As suggested earlier, the perception that a substantive quality 'turn' is underway in contemporary food practices is a key premise of the Western European literature on AAFNs. We have already mentioned the catalytic effect of food 'scares' and the unease among more affluent consumer groups and food activists about the safety, nutritional value, and environmental consequences of the standardised foods produced by industrial agriculture, and processed and distributed by highly concentrated, globalising agro-industrial corporations. Observers have linked this 'turn' to the innovative development of 'new economic spaces' within local and regional economies that can better resist the forces of globalisation. The negative association between food safety and globalisation has given new salience in consumer knowledge practices to transparency, a criterion met by schemes to assure quality, provenance and traceability, organic agroecological production practices, and forms of direct marketing. Thus Marsden [et al.] (1999: 299) situate demands for greater transparency in food practices in 'the conflict between globalised aspatial systems of production and locally situated ecological systems'.

Several contributors recently have made preliminary attempts to 'map' the incidence and economic significance of quality food production in Western Europe. Thus in Van der Ploeg and Renting's (2000) analysis of 30 case-studies from the IMPACT programme, quality production constitutes one of three overlapping, often synergistic, clusters of what they define as 'rural development' practices. These clusters are pluriactivity, which has strong historical roots in many countries, cost-reduction by 'farming economically' using on-farm resources and, thirdly, high quality food production, a cluster that encompasses Parmigiano Reggiano cheese in Italy, organic dairy and meat, agro-tourism, and farmers' markets. As these authors stress, the socio-economic impact of these practices depends on 'synergy at farm enterprise level' (p. 533). For example, 'Organic farming in Tuscany results on the average in some 20 percent increase of total farm income' but, when combined with direct marketing and agro-tourism, 'the contribution of rural development practices to total farm income rises to 84 percent' (p. 533–534). In the case of the Dutch Wadden islands, Roep (2000) demonstrates that 'on one of the '*avant garde*' farms, the extra value added of a range of interconnected activities' is more that four times 'the value added on a comparable conventional farm…' (p. 534, their emphasis).

The question of synergy is also important in assessing the regional socio-economic impact of rural development practices, which Van der Ploeg and Renting (2000) insist should not be analysed in isolation. In the West of Ireland, for example, 'Taken *together* pluriactivity, cost reduction and quality production account for a *total* contribution of 29+5+1=35 percent to regional agricultural income' (p. 538, their emphasis). The authors add that 'Similar examples can be found in other countries. In the Dutch context, nature conservation, quality production, farming economically and pluriactivity contribute at least 50 percent to regional agricultural income' (p. 538).

Van der Ploeg and Renting (2000) usefully reveal the range and diversity of rural development practices and the significance of synergies at the farm and regional levels. However, their analysis can do little more than illustrate the potential socio-economic impacts of the quality 'turn' and paradigmatic rural development practices since comprehensive data are lacking and because these practices are so broadly defined. A second review of the IMPACT research studies identifies 'new' rural development practices with SFSCs, defined empirically as organic farming, quality production and direct marketing (Renting, Marsden and Banks 2003). Some kind of SFSC activity is found on 20–30 percent of all farms in France, Germany, Italy and Spain, 5–10 percent of farms in the Netherlands and UK, and under 1 percent in Ireland, but economic impact figures are described here as 'best educated guesses'. Even on this narrower definition, the authors acknowledge that far more detailed investigation of 'the temporal, spatial and evolutionary dynamics involved in SFSCs' is required before we can 'gauge whether they are economically, socially and environmentally stable over the long-term'. Moreover, as we argue below, these differences in the contents of the rural development 'box' call into question the notion of paradigm change.

III.1. Adding value, 'repeasantisation', and multiplier effects

The proposition that a paradigm shift is underway depends, in substantive terms, on the economic calculation that 'alternative' rural development practices provide secure new bases of farm income growth and territorial value-added. In the lexicon of mainstream regional economic analysis, what are the likely regional income and employment multiplier effects of AAFNs, and to what extent will 'leakages' reduce these coefficients, to the detriment of regional gross product? (Van der Ploeg and Renting, 2000; Knickel and Renting, 2000). Although regional multiplier effects are difficult to quantify at this stage, the emphasis in the AAFN/agrarian-based rural development literature on innovation at the farm enterprise level and internal, multi-functional synergies provides some tentative clues.

This literature repeatedly affirms the 'strategic' or 'central' role of farmers and their privileged ability to appropriate the higher income flows generated by rural innovation (cf. Van der Ploeg and Renting, 2000). That is, innovation often is a question of re-allocating existing resources within the farm enterprise since 'they have the land, space, craftsmanship, buildings, animals, products and the capacity to *recombine* and *reconfigure* the resources at their disposal. It is important to note that these resources are 'already paid for' and that they are multi-purpose (...) By building primarily upon resources that are owned and controlled by the farm family (...) transactions costs can be kept low,' as well as new investment (Van der Ploeg and Rening, 2000: 531, original emphasis). In a separate paper, Van der Ploeg (2000) notes that 'Farming economically is basically a strategy to *contain monetary costs...*' (499, original emphasis) by reducing resource flows mobilized through markets in favour of those mobilised in non-commoditized circuits. Clearly, such practices raise doubts about the magnitude and distribution of local multiplier effects.

However, despite the relevance of these technicalities, this literature fails to integrate farm-level innovation and AAFN-centred strategies into a wider political economic assessment of rural development. That is, how will these strategies address such long-standing rural problems as income inequality, low paid employment, rural poverty, social exclusion, gendered family and property relations, and more general questions of uneven development? In some formulations, advocacy of paradigm change, with AAFNs as new vectors of farm income diversification embedded in the cost-containment logic of non-commoditized production circuits, seems to evoke nostalgia for an earlier, romanticised vision of rural Europe.

Thus, against the modernisation paradigm's 'farmer entrepreneur' oriented by 'the logic of the market,' Van der Ploeg [et al.] (2000: 403) call for the reconceptualisation of the farmer: 'the interrelated movements away from the 'script' of agricultural entrepreneurship, reflected in...newly emerging rural development practices can be understood as a kind of repeasantization of European farming'. The expansion of AAFNs/SFSCs 'on the bases of new, non-commodity circuits' and, it needs to be said, non-commoditised gendered domestic labour relations, certainly offers farm households one type of livelihood defence mechanism in response to the income 'squeeze on agriculture' (p. 395). However, it is difficult to envisage the revival of agrarian populism as the foundation of equalising trajectories of rural development. As Shucksmith (2000: 215) observes of efforts to build 'cultural-territorial identity, rurality, sustainability or indeed endogenous development itself'...'the greatest challenge will be...to reflect on who gains and who loses in the process'. Advocates of the new rural development paradigm and 're-peasantisation' need to take this challenge to heart.

III.2. Territorial quality schemes and spatial valorisation

Even on its own terms, the territorial value-added approach to agrarian-based rural development has significant potential weaknesses, which throw its longer term political economic viability into question. These limitations are discussed briefly under the broad headings of competition, replicability, and uneven development. The ability of quality food products to secure premium prices and so generate excess profits is a central plank of the market-led, value-added model. These excess profits represent economic rents, which may arise as short-lived gains from product differentiation or more permanently as monopoly rents created by bureaucratic and regulatory barriers to entry, as in the case of organic certification and production quotas, and by the scarcity value of certain speciality foods, such as A.O.C. products.

Since rents attract rent-seekers, the durability and magnitude of these income flows and the location of the actors who capture them become key issues. The competitive erosion of excess profits can reflect widespread adoption of product differentiation strategies, including territorial identity labels and 'traditional' 'local' and 'organic' designations that are broadly generic in character. The rapid growth of food products claiming territorial identity following the introduction of EU Regulation 2081/92 and PDO designations is a case in point. Thus, Valceschini [et al.] (forthcoming: 24) observe that regional quality promotion by public authorities is undermining the criteria and intent of this regulation by placing 'the emphasis on the reputation of the product's 'native' region, rather than on specific production conditions'. The authors add, 'The possible regulatory authorization of regional brands could lead to sweeping changes in the distribution of the economic rent that is derived from such differentiation among the players in the agrifood chain across Europe'.

This combination of imitative expansion and strategic convergence also is open to corporate food interests, accentuating downward pressures on price margins and threatening to shift economic rents away from the farm and local level. Corporate food retailers have responded quickly to food safety concerns by developing a variety of product quality strategies, including supply chain management to enforce quality assurance standards, such as HACCP, product traceability norms, the sourcing and labeling of 'local' foods, and the introduction of own-label territorial identity foods. Examples here are Tesco's sale of over 100 locally-sourced products with some form of Welsh label in their supermarkets in Wales, the development by Waitrose of 'Welsh Organic

Lamb' (Banks and Bristow, 1999), and 'Carrefour's filières de qualité... and another major distributor's Reflets de France product line, which banks on the regional tie' (Valceschini [et al.], forthcoming: 25). The market power of these corporate networks co-opts and subverts agrarian imaginaries and the territorial or endogenous value-added logic of SFSCs and AAFNs. In the UK the current struggles between Tesco, Sainsbury and Asda/Wal-Mart, exemplifies the concentration of retailer buying power. These retail struggles arouse fears of even greater oligopsonistic control over supply chains, intensifying the squeeze on the production costs and profit margins of farmers and other actors upstream.

A focus on supply chain relations also warns against the simple conflation of AAFNs/SFSCs and the local embeddedness of the main actors. As Marsden, Banks and Bristow (2000: 426) observe about industrial and alternative food provisioning, 'Types of speciality, quality, region specific, or organic foods are by no means solely the preserve of the alternative mode...' (and) 'This is producing some interesting mutations with regard to supply chains... in terms of the types of relations and organizational features they display'. These authors point, for example, to local product sourcing by corporate retailers and, on the other hand, to the international reach of some alternative quality networks, as in the case of Parmigiano Reggiano and other AOC/PDO products. In this respect, the 'key influences upon the attribution and allocation of economic value across the different actors in the supply chains...' (is) '...a significant research gap in recent literature' (ibid.: 26).

Strategic imitation and convergence on the various modalities used to represent territorial identity raise the very real prospect that quality differentiation by AAFNs will be trivialised and economic rents redistributed from the farm level and other local actors. As Valceschini [et al.], (forthcoming: 26) warn, territorial quality schemes then 'would just be one of a series of marketing tools'. Similar limitations come into play as AAFNs scale-up from local niche markets and rely on identity labels to make relational connections between 'local' sites of production and more distant spaces of consumption. A.O.C. labels typify these material and symbolic exchanges between worldwide consumers and situated terroir, entangled in the histories of people and place – 'cohabited nature,' 'cohabited landscape (Barham, 2003). The effectiveness of these symbolic mediators of quality, commitment and certification inevitably suffers from processes of abstraction as supply chains are extended.

The critical question, therefore, is whether the logic of territorial valorisation governing AAFN/SFSC development will produce a bewildering and counter-productive proliferation of competing quality schemes, labels and logos. This scenario of market-led 'competitive territoriality,' to paraphrase Henry Buller (2000), invites parallels with neo-classical trade models of comparative advantage (Ray, 2000), and the attendant dangers of the fallacy of composition. That is, in certain individual cases, regions may succeed in appropriating the gains from trade, but this condition may not hold in the aggregate as competition intensifies.

These considerations pose serious questions about the replicability and durability of the AAFN/SFSC model of spatial valorisation, and stress the importance of assessing powerful disembedding forces (Sayer, 1997). Spatial uneven development certainly is a possible corollary of competitive territoriality insofar as only the most distinctive food products are likely to have the capacity to resist pressures to redistribute economic rents to extra-regional agents as production expands and supply chains are extended. Marked differences can be expected to emerge between new, embryonic AAFNs and those with well-entrenched institutional defences of 'captured' markets, and which can operate as '"semi-oligopolies", such as the Parmesan cheese cluster' (Van der Ploeg and Renting, 2000: 536). Even in this emblematic and 'mature case of rural development' (p. 446), producers face a cost-price squeeze and must continuously reduce costs in order to prevent too great a price differential emerging between Parmigiano Reggiano and its close industrial substitutes (Roest and Menghi, 2000).

IV. Paradigm shift or continuity in change?

Recent programmatic statements on AAFNs and agrarian-based rural development press the case for paradigm change in vivid binary contrasts: old and new, crisis and rupture, modernisation and alternative models (Van der Ploeg [et al.], 2000; Van der Ploeg and Renting, 2000; Marsden, Banks and Renting, forthcoming). These oppositions overlook the complexities of transition, its uneven spatial and temporal intensity, and the possibility that change may accentuate dualisms, as between highly intensive industrial agriculture in East Anglia and the Paris Basin, for example, and other rural areas of regionally-embedded, multi-functional agriculture. Moreover, irrespective of the developmental potential of AAFNs, these new economic forms are mainly embryonic at this point, suggesting that claims of binary transition are unnecessary and premature. We have also seen that the spatial incidence and economic importance of alternative rural development activities in national farming structures varies very significantly.

In its binary formulation, the crisis of industrial agriculture is absolutely central to theorisations and prognoses of a new European rural development paradigm. Yet this crisis, its social and spatial patterns, and ways in which the 'old' might shape the 'new' receive little analytical attention. Indeed, since the eclipse of the modernisation paradigm is taken as a foregone conclusion, the unexamined assumption is that farmers and other rural actors face an unyielding imperative to embrace the new model. This recalls earlier characterisations of agricultural and rural change in terms of post-productivism and endogenous development, which are the theoretical antecedents of alternative rural development (Wilson, 2001). Van der Ploeg and Renting (2001) draw this parallel when they refer to tertiarisation, the integration of service activities into farming, and the well-endowed capacity of 'traditional farming' to enter the post-Fordist era. Commenting that 'L'histoire se répète' (p. 530), they suggest that 'Through rural development, agriculture is increasingly moving 'beyond modernisation'' (p. 531).

However, recent reviews of the production/post-productivism debate have questioned both the empirical validity of this representation of farm output diversification (Evans [et al.], 2001), and whether the values, attitudes and behaviour of rural actors actually correspond to this uncompromising binary definition of their alternatives (Wilson, 2001). Studies of the adoption of post-productivist farm management practices, including agri-environmental schemes, reveal a plurality of action and thought, suggesting that agricultural and rural change 'on the ground' is complex and often contradictory (Morris and Potter, 1995; Morris and Evans, 1999; Buller [et al.], 2000). A further complication is that some countries may be more resistant to such practices than others. As Wilson (2001: 91) observes, 'Spain, Portugal and Greece, in particular, have criticized the EU for imposing policies that aim at the *extensification* of agriculture at a time when they are still mostly concerned with 'catching up' with their Northern counterparts through the *intensification* of commodity production' (original emphasis).

While accepting that the modernisation paradigm does indeed have an uncertain future, such actor-oriented and behaviourally-grounded research is needed to clarify the multi-faceted nature of this crisis and its likely evolution in time and space. In this light, a more modest approach to the contours of a successor model and the role of AAFNs might first see their present efflorescence as innovative responses to the ongoing struggle for rural livelihoods.

In this weaker formulation, AAFNs create 'new spaces of possibility' for farm reproduction and rural livelihoods, building on the heterogeneity and polyvalence that are such distinctive features of contemporary European food practices (Goodman, 2002). To use a now less fashionable term, AAFNs creatively extend the repertoire of rural 'survival strategies.' Interestingly, many leading contributors to current debates on rural change were instrumental in giving wider prominence in European rural sociology to the diversity of 'farming styles' and rural livelihood strategies, well before multi-functionality became ubiquitous in the lexicon of EU rural development policy (Van der Ploeg, 1993; Marsden, 1990; Whatmore, 1991; Pugliese, 1991).

Moreover, this earlier body of research and the continuities it evokes also are more relevant to the 'strong' AAFN/SFSC position than might be apparent initially because it is unclear exactly which activities belong in the rural development 'box.' This suggests that advocates of rural paradigm change are torn between a focus on innovative but embryonic organisational forms and a more inclusive position that plays down the 'rupture' with conventional intensive agriculture. Thus, as we have seen, Renting, Marsden and Banks (2003) identify this 'box' only with SFSCs (organic farming, quality production, and direct selling), whereas other protagonists include low external input styles of 'farming economically' and pluriactive, part-time farming. For example, Van der Ploeg (2000: 497) remarks that 'a considerable proportion of what is vaguely referred to as 'normal agriculture' should also be situated within that other model of rural development,' and argues for the inclusion of 'pluriactivity and part-time farming, from which sources the majority of people living in the countryside derive their livelihood'. Analytically, this re-labelling of long-standing or 'old' practices shifts the centre of gravity away from 'vanguardism,' 'rupture' and paradigm change towards continuity and incrementalism.

The continuities implicit in this wider conceptualisation of rural development are suggested by the case of Ireland, where 47 percent of farm households engaged in off-farm work in 1998, which accounted, on average, for 40 percent of total farm household income (Kinsella [et al.], 2000: 485). Furthermore, in the late 1990s, Ireland stood only in fourth place behind Finland, Germany and Austria in terms of the proportion of pluriactive farm households. As these authors observe, 'Since pluriactivity appeared on the research agenda in the late 1970s, it has become accepted as a structural phenomenon… that is prevalent throughout the European countryside' (ibid.: 481).

These additions to the rural development 'box' significantly broaden its social and economic base. At the same time, however, the radical or distinctive identity of the new rural development, with AAFNs/SFSCs as its dynamic, innovative expression, becomes less pronounced. This loss of conceptual clarity and empirical understanding of what distinguishes the 'alternative' from the 'normal' is offset by a fuller appreciation of the complexities and continuities of contemporary rural change.

V. By way of a conclusion: the missing guests

Finally, returning to an earlier theme, both the strong formulation of paradigm change and its weaker counterpart of continuity in change 'see' rural development through the prism of production and supply. Demand for AAFN/SFSC production is treated as an exogenous datum, as if these new organisational forms were the principal determinants of rural change and its magnitude. Apart from this dubious neglect of the active, relational and political role of consumers in the genesis and reproduction of these forms, quality food production is destined to remain as a narrow 'class diet' of privileged income groups in the absence of consumer price subsidies and related institutional changes. In the case of organics, for example, the premiums paid over prices for conventional products vary from a low range of 20–50 percent to 100–200 percent at the other extreme (Miele, 2001: 37–42). In addition, 1998 market survey data indicate that organic products, including imports, account for less than 2 percent of total food sales in the EU, with projections of 6–7 percent by 2005.

For all its well-known limitations, the industrial food system arguably has attenuated income-related class differences in food consumption by democratising access. With AAFNs and direct selling schemes as presently structured, this process is in danger of being reversed and further fragmented by the emergence of a new multi-tiered food system differentiated by income and class. In the flight to quality, upper income groups will be provisioned increasingly by the AAFNs/SFSCs associated with 'repeasantization' and the new rural service economy. In short, the provision of safe, nutritious food for all is the glaring omission in characterisations of alternative food practices as the new rural development paradigm. Without close attention to the consumption side of the development equation, innovative food quality networks will continue to represent socially exclusive niches rather than the future of European rural economy and society.

Much of the literature to which we have referred has paid little attention to the processes through which the mainstream food system has made partial – and often largely 'cosmetic' – accommodations to the challenges represented by alternative food networks. Not far from most community-based 'farmers' markets' in the United Kingdom, are major food retailers, like Tesco, Sainsburys and Asda (Wal-Mart) which not only sell authenticated organic produce but, equally importantly, repackage major food items within a consumer discourse linked to health and diet. New products are launched which incorporate health and 'fitness' elements, within a corporate marketing strategy and under a branded logo (Lean Foods, Healthy Alternative, Natural Choice). These sometimes, but not invariably, point to the sourcing of food (extensively-reared pigs, organic produce) but this re-packaging and labelling is linked fundamentally to human health and nutrition. Products that are low-calorie, for example, are assumed to be more 'healthy'.

There is a confusion of messages here that may reflect confusions on the part of consumers, but also suggests retailing strategies to identify different markets among consumers, some of them geared to animal welfare or vegetarian preferences, but overwhelmingly linked to human health, fitness and risk aversion. In recent years there has been a concerted reconfiguration of elements in the mainstream retail market, partly prompted by the growth of alternative networks and social movements to which we have referred. Whether such responses are more than retail marketing strategies, and have the potential to alter broad consumer diets, remains an open question.

In fact, the closer one looks at the spatial distribution of the alternative food networks described above, the more one witnesses some major temporal disjunctions in the advancement of the process of 'substitution' of greener food, through which challenges to the agri-food system are established. We noted above that there are significant differences between parts of southern Europe, where some producers are willing to embrace more rather than less integration (even at the cost of losing the identity of local provisioning), and northern Europe, where growing numbers of farmers are challenging the mainstream food system themselves.

What we are apparently seeing is a process of food system 'advancement' that is very incomplete, in terms of geographical regions as well as types of producers, and where major disjunctions occur. Where the transition towards an integrated modern food system is most advanced, and the contradictions for diet (and the environment) are most visible, a process of social resistance is underway, on the part of some consumers, as well as producers. In areas of more inclusive and 'traditional' farming, where consumers still make daily use of small, local food markets and less processed food products, the earlier 'stages' of the agro-food system assault are still being played out. The considerable media attention given to the widespread closure of family-run brasseries and restaurants in France, together with evidence that French women are spending less time on food preparation and cooking in the home, may be indications of the very uneven spread of apparently contradictory processes. The growing attraction of new locations for consuming food and drink, such as the themed bars and restaurants owned by Guinness Group in southern Europe, are indications that these processes are also reflected in business decision-making.

Another avenue well worth exploring is that charted by the 'performance' of the body. Challenges to the agri-food system at the production end are represented, as we have argued, by new nexuses of producers, many of them part-time, or linked to livelihood aspirations that are every bit as 'post-productivist' as their business strategies. From the consumption end the real challenges to the ubiquity of the system lie in public health, diet and nutrition, domains in which the consumer reflects not on the social consequences of their practices, but their own internal implications. The body has taken hold of much of the discursive territory surrounding food, reflected in popular magazine publishing, Web-based communication and local activity (groups like 'Weight-Watchers' involve many times as many participants as farmers' markets). In a sense the new terroir is that of the human body, rather than the external environment (Redclift, 2001).

There is little evidence that the diet of most Europeans has been reversed in ways that are consistent with major advances for local, alternative food networks; although niche markets of consumers are significant, and hold sway among the urban 'chattering' middle classes. The overall figures for diet (and more worryingly, for health) still suggest that the agri-food system has replaced carbohydrates with fats, in the mass popular diet, and increased *per capita* consumption of meat (Lang, 2003). The European consumer is urged to adopt a diet featuring at least 'five' items of food and vegetables a day (or seven or nine). However countervailing forces are at least as important, including the changing demographic profile of 'Europeans', and new forms of vulnerability and insecurity. The implications for public health of an ageing European population, living off pensions of diminishing value, and plagued by increasing ill health, have not been addressed by public policy. Perhaps this demonstrates that the parallel importance of consumer-driven changes in the agri-food system, although acknowledged in principle, have hardly begun to alter practices. To redesign diet, and to identify the 'external' costs of mass food production, for health and nutrition, as well as the environment, would require a new project of social engineering, at the very moment when positions on personal 'choice' are inspired largely by neo-liberalism, rather than regulative fiat.

To inform such an ambitious project requires a serious consideration of factors behind current consumer habits. It requires that we consider the time and space equations, which operate for households, within specific social formations. Despite increased political activity linked to more sustainable food production, and healthier and less wasteful food consumption, most consumers in the North (and many in the South) make increasing use of an industrial food system. This system is more concentrated, in retail terms, than ever before, and has developed levels of horizontal and (less clearly) vertical integration, that link food production with its processing, marketing and sale. When food is delivered to the consumer it is at locations which serve as the locus of 'community' (or what substitutes for it) particularly supermarkets and hyper-markets. The cultural hold of the modern food system is, therefore, not only central to the semiotics and communication of the whole society, but it is spatially linked to peoples' homes and work. 'Getting and spending' does not exist in a cultural vacuum, buttressed by advertising and the media; it is increasingly where taste and 'choice' and modernity are celebrated, and sometimes enjoyed. Future research needs to identify the changing demographics of our societies, by gender as well as age, and geographically as well as socially. These parameters of what we have called 'time and space equations' help explain why the food system has such ubiquity, even while it is being challenged from a number of sources.

As well as an increasingly diverse food system, the revival of interest in alternative foods and the social networks that lie behind them, is not conclusive evidence of resistance to globalisation. Indeed, in some respects, alternative food networks are themselves evidence of the influence of globalisation: bringing diverse ethnic and specialised foods to a broader market. Our point is that individuals as consumers (and households are increasingly made up of single individuals in Western Europe) continue to participate in both 'mass' and 'alternative' food and environment systems, in an instrumental and opportunist way. They are able to combine different components of both mainstream and alternative systems, linking actual consumption practices to their own different identities as consumers, citizens, and family or household members. As identities change so, in a sense, does the food system: as both their mirror and their material embodiment.

We are witnessing partial, and politically accommodated, European Union regulation (and re-regulation) to reduce the costs of poor environmental health, and the accompanying risks to European 'publics'. But such limited measures, themselves invariably reactive responses to 'food scares', are unlikely to shift the main parameters of diet and food consumption. These seem to be broadly consistent with the model we set out at the beginning of this chapter. What may be required, from the consumption side, is a new 'social compact', like that heralded in post-war Europe, which envisaged the development of a European industrial base on the foundations of a reinvigorated agriculture. The answer to 'Europe's farm problem' was to emphasise the systemic importance of agriculture to accumulation, as we have argued elsewhere. The challenge now, however, is to introduce new political and social priorities into the debate about European integration and enlargement, and to do so in the face of a well advanced coalition of agri-food capitals, and a fading European momentum towards integration. In our judgement, the next generation of political conflicts over food will take us into rather unfamiliar territory. The performance of the body, as well as the environment, is likely to be the stage for any consumer assault on the continued progress of the modern food system.

Bibliography

Banks, J. and Bristow, G. (1999) 'Developing quality in agro-food supply chains: a Welsh perspective', *International Planning Studies,* 4, 3, pp. 317–331.

Barham, E. (2003) 'Translating *terroir*: the global challenge of French AOC labeling', *Journal of Rural Studies*, 19, 1, pp. 127–138.

Beck, U. (1992) *Risk Society: towards a new modernity*, London.

Buller, H. (2000) 'Re-creating rural territories: LEADER in France', *Sociologia Ruralis*, 40, 2, pp. 190–199.

Buller, H., Wilson, G.A. and Holl, A. (eds) (2000) *Agri-environmental policy in the European Union*, Aldershot.

Evans, N., Morris, C. and Winter, M. (2001) 'Conceptualizing agriculture: a critique of post-productivism as the new orthodoxy', *Progress in Human Geography*, 26, 3, pp. 313–332.

Goodman, D. (2002) 'Polyvalent foodways and alternative foodways: Some reflections on paradigm change in European rural development', Keynote lecture, Final Conference COST A12 Rural Innovation, 5–7 April, Budapest.

Goodman, D. and Redclift, M.R. (1991) *Refashioning nature: Food, ecology and culture*, London.

Goodman, D. and DuPuis, E.M. (2002) 'Knowing food and growing food: Beyond the production-consumption debate in the sociology of agriculture', *Sociologia Ruralis*, 42, 1, pp. 6–23.

Hobsbawm, E. (1979) *The age of capital, 1848–1875*, Harmondsworth.

Kautsky, K. (1988[1899]) *The agrarian question*, 2 vols, London.

Kindleberger, C.P. (1951) 'Group behaviour and international trade', *Journal of Political Economy*, 59, 1, pp.432–469.

Kinsella, J., Wilson, S., Jong, F. de and Renting, H. (2000) 'Pluriactivity as a livelihood strategy in Irish farm households and its role in rural development', *Sociologia Ruralis*, 40, 4, pp. 481–496.

Knickel, K. and Renting, H. (2000) 'Methodological and conceptual issues in the study of multifunctionality and rural development', *Sociologia Ruralis*, 40, 4, pp. 512–528.

Lang, T. (2003) 'Battle of the food chain', *The Guardian*, 17 May.

Lowe, P., Buller, H. and Ward, N. (2002) 'Setting the next agenda? British and French approaches to the second pillar of the Common Agricultural Policy', *Journal of Rural Studies*, 18, 1, pp.1–17.

Marsden, T. (1990) 'Towards a political economy of pluriactivity', *Journal of Rural Studies*, 6, 4, pp. 375–382.

Marsden, T. (1999) 'Rural futures: The consumption countryside and its regulation', *Sociologia Ruralis*, 39, 4, pp. 501–520.

Marsden, T., Murdoch, J., and Morgan, K. (1999) 'Sustainable agriculture, food supply chains and regional development: editorial introduction', *International Planning Studies*, 4, 3, pp. 295–301.

Marsden, T., Banks, J. and Bristow, G. (2002) 'Food supply chain approaches: exploring their role in rural development', *Sociologia Ruralis,* 40, 4, pp. 424–438.

Marsden, T., Banks, J. and Renting, H. (forthcoming) 'Alternative food networks and institutional innovation: Exploring the role of short food supply chains in rural development', *Environment and Planning A*.

Miele, M. (2001) *Creating sustainability: The social construction of the Market for Organic Products*, Wageningen.

Morris, C. and. Potter, C. (1995) 'Recruiting the new conservationists', *Journal of Rural Studies*, 11, pp. 51–63.

Morris, C. and Evans, N. (1999) 'Research on the geography of agricultural change: redundant or revitalized?', *Area*, 31, 4, pp. 349–358.

Murdoch, J. and Miele, M. (1999) '"Back to Nature": changing "worlds of production" in the food sector', *Sociologia Ruralis*, 39, 4, pp. 465–483.

Murdoch, J. and Miele, M. (2002) 'The practical aesthetics of traditional cuisines: slow food in Tuscany', *Sociologia Ruralis*, 42, 4, pp. 312–328.

North, D.C. (1958) 'Ocean freight rates and economic development', *Journal of Economic History*, 18, 4, pp. 87–103.

Ploeg, J.D. van der (1990) *Labor, markets and agricultural production*, Boulder, CO.

Ploeg, J.D. van der (1993) 'Rural sociology and the new agrarian question: a perspective from the Netherlands', *Sociologia Ruralis*, 32, 2, pp. 240–260.

Ploeg, J.D. van der (2000) 'Revitalizing agriculture: Farming economically as starting ground for rural development', *Sociologia Ruralis*, 40, 4, pp. 497–511.

Ploeg, J.D. van der and Renting, H. (2000) 'Impact and potential: A comparative review of European development practices', *Sociologia Ruralis*, 40, 4, pp. 529–543.

Ploeg, J.D. van der, Renting, H., Brunori, G., Knickel, K., Mannion, J., Marsden, T., Roost, K. de, Sevilla-Guzman, E. and Ventura, F. (2000) 'Rural development: From practices and policies towards theory', *Sociologia Ruralis*, 40, 4, pp. 391–408.

Pugliese, E. (1991) 'Agriculture in the new division of labor', in W.H. Friedland [et al.] (eds), *Towards a New Political Economy of Agriculture*, Boulder, CO, pp. 137–150.

Ray, C. (2000) 'Further thoughts about local rural development: Trade, production and cultural capital', Working Paper 49, University of Newcastle-upon-Tyne.

Redclift, M.R. (2001) 'Environmental security and the recombinant human: sustainability in the twenty-first century', *Environmental Values*, 10, pp. 289–299.

Renting, H., Marsden, T. and Banks, J. (2003) 'Understanding alternative food networks: exploring the role of short food supply chains in rural development', *Environment and Planning A*, 35, 3, pp. 393–411.

Roep, D. (2000) 'The Waddengroup Foundation: the added value of quality and region',

IMPACT Working Paper Series NL3, Wageningen.

Roest, K.D. and Menghi, A. (2000) 'Reconsidering "traditional" food: the case of Parmigiano Reggiano cheese", *Sociologia Ruralis*, 40, 4, pp. 439–451.

Salais, R. and Storpor, M. (1992) 'The four "worlds" of contemporary industry', *Cambridge Journal of Economics*, 16, pp. 169–193.

Sayer, A. (1997) 'The dialectic of culture and economy', in R. Lee and J. Wills (eds), *Geographies of Economics*, London, pp. 16–26.

Schucksmith, M. (2000) 'Endogenous development, social capital and social inclusion: perspective from leader in the UK', *Sociologia Ruralis*, 40, 4, pp. 208–218.

Tracy, M. (1982) *Agriculture in Western Europe: Challenge and Response*, London.

Tubiana, L. (1989) 'World trade in agricultural products: From global regulation to market fragmentation', in D. Goodman and M.R. Redclift (eds), *The International Farm Crisis*, London, pp. 23–45.

Valceschini, E. [et al.] (forthcoming) 'Agriculture and quality in 2015: the outlook based on four scenarios', *Environment and Planning A*.

Whatmore, S. (1991) *Farming Women: Gender, Work and Family Enterprise*, London.

Wilson, G. (2001) 'From production to post-productivism... and back again? Exploring the (un)changed natural and mental landscapes of European agriculture', *Transactions of the Institute of British Geographers*, NS 26, pp. 77–102.

8 *Kulturkampf* in the countryside. Agricultural education, 1800–1940: a multifaceted offensive

Leen VAN MOLLE, University of Leuven

The word 'agriculture' derives from the Greek *agros*, meaning 'acre', and the Latin *cultura*, meaning both 'cultivation' (in the sense of tilling the soil) and 'civilisation'. Cultivation takes place with the use of techniques, themselves the evidence of mankind's intellectual or cultural progress. Agriculture and technology (the latter being the key word in the 'Tensions of Europe' project – hereinafter referred to as ToE) have been inseparable since the time when mankind turned for food from hunting, gathering and fishing to arable farming and animal husbandry. It is precisely with the introduction of techniques – in the sense of methods and the use of technical tools – that *agricultura* arose. The gradual increase in the harvest since prehistoric times is thus explained chiefly by technological innovation, more particularly the improvement in tools and in the methods of tilling the soil.

Over time, the process of agrarian innovation has both multiplied and accelerated. During the past century, in particular, Western agriculture underwent spectacular change, as the following extract from a recent survey illustrates: 'If a farmer from Old Testament times could have visited an American farm in the year 1900, he would have recognised – and had the skill to use – most of the tools he saw: the hoe, the plow, the harrow, the rake. If he were to visit an American farm today, he might think he was on a different planet' (Paalberg and Paalberg, 2000: xiii).

This transformation process is easily viewed as the combined effect of science and professional training, the result of the dissemination of scientific knowledge in rural areas. Nineteenth–century authors, including Karl Marx, were already convinced of the impetus provided by agricultural science: 'dass sie (...) die Agrikultur aus einem blossen empirischen und mechanisch sich forterbenden Verfahren des unentwickelten Teils des Gesellschaft in bewusste wissenschaftliche Agronomie verwandelt (hat)' ('that agricultural science has transformed what was a purely empirical and mechanical process practiced by an uneducated section of society into a consciously scientific agronomy').[1] Since the time of the Enlightenment, it has been agricultural science that has enjoyed all the plaudits, whereas the farmer has been regarded chiefly as the obstinate subject of scientific and professional education. In this respect, definitions in encyclopaedias are very revealing, with such words as 'art', 'science' and 'industry' being repeatedly and respectfully used to designate agriculture, as in the following French definition from the end of the eighteenth century: 'Agriculture: c'est l'art de cultiver la terre pour la rendre fertile, c'est la science de faire produire à la terre des plantes et des fruits' (*Encyclopédie,* 1770, I: 595–601) ('Agriculture is the art of cultivating the soil to make it fertile, the science of producing plants and fruit from that soil'), a definition that, notably, barely differs from the present-day one, i.e. that 'Agriculture (is) the science or art of cultivating the soil, growing and harvesting crops, and raising livestock' (*Encyclopaedia*

[1] From the third part of *Das Kapital*, quoted in Klemm, 1992: 270.

Britannica, CD-Rom 1999). However, there are two divergent definitions of the practitioners of this art or science. On the one hand they are described as 'farmers' (*agriculteurs, Landwirte, landbouwers*), regarded as experts and evoking positive connotations: 'Agriculteur: celui qui cultive la terre, un bon agriculteur' ('Farmer: a person who cultivates the soil, a good farmer'); on the other, they are 'peasants' (*paysans, Bauern, boeren*) who conjure up a picture of backwardness: 'un pauvre paysan', 'c'est un homme rustre, impoli, grossier dans ses manières et dans son langage'[2] ('a poor peasant, a bumpkin, impolite, coarse in manner and speech'). Today, the term 'peasant economy' is still applied to small, poor farmers and the use of relatively simple technology (*Encyclopaedia Britannica*, CD-Rom 1999).

Closing the real or imagined gap between the (few) scientific farmers – i.e. those that practice farming as a rational science – and the (numerous) unschooled peasants was and remains the purpose of agricultural education. That education, in short, is presented as the necessary link between agricultural science and the business of farming, as the lever in the dissemination of know-how and innovation, and consequently as the key to modernising farming and increasing productivity. In this way, via education, science and technology are coupled to economic return and social progress.[3] This presentation of matters is not unproblematic, as both the top-down logic and the one-sided economic reasoning are simplifications of reality.

The educational offensive in rural areas began around 1800, rapidly assuming the guise of a veritable *Kulturkampf* or struggle against so-called peasant conservatism and ignorance. What is known of this offensive? How did it progress in time and space? Who set it in motion, who did provide the instruction and how many of the farming community did they reach? What was taught? What forms did the teaching take and what teaching methods were used? What was the effect? Above all, how is the totality of the educational offensive to be interpreted? This chapter does not attempt to provide a definite answer to all these questions, but rather to explore their contours through the medium of available writings on developments in the agricultural sector in Belgium, The Netherlands, France, England and Germany, the countries of the CORN group (the Ghent-Louvain research group 'Comparative Rural History of the North Sea Area'). Additionally, I have drawn substantially from my own research on Belgium, a small and relatively young European State that, prior to the First World War, presumed to be a model in matters of agricultural education. As Julien Vander Vaeren, Inspector at the Ministry of Agriculture and Belgian delegate at the World Exhibition at Paris in 1900, put it, 'La comparaison entre le développement de l'instruction professionnelle agricole en Belgique et à l'étranger est certes à notre avantage' (Vander Vaeren, 1902: 172; Vander Vaeren, 1913: 36). ('The comparison between the development of professional instruction in Belgium and other countries is certainly to our advantage'). The chapter falls into five sections, each covering a particular aspect: the place of education in agrarian historiography, the problem of periodisation, the production and broadcasting of the agronomic message, the reception of the message, and, finally, the question of the hidden curriculum of agricultural education.

---

2 *Dictionnaire*, 1798, I: 33; II: 253; also 6th edition, 1835, II: 409. See also Barral, 1966: 72–80; Muth, 1968.

3 See, for example, Boulet, 2003: 7–10.

I. A missing link in historiographical traditions

European historiographical traditions have not been conducive to the study of agricultural education, and even less so to the relationship between agricultural science, professional education and farming. In agricultural history, the focus has been chiefly on the production side. In history of science, the preference has been for the 'pure' sciences, such as physics and chemistry, biology and geology, which, it is true, have in turn been a breeding-ground for agricultural science (Halleux et al., 1998–2001). In educational history, most attention has been given to general education. None of the countries mentioned here has produced a satisfactory survey of the development of its agrarian vocational education over the past two centuries.[4] It is at least symptomatic of the lack of interest that only four volumes of the French periodical *Les Annales d'histoire des enseignements agricoles* have appeared, the first in 1980 and the last in 1992.

After the Second World War, the British agrarian historians, with the Leicester School and Reading in the van, concentrated on research into farming practices and regional differences in agricultural systems: crop selection, crop rotation, tractive force, the development of the plough, soil productivity, labour productivity, etc. The orientation was technical and economic, with a bias towards long-term developments and quantification. Dovetailing with this was also the research into population movements, which was begun in 1964 by the Cambridge Group for the History of Population and Social Structure. A comparable technological, economic and demographic approach predominated in the Dutch School of Wageningen, whose pioneers were B.H. Slicher van Bath and E.W. Hofstee. In recent years, especially in Great Britain, The Netherlands and Belgium, a great deal of time and energy has gone into the macroeconomic reconstruction of the agricultural product.[5] The social and institutional approach to the agricultural sector focused, in the Marxist tradition, on such themes as the loss of common rights and the enclosure process, on shifts in respect of private property and leasehold (cf. the investment policy of the old aristocracy and the new capitalist bourgeoisie), farm size and the splitting-up of land, social stratification of the labour force (farmers, servants and day labourers), market organisation and market regulation (sellers, buyers, traders, consumers, credit), class conflicts in the countryside and poverty. This research, too, followed much more of a structural and macro approach, the major explanatory background being the imperatives of the capitalist market economy (Campbell and Overton, 1998). Jean Gadisseur reduced the new perspectives for Belgian agricultural history a decade ago to the quantitative study of the structural changes in the relationship between what he termed 'the agrarian sub-structure', namely the farming people/proprietor connection, and the supplying and purchasing industrial and urban sub-structure (Gadisseur, 1993). In the predominating historiography, agriculture appeared mainly as a means of subsistence, less as a multifaceted way of existence in which social and cultural determinants were present side by side with economic. Dealt with much less explicitly were why, by which information channels, in which way and with

[4] The lively book by Goudswaard (1986) covering the Netherlands is primarily informative, but lacks depth.

[5] Buyst, Smits and Van Zanden, 1995. In respect of Belgium, this resulted in the following publications: Gadisseur, 1990; Dejongh, 1999 and 2000; Goossens, 1993; Blomme, 1992.

what consequences new methods were introduced into farming and thus also within the farming families involved.

Until recently, little thought was given to the role of teaching and education in the process of agrarian transformation. This had primarily to do with the preference of researchers for the Middle Ages and Modern Times: in the publications of the 1950s and 1960s, agrarian history appears to stop around 1850, when the so-called 'process of modernisation' got under way and shape began to be given to agricultural education.[6] Thereafter, however, the time perspective shifted, though without any systematic broadening to take in agricultural education. It is true that one volume of the five-volume *Deutsche Agrargeschichte* of the 1960s was devoted to the *deutsche Landwirtschaft im technischen Zeitalter* from 1815 to 1940, but it made only minimal mention of agricultural education (Haushofer, 1963: 77–84, 162–163 and 198). In the four-volume *Histoire de la France rurale* (1976), no perceptible place is given to nineteenth- and twentieth-century agricultural education, which is also the case in the eighth part, covering the 1914–1940 period, of *The Agrarian History of England and Wales*, where education is tucked away in the chapter on agricultural policy and in a few pages on science and farming practice (Duby and Wallon, 1976: III and IV; Wetham, 1978: 199–204, 282–284). Where agricultural education gained more attention, it was separately, i.e. in books and articles on the history of agricultural science and biographies of great men, on the one hand, and histories of various agricultural colleges and farmers' associations, on the other. Much of the writing was commemorative in character; indeed, anniversaries appear to provide good pickings for research into agricultural history.[7] The relationship between science, education and farming was dealt with more implicitly than explicitly as being self-evident.

The 'new agricultural history' that has emerged during the past decade and that is influenced by the post-modern cultural turn offers a much broader prospect, in which there is scope for the interaction among economic, political, social and cultural factors. This new agricultural history is much more a history of farming people as a social group, including the strong differentiations within that group. The new approach is evident, for example, in the most recently published part of *The Agrarian History of England and Wales* (Collins, 2000), covering the 1850–1914 period. Of the forty-five chapters, there are two in which the educational offensive is given prominence: Chapter 8 'Agricultural Science and Education' by Paul Brassley, and Chapter 9 'Agricultural Institutions: Societies, Associations and the Press' by Nicholas Goddard. The new direction being taken in research was made clear in Great Britain with the new journal *Rural History: Economy, Society, Culture* (1990–) and in France with *Histoire et Sociétés Rurales* (1994–).

In Germany, historical research into agriculture went into decline after the post-war generation that included Günther Franz, Wilhelm Abel and Friedrich Lütge, whose work, as was the case elsewhere, was strictly quantitative, macroeconomic and demographic.

---

[6] An example of this is the successful synthesis of Slicher van Bath, 1960; another example is Lindemans, 1952.

[7] In this respect, tertiary-level agricultural education came particularly into the picture. For Belgium, the following publications deserve to be mentioned: Antoine and Hennebert, 1985; *125 jaar tuinbouwonderwijs*, 1973; *Institut agricole de l'Etat*, 1910; Moerman, 1970; Pastoret et al., 1986; *50 jaar Nederlandstalig diergeneeskundig onderwijs*, 1984.

Recently, according to Rita Gudermann (2001), a revival has been apparent, whereby in Germany, too, the path of the 'new agricultural history' has been taken. Increasingly, the agricultural sector is coming to be seen as a part of a complex social environment in which self-perception and perception by others are factors in the *status quo* or in change. In Clemens Zimmerman's view (1999; see also Trossbach and Zimmerman, 1998), research is rightly abandoning the 'traditional/modern' dichotomy, as, for example, technological modernity does not exclude political conservatism and religious traditionalism. Transformation processes are not a linear given; in order to understand them, we have to learn to regard rural society as being multi-stratum in both time and space, and to break those processes down into human dimensions, meaning into terms of households of men, women and children, and their life cycles, as well as into terms of village life, including the commercial, ideological, political and cultural relationships that play a role in it. It does not need much imagination to realise that the introduction of agricultural education to boys and girls, men and women could have cut across, changed and distorted those complex family and village structures.

In France, a stimulating initiative has recently been set in train that is prompting thought about the significance and function of agricultural education in its most divergent forms: model farms, agricultural orphanages, schools for secondary and tertiary education, correspondence courses, lectures for adults, textbooks and agricultural periodicals, agricultural weeks for females, etc. The starting-point was the 150th anniversary in 1998 of the decree of October 1848 concerning the establishment of a network of farming schools, which prompted two great colloquia on the theme of 'les acteurs de l'agriculture' ('the protagonists in agriculture'), the first covering 1760–1945 and the second 1945–1985. Although the disturbingly uneven quality of the total of eighty-nine contributions tests the reader's patience, both colloquia served to prompt international comparison and stimulate further research (Boulet, 2000; Boulet, 2003).

II. The problem of periodisation

This brings us to the question of periodisation, or to what the chronological sequence is of the professional education of farmers. For example, is 1945 a logical cut-off point, as suggested in the French volumes? Agricultural history uses economic output and the techniques for achieving it as chronological milestones, of which agricultural education appears logically to be an adjunct. Historians distinguish a 'first' agrarian revolution, which, depending on the writer and the region under review, would have taken place at some time between the Late Middle Ages and the nineteenth century (Overton, 1996: 1–9). Partly as a result of demographic pressure, there was an acceleration in agrarian development – or revolution – during particularly the eighteenth and nineteenth centuries, in which several technological innovations played a key role: the reclamation of wasteland and the draining of fenland, the spread of Dutch husbandry and consequently the abandonment of the system of leaving land fallow in favour of that of crop rotation (the oldest traces of this in Flanders date back to the thirteenth century), the switch from ox- to horse-traction, the spread of potato cultivation, the improvement of the plough (cf. the distribution of the 'Flemish' plough), seed improvement, greater attention to cattle-breeding, and the increased use of natural and – later – chemical fertilisers. This process was accompanied by the escalating integration of agriculture into the national and international market economy and of the farming community into the urbanised and industrialised society.

The technological innovations that have occurred since the 1950s have been hailed as the 'second' agrarian revolution, the tractor, in particular, being accorded a great symbolic value, having changed the entire face of agriculture (see Figure 8.1); in fact, the farm has become a high-tech enterprise that has achieved an unprecedented level of soil productivity. Post-war agriculture has been characterised by mono-cultures, intensive cattle-farming not linked to land, glasshouse horticulture and far-reaching mechanisation, though also very substantial dependence on the ancillary industry (for example, agricultural machinery, commercial fodders, chemical fertilisers, crop protection products) and the food industry (including mills, creameries, canned food and deep-freeze industry).

Figure 8.1 Kilograms of cereals per head and per year

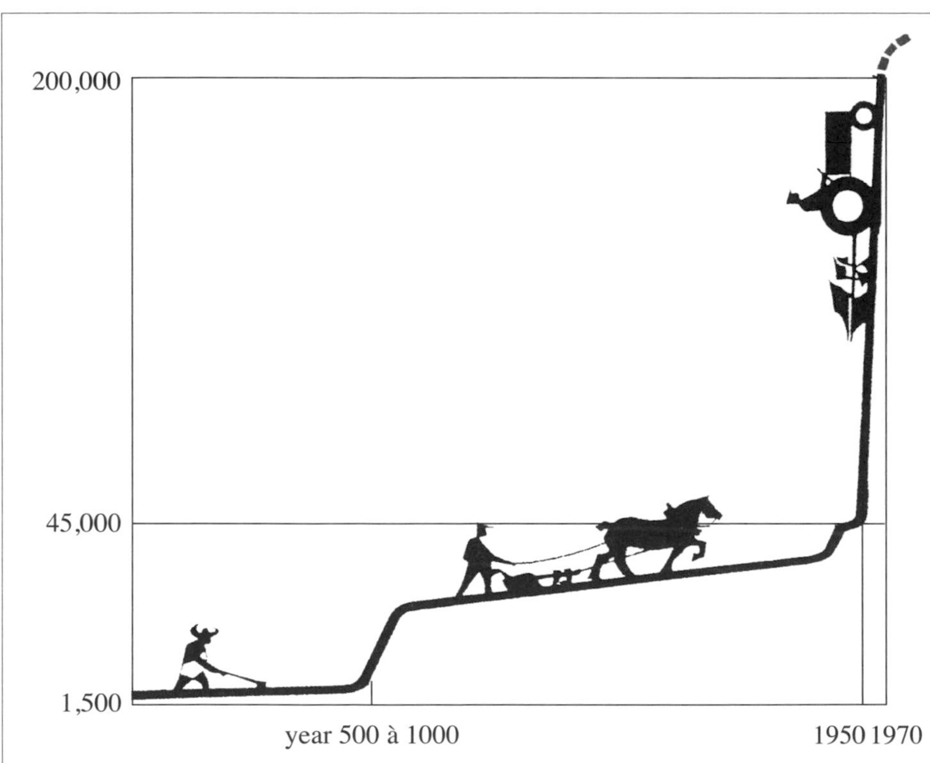

Agricultural education has played its part in this revolutionary process of transformation, and not just during the twentieth century; as regards the questions that arise here, the *ToE* project's focus on the twentieth century is consequently artificial, or even of little relevance. The beginnings of the educational offensive in the countryside ought to be sought earlier, at least during the first half of the nineteenth century when agricultural chemistry began to chalk up successes and the first agricultural schools were set up in a number of countries. They might even be sought earlier during the second half of the eighteenth century, when the first learned agricultural societies were established whose stated purpose was to introduce 'enlightenment' to agriculture and which, according to Jean Boulaine, coined the vocational name 'agronomist' during the 1790s in France. Sir John Russell, the former director of the British research station at Rothamsted actually places them in the seventeenth century when the basis of agronomic science was being laid and the first agronomist journals appeared. In support of all this are that Francis Bacon (1561–1626) was already experimenting with ways to improve soil fertility and that the problem of plant nutrition was exercising numerous scholars before 1800 (Russell, 1966; Klemm, 1992; Boulaine, 1992: 202; Buttress, 1950).[8]

On the other hand, there are those who claim that it was only from the late twentieth century on that any real importance was attached to agricultural education in terms of effect on agricultural practice. This, at least, was the conclusion of a French colloquium in 1985: 'On peut se demander de quoi on parle quand on parle d'enseignement agricole avant 1960. C'est peu de chose et cela touche peu de monde' (Boulet, 2000: préface) ('The question could be asked of what is meant when one speaks of agricultural education before 1960. There was nothing much to it and it reached few people'). Around 1960, indeed, against the background of the establishment of the European Economic Community, a number of countries set about a complete or partial reform of their agricultural education: France in 1960, Belgium and the Netherlands in 1968. However, a problem arises if it is only at so late a date that real weight can be given to the efficacy of agricultural education. What, in fact, brought about the radical transformation of the agricultural sector? Can it be that something is lacking in the definition of agricultural education? Only if 'education' is understood in the broadest sense – i.e. the entirety of the channels of information and advice to the farming community within the complex chain linking agricultural research and practice – can insight be gained into the process of transformation. Presenting that chain solely in top-down terms, as a vertical communication process, and to ascribe the key role in that to agricultural education at school is problematic (I shall return to this below). To put it another way, farming people do not in any way have to have attended an agricultural school in order to be abreast of the technical aspects of modern agriculture. A broader definition reveals a substantial degree of continuity in the development of agricultural education between about 1800 and the present day and my perception is that, since the Second World War, there has been principally a further increase in scale, stricter public regulation and a stronger move to commercialise the professional education of farmers.

In the remainder of this chapter, I have limited myself to roughly the years from 1800 to 1940, the period of the constitution and initial expansion of the agrarian educational system. It is striking to see how strong the similarities in institutional set-up were during this period in England, France, Germany, Belgium, The Netherlands and even further away in the Scandinavian and Southern European countries. At the same time, however, one ought not to be blind to the differences in the timing and intensity of the educational initiatives. How does it happen, for example, that countries with a similar (but not identical) system of agricultural education record very divergent economic results? Around

[8] The oldest definition of agronomy is attributed to the Swedish chemist Wallerius and dates from 1764, see Schling-Brodersen, 1989: 4–8.

1870, a date that in England marked virtually the culminating point of the Golden Age of high farming, Belgium, The Netherlands, Denmark and England were posting the highest yield per hectare and per capita; thereafter, they were overtaken by Germany. By about 1900, indeed, agricultural productivity in France and England had fallen behind that of not only Germany, but of Belgium and The Netherlands, too. In 1910, the use of chemical fertilisers in England and France was about 10kg/ha; in Belgium, The Netherlands, Germany and Denmark, about 30 kg/ha (Van Zanden, 1991: 219; Collins, 2000: 138–140). The question thus arises of to what extent these differences can be explained by variations in agricultural education, but, as yet, specialist writing has provided no adequate answer.

III. Formulation and dissemination of the agronomic message

According to the way matters are traditionally presented, scientists are the 'inventors' of agrarian modernity, those teaching agriculture the 'messengers' and farming people the 'receivers' of the message. This knowledge chain, within which three homogenous groups appear at first sight to be interacting is easily presented as being top-down, something that of itself already indicates a lack of respect for farmers (see Figure 8.2).

It is perhaps not surprising that contemporary, university-trained teachers of agriculture were not particularly flattering in portraying their target group, but the situation becomes more problematic when such a representation is regurgitated uncritically in historical

Figure 8.2 The knowledge chain

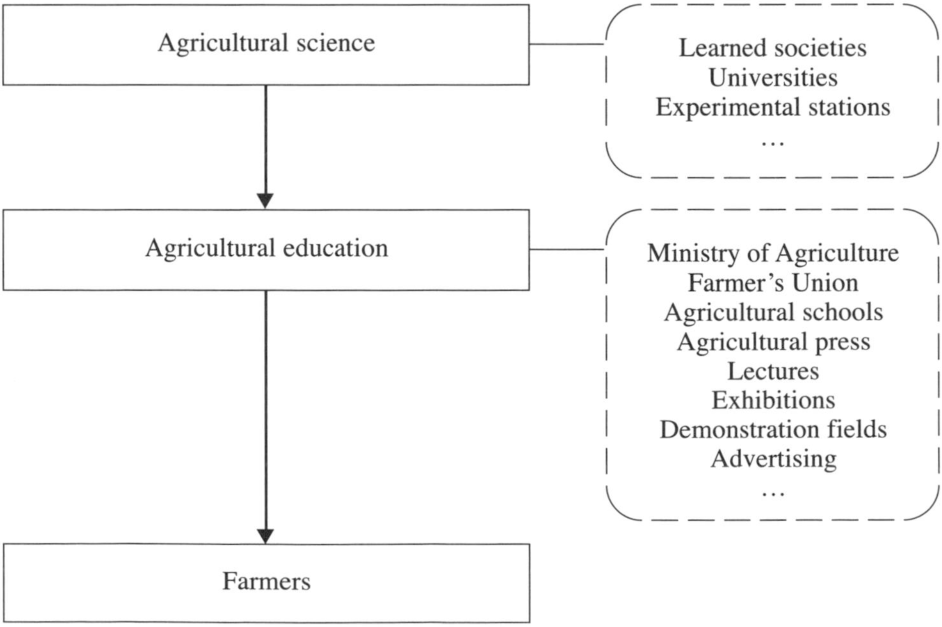

Agricultural science

Learned societies
Universities
Experimental stations
...

Agricultural education

Ministry of Agriculture
Farmer's Union
Agricultural schools
Agricultural press
Lectures
Exhibitions
Demonstration fields
Advertising
...

Farmers

research. I offer two examples of this that are also notable in their one-sided association of knowledge with book learning. Caesar De Bruyker (1878–1924), highly appreciated in Flanders as a lecturer on botany, stated that around 1850, when agricultural education began to take shape in Belgium, 'the proverbially distrustful Flemish farmer' ought first 'to attend nursery school to learn his abc'.[9] In 1963, the agricultural historian Haushofer shamelessly described the German farmers of the early nineteenth century as the illiterate '*Volk vor der Schrift*' that had to learn to read and write before thought could be given to any 'fachliche Ausbildung in der modernen Landbautechnik' (Haushofer, 1963: 77–78) ('professional training in modern agricultural techniques').

III.1. Agromania

The modernity that Haushofer alluded to was rooted in the 'agromania' that made its successful appearance from the mid-eighteenth century. The mercantilism and physiocracy that were in their heyday in France and England, and likewise the German *Kameralwissenschaft* (a forerunner of political economy) heightened respect for the productive power and economic value of the primary sector. It became fashionable to come up with agricultural innovations and to propagate them, examples being the competition for the best plough organised by François de Neufchâteau, French Minister of Home Affairs during the Directory, and the launching in 1808 of 'his' best plough by the American President Thomas Jefferson.

The active search for new agricultural methods was conducted against the background of the growing regard of the 'enlightened' man for natural sciences. The fascinated middle-class public expected the natural sciences to provide it with answers to its questions about the origins of mankind and the working of nature.[10] The particular interest in agricultural science has also to be seen in combination with the Malthusian tensions that, during the mid-1840s, led to the last great famine in Europe, with industrialisation and the process of urbanisation, and with the political changes that brought about the admission of farmers to the electorate (viz. the revolutions of 1789, 1830, 1848 and 1870). Successive events served again and again to place 'the agricultural question' onto the political agenda: the Agricultural Invasion of the 1880s that appeared to herald the end of European agriculture; the economic depression of the 1930s, which severely squeezed farmers' incomes again; the First and Second World Wars, which exposed the inability of national agricultural economies to meet domestic food requirements; and, lastly, European economic unification. In all of these events, it is noticeable how, time and again, politicians and public alike have turned to science for the answer, as was the case, for example, in France in 1891: 'La solution définitive du problème agricole n'est pas dans la douane, elle est dans la science' ('The definitive solution to the problem lies not in customs and excise, but in science').[11]

During an initial phase, i.e. the eighteenth and the early nineteenth century, scientists, practising agronomists and learned societies played a leading role in the development of agronomy. Right up to the present day, with barely disguised pride, each country has

[9] De Bruyker, physician and Doctor of Science (Botany), was also a convinced advocate before the First World War of agricultural education in Flemish (De Bruycker, 1911: 72–73).
[10] For the popularising of the natural sciences, see the ground-breaking study of Daum, 1998.
[11] Quoted from a political speech of 1891, in Duby and Wallon, 1976, III: 407.

lauded the names of its own spiritual fathers of agricultural innovation: the British Jethro Tull (1674–1740), Arthur Young (1741–1820) and Humpry Davy (1778–1829), the Germans Albrecht Thaer (1752–1828), Johann Schwerz (1759–1844) and Johann von Thünen (1783–1850), the French Mathieu de Dombasle (1777–1843) and Jean-Baptiste Boussingault (1802–1887), the Swiss Nicholas Th. de Saussure (1767–1845), the Dutch Hermann C. Van Hall (1801–1874) and Winand C. Staring (1808–1877), the Belgian horticulturalists Charles Van Hulthem (1764–1832) and Charles Morren (1807–1858) and certainly the German agricultural chemist Justus von Liebig (1803–1873), the Austrian geneticist Gregor Mendel (1822–1884) and the French bacteriologist Louis Pasteur (1822–1895). These pioneers raised branches of empirical agriculture to respected agricultural science. However, it may be that the focus has been directed too one-sidedly at them, as there are reasons to be cautious about mythologising great men as pioneers. The elite has always been good at publicising its own role and in that publicising there has been little, if any, room for the role played by farming people. The history of science is constructed in the first place by scientists themselves and it has become known in the meantime that Liebig, for example, was a very accomplished self-publicist (Schling-Brodersen, 1989: 20–27). Besides the big names, there were numerous and often anonymous landlords and gentlemen farmers who experimented with innovations on their own lands and supported their tenants in making improvements, as it was in their interest to encourage good farming practices. Here, too, however, one has to be cautious: numerous estates were encumbered with heavy mortgages, so that the proprietors were not in a position to make capital improvements and it is not unlikely that most landlords played safe, encouraging their tenant farmers to stick to well-tried and established practices (Overton, 1996: 4, 184, 205).

Eminent persons with a scientific bent for or financial interests in agriculture, or both, often took the lead in the first agricultural societies. These private, learned societies were meeting places for prominent agronomists and well-meaning landlords. The members were interested especially in experiments and also acted as catalysts and diffusers of the modern agriculture. As early as 1757, Rennes in France was already boasting its *Société d'agriculture, du commerce et des arts de Bretagne*, whose aim was to disseminate 'enlightened ideas'. Among the oldest and most renowned agricultural societies were the *Société royale d'Agriculture de la Généralité* in Paris (1761), the *Maatschappij ter bevordering van den Landbouw* in Amsterdam (1776), the *Royal Horticultural Society* in London (1804), the *Société royale d'Agriculture et de Botanique* in Ghent (1808) and the *Royal Agricultural Society of England* (1838). The motto of this last was the eloquent 'Practice with science', which could apply to all these societies: indeed, their field was not pure, explicative science, but rather the description, the practical experiment, the patient observation and the demonstration to encourage imitation. But, as contemporary criticism also had it, their attention was sometimes drawn more to prestigious exhibitions of ornamental plants and to competitions for the fattest cattle than to routine agricultural practice.

III.2. A science for universities

The high esteem in which the burgeoning agricultural science was held found expression in the relatively early development of agricultural education. It is typical of the fascination for the science that agricultural education first became firmly established at a higher, even university level and only thereafter at the lower vocational level. The academic institutionalisation of the young science was crucial for its continued existence and professionalisation; its practical applicability and the profit angle were vital to its spread, social legitimacy and respectability. Nevertheless, it was precisely this institutionalisation that also created the tension between schooled and unschooled, between professionals and amateurs; henceforth intermediaries would be necessary to interpret agricultural science for farmers.

Leading the way was veterinary science, a field of study that was a mixture of medicine, zoology and animal husbandry. It owed its early recognition as an academic discipline to its affinity with medicine and the fact that it could be mobilised in the struggle against epidemic animal diseases. The oldest, initially private, veterinary schools were established at Lyon (1762) and Alfort (1765) in France. Thereafter, schools were set up at, for example, Copenhagen (1773), Vienna (1777), Dresden (1778), London (1792), Utrecht (1821) and Liège (1830). During the nineteenth century, the schools obtained official recognition, were attached to universities or were subsidised or taken over entirely by the State (Mammerickx, 1967). Certain universities were already offering a measure of agricultural instruction at the height of the eighteenth century, examples of this being agrarian asset management within the framework of the *Kameralwissenschaft* at a number of German universities (starting with Halle and Frankfurt a/d Oder from 1727 on), agrarian law at Franeker (Friesland), agricultural statistics at Leiden and agricultural science at Edinburgh (1790) (Klemm, 1992: 6–12; Brassley, 2000: 624–625).

Separate, tertiary agricultural colleges or fully fledged academic agricultural faculties followed only during the nineteenth century. Thaer in particular judged that the *rationelle Landwirtschaft* (rational agriculture) ought to be detached from the study of asset management and be taught in autonomous research and educational centres. He himself, at Celle near Hannover (1802) and Möglin near Berlin (1806), set up the first agricultural colleges that he called agricultural academies, which concentrated on the middle ground between theoretical (university) and practical instruction and aimed their efforts at the sons of large-scale farmers, the reasoning behind this being that it was sufficient to form an elite that, in time, would set an example to smaller-scale farmers. Between 1818 and 1858, twenty or so such private academies were established in German-speaking States, that of particularly Hohenheim (1818) in Württemberg becoming well known; some of them worked closely with a nearby university. In France, the oldest agricultural college was the *Institut agricole* of Roville near Nancy, established in 1819 by de Dombasle and boasting its own journal, the *Annales de Roville*. In turn, old students of de Dombasle set up agricultural colleges at Grignon near Paris (1826), at Trois-Croix near Rennes (1832) and others. In England, the doors of the *Royal Agricultural College* opened in 1845 at Cirencester. The integration of agricultural science as a separate discipline within the universities dates from the second half of the nineteenth century when, encouraged by the Prussian government, the university of Halle established an agricultural institute in 1862. This example was followed by, among others, the university of Leipzig in 1869 and by that of Louvain in 1878. It was only during the last years of the nineteenth century that it became possible to graduate in agronomy at English universities. In certain instances, the State itself set up institutions for tertiary agricultural education: in 1848, France created the *Institut national agronomique* (INA) at Versailles and, in 1860, Belgium the *Institut agricole de l'État* at Gembloux. However, few, if any, real farmer's sons could be found at these institutions.

Meanwhile, private agronomic research centres had become the hub of fundamental and applied scientific research, of the promotion of new agricultural methods and above all of the battle against the adulteration of fertilisers and cattle feed. In 1836, Boussingault established an experimental farm at Bechelbronn (Alsace). The first, and for a long time the only, English experimental station was founded at Rothamsted in 1843, the first German at Möckern in 1851. For Germany, this represented the beginning of a long series: by 1863 there were seventeen and by 1871 fifty-two; in 1900 what were by then a good 500 German experimental stations were employing more than 1 500 scientists. France began to set up comparable experimental stations in 1867, Belgium in 1871 and the first Dutch one, at Wageningen, dates from 1877. Certain stations specialised in one specific and limited area of research: for example, seed improvement, phytopathology, dairy produce, forestry or sugar beet growing. Others worked chiefly as service providers and conducted quality controls on fertilisers, animal feeds, butter and other foodstuffs. Until the First World War, the German experimental stations in particular were – rightly or wrongly – held up internationally (and as far away as the USA) as being exemplary. The subsidisation or take-over of agricultural stations by modern Nation States made them emblems of those States' progressive agricultural policy, as was the case in Germany and Belgium. Prior to 1914, the Belgian State had already systematically mobilised seven of its fifteen analytical laboratories into the combating of fraud, but France and especially England lagged behind in this respect.[12]

III.3. In the national interest

It is remarkable how rapidly some Nation States came to play a guiding role in the institutionalisation and diffusion of the new agricultural science; indeed, the development of modern agriculture came to be an aspect of the policy of furthering their national prestige. Itinerant agronomists contributed to mutual comparison and competition, and this involved agricultural science being drawn into a dialectic argument about progress and backwardness. Around 1800, the Dutch were looking jealously at the development of agricultural science in England, France and Germany.[13] During the early decades of the nineteenth century, what in 1830 became the Belgian State basked in the complacency of knowing that its intensive agriculture – particularly that of Flanders and Brabant – was being taken as an example in neighbouring countries. Such authorities as Arthur Young, John Sinclair, André Thouin, Johann Schwerz, Samuel von Grouner and many others sent enthusiastic reports about Flemish agriculture to Great Britain, France and the German States. Around 1820, responding to a price request from the English *Board of Agriculture*, the Fleming J.L. Van Aelbroeck penned an exhaustive book about Flemish

[12] Schling-Brodersen, 1989: 136–176, with a chronological list of the German stations until 1871 on 248–249; Finlay, 1988; Jas, 2000; Van Molle, 1984: 32–34; Van Molle, 1989: 280–281; Boulaine, 1992: 300; Brassley, 2000: 617–619.

[13] Thus, Jan Kops in the first number of his journal *Magazijn van Vaderlandschen Landbouw* (1803): 'In Engeland, Vrankrijk en Duitsland wedijveren de beroemste en kundigste mannen, om den Landbouw in deszelfs wijden omvang te doen kennen' ('In England, France and Germany the most famous and competent men strive for the diffusion of agricultural science in their surroundings'), quoted in Goudswaard, 1986: 40.

agriculture that was ultimately published in both Dutch and French.[14] Around the mid-nineteenth century, many French people were gripped by a veritable Anglomania and cast envious eyes at the well managed large estates that abounded in Great Britain (Duby and Wallon, 1976, III: 116–118; Sigaut, 2000: 489). For their part, from the 1830s on, the British complained increasingly of their agricultural science and education falling behind the levels being set in Germany (Sykes, 1981: 267–270; Brassley, 2000: 594–595, 616–622, 647). For her part, from 1848 onwards and certainly after the unification in 1870, Germany appeared no longer to have much, if any, fear of comparison: indeed the pursuit and popularisation of science, and formed an integral part of the *Volksbildung* (national education) and of nation-building (Daum, 1998: 458–471). The belief in progress embraced both belief in the possibilities of agricultural science and belief in the capacity of that science to make all farmers aware of its inventions via an appropriate network of communication channels. Consciously, farming became the subject of State concern, farmers being required to transform their agricultural practice in their own interest, in the interest of consumers and thus in the interest of the whole nation.

The inclusion of the agricultural sector in the political debate on the general interest was naturally not exclusive to Germany. Less unequivocal was the moment when and the degree to which that debate led to State intervention in the organisation of the agricultural sector and to stimuli in respect of agricultural education and research. It is notable how soon, via the spoken and written word and via illustration and demonstration, various educational methods were applied to encourage farmers to innovate. As early as 1783, the Dutch Republic launched a competition to encourage the publication of 'agricultural schoolbooks' for primary education (Goudswaard, 1986: 51–52). In 1808, Jan Kops, Commissioner for Agriculture in the Northern Netherlands, established a 'Tools and Implements Office', first at Utrecht and thereafter at Amsterdam, where tools and implements could be inspected, borrowed, tested, ordered or copied (Goudswaard, 1986: 42–48). Bearing witness to the bureaucratic propensities of Revolutionary France's central government, the Directory and the Empire launched one agricultural census after another as a means of following up developments in the sector closely and, where necessary, adjusting them. During the first half of the nineteenth century, France set up a hierarchical structure for the official representation of agricultural interests and the exchange of information between the government and farmers. The embryonic structure already in place during the Napoleonic regime was expanded under Louis-Philippe. It consisted of a Central Agricultural Society, departmental agricultural committees, *comices agricoles* (district agricultural committees) and agricultural inspectors. At the top of the structure stood the relevant minister who, after the Agricultural Invasion, headed a separate Ministry of Agriculture (1881). Between 1830 and 1848, the French system was copied in full in the young Belgian State (Central Agricultural Council 1834, provincial agricultural committees 1845, district committees 1848, Ministry of Agriculture 1884) and partially in The Netherlands (from 1815, provincial agricultural committees; from 1884, the National Agricultural Committee). Certain of the German states, too, set up a comparable structure: among them was Saxony, where, beginning in 1834 and partially on the basis of existing agricultural associations, a hierarchically represen-

[14] A list of forty-three foreign publications concerning Belgian agriculture that appeared between 1650 and 1850 (including Van Aelbroeck, 1830) may be found in David, 1975: 85–87.

tative system was put in place, ranging from local level to an *Landeskulturrat* (agricultural board) with a brief to advise the government. There was no such system in England. It is true that the central and regional Chambers of Agriculture (established from 1865 on) assumed the role of political spokesmen for agricultural interests, but they lacked any official mandate and were in turn eclipsed by other private, farmers' organisations (Schling-Brodersen, 1989: 218–221; Goddard, 2000: 662–664).

Particularly able – at least in principle – to respond to the needs of and opportunities for farming people were the French and Belgian *comices agricoles* and other small-scale agricultural associations that were firmly entrenched in the local community. It was no coincidence that they enjoyed considerable expansion precisely during the 1840s, a period of failed harvests and food scarcity. In England and Wales, farmers' clubs began to be set up from 1837 on, which, via regular meetings, attempted to spread the 'enlightened' agricultural science among their members, chiefly tenant farmers. By the early 1850s, there were already nearly 700 of them, but thereafter many of them folded or continued an uncertain existence (Goddard, 1989: 378–379; Goddard, 2000: 669). More continuity and uniformity was to be seen on the European continent. In certain German States, there was an early proliferation of local *Landwirtschaftliche Vereine* (agricultural associations): a first *Verein praktischer Landwirte* (association for practical farmers) was created in Saxony as early as 1810 and by 1864 there were 242 of them in the State. By 1870, Prussia had 865, with more than 100,000 members (by comparison, the English *Royal Agricultural Society* then had around 5,000 members). 1862 saw the creation of the *Westfälische Bauernverein* (Westphalian Farmers' Union), a co-ordinating structure that from 1880 on was imitated across continental Europe (Cocaud, 2000: 199-206; Schling-Brodersen, 1989: 213-231; Klemm, 1992: 204-206; Goddard, 2000: 650-2 and 668-71). In France and Belgium, the *comices* continued their activities, albeit with varying intensity. Although they have often been accused of reaching only the local elite, recent research has shown that the part they played in disseminating innovation was greater than hitherto recognised (Bourrigaud, 2000; Mayaud, 2000). Via contests, exhibitions, lectures, demonstrations and the like, they created a forum for information, inspiration and debate, and with fine-looking diplomas, medals, ribbons, cups, premiums and cash prizes, they encouraged the farmer's ambition and pursuit of profit to good advantage. An instance of this was France's introduction in 1857 of *primes d'honneur* (medals of honour) for innovative farmers who could act as examples for others in their region; these premiums were supplemented in 1883 with the prestigious *Mérite agricole* (for agricultural merit) award (*Terre récompensée,* 2002). In this way, a collective, psychological climate was created within which innovation was a 'must'. France also introduced a channel for the dissemination of innovation, the city council of Bordeaux leading the way in 1836 with the establishment of a departmental chair of agriculture, a sort of itinerant professorship, the purpose of which was to fill the gap between agricultural research and practice. The professor's task was to establish experimental fields, to give lectures to the *comices*, to organise courses, to teach notions of agriculture in training colleges, etc. In short, he was to usher in modernity. In 1879, the French government made the creation of such a post obligatory in every department of the country.

III.4. Teaching farming at school

While adult instruction was being given shape in this way from the first half of the nineteenth century on, attention was also being paid to the training of future farming generations. In France and The Netherlands, and possibly in other countries, too, the idea had already been floated early during the nineteenth century of introducing the principles of agriculture into the primary school, though little or nothing came of this in practice. However, because existing tertiary agricultural education was directed solely towards elitist circles, more practically oriented secondary agricultural schools began to be established from the 1840s on, States being prompted to act because of pauperism, the exodus from the countryside, threatening famine and the fear of revolutionary uprisings. In German-speaking countries, numerous lower-secondary *Ackerbauschule* (arable farming schools) were set up for boys from 14 years of age and, above them, the *Landwirtschaftsschule* (agricultural colleges). From 1875, those graduating from these last were granted a remission of half the term of their military service. Germany herself was alone in being able to boast an established system of agricultural education already in place during the 1870s: by 1878, Prussia had nine tertiary agricultural institutions (five of which within universities and two of which veterinary schools), seventeen upper-secondary agricultural schools, twenty-six lower-secundary agricultural schools, ten horticultural schools and three schools for arable farming. Bavaria, Saxony, Baden, Hesse and other German *Länder* could also take pride in similar results. Furthermore, the Rhineland began in 1861 with *Wanderlehrer* (itinerant teachers) and Prussia in 1866 with agricultural winter schools for adults (Klemm, 1992: 204).

France, Belgium and The Netherlands lagged behind in comparison to all this. The French decree of October 1848 previewed the creation of *fermes-écoles* (farm schools) at departmental level, a few *écoles agricoles régionales* (regional agricultural schools), where theoretical and practical instruction would be combined, and above that the INA. In addition, a French law of 1850 encouraged the setting-up of *orphelinats agricoles* and *colonies agricoles* (agricultural orphanages and colonies). In 1849 and 1850, the Belgian State decided upon the establishment and subsidization of twelve secondary agricultural schools, two horticultural schools and a school for agricultural mechanics, in each case for young men of between sixteen and twenty years of age; some of these schools were private agricultural enterprises, in which instruction was henceforth also to be given. A Dutch law of 1863 likewise previewed the creation of secondary agricultural schools. Reality, however, disappointed, the first wave of the initiatives of the 1840s exhibiting little permanence. Most French, Belgian and Dutch secondary agricultural schools or *fermes-écoles* functioned poorly or gave up the ghost after just a few years. Financial and management problems, coupled with a lack of pupils and demonstrable results, served literally to close the doors. In the *fermes-écoles*, difficulties were created by the combination of private interest and State assignment. All that remained in Belgium after 1860 was the tertiary *Institut agricole de l'État* (agricultural institute) at Gembloux, the veterinary school at Cureghem and the horticultural schools at Vilvoorde and Ghent. The French INA was closed down in 1852 and reopened only in 1876. According to a report of 1884, England was the most lacking in secondary agricultural education; it was only with the Technical Instruction Act in 1889 and the ensuing levies on spirits (the so-called 'whisky money') in 1890 that substantial financial resources were put into agricultural education by the British government (Brassley, 2000: 625–629).

This last-mentioned piece of legislation was part of the second educational offensive that, in both Great Britain and continental Europe, was set in train during the 1880s, at the height of the Great Depression and the Agricultural Invasion. Thereafter, government ambitions in respect of education blossomed. In the attempt to reach all farming people, the range of educational resources was broadened and there was an increase in the number of persons involved; agriculture came to be regarded as a vocation that, just like all others, formed a subject to be taught and regularly leavened via schools and post-school instruction. School education was chosen to spearhead the effort: the rudiments of agriculture and horticulture were introduced into the rural, primary school curriculum and numerous secondary agricultural schools for boys were established, most with the accent on arable farming, others – depending on the needs of the region – with the accent on cattle rearing, dairy, agricultural mechanics, horticulture, forestry or wine growing. For girls, there followed, beginning during the 1890s, the establishment of secondary and tertiary agricultural housekeeping instruction. There was also a brisk expansion in tertiary agricultural education: around 1910, the two Belgian agricultural faculties were together sending around ninety agricultural engineers a year into the world; immediately before the Second World War, France was producing yearly around 200–300 (Boulaine, 1992: 335).

The sum of all these efforts on the agricultural front was impressive and even England was catching up with the Continent by 1940. Belgium, for example, had recognised countrywide by 1934 four tertiary agricultural institutions (the State institute at Gembloux, the agricultural faculties at Louvain and Ghent and the State veterinary school at Cureghem), sixteen secondary agricultural schools and seventeen secondary agricultural departments, seventeen secondary horticultural schools, two schools for agricultural mechanics, eight tertiary agricultural housekeeping institutions and no less than seventy-five secondary agricultural housekeeping schools and departments (*Code van het landbouwonderwijs,* 1935: 256–264). In all countries, meanwhile, there was a virtually endless proliferation in the extent and variety of adult education: lectures and courses, itinerant dairy schools (from the 1880s in England), short courses of instruction for soldiers, winter schools, Sunday schools, agricultural weeks for females (from 1919 in France), correspondence courses in agriculture (from 1927 in France), etc. By way of further example, 1905 saw the Belgian government organising 1 350 courses of five to fifteen hours of lessons for adults (De Bruycker, 1911: 119). Over and above these somewhat bookish forms of education, governments made use of experimental fields, demonstrations, exhibitions and competitions to intensify demonstrative instruction. Prior to, though chiefly after, the First World War, they added new forms of communication to their educational offensive, including wall charts for schools, posters, postcards, radio talks and educational films, as was the case in the battle against the Colorado beetle.

III. 5. A complementary circuit

To all this, the free associations of farmers (be they *Farmers' Unions, Syndicats agricoles*, *Bauernbunde*, *boerenbonden*) added their own form of professional instruction and additional training, complementary to that of the government, but nevertheless in competition with it; however, this last was not necessarily a disadvantage in terms of the amount of initiatives taken, the quality and the number of persons reached. The free associations functioned as intermediary bodies between the individual farmers, science, the supplier and consumer market, and generally, too, the political world. From the final quarter of the nineteenth century on, i.e. again within the framework of the Agricultural Invasion and the political wrestling to gain the electoral allegiance of the farmers, they recruited massively in all countries discussed here. The divisions between them – reflecting their ideological identity (conservative, confessional, liberal, republican, the left and, during the 1930s, also the extreme right) or target group (proprietors, tenants, agricultural workers) – served perhaps to compromise their political effectiveness, but took little away from the similarity of their educational function. Livestock insurance companies, livestock-breeding syndicates, dairy co-operatives, horticultural syndicates, etc. also saw themselves as having an educational function. The vehicles they used to impart professional information were largely those that have already been mentioned, i.e. exhibitions, competitions, lectures, courses, etc., but also included individual business consulting. It is not improbable that, whether or not prompted by their members, they in turn produced agricultural innovations and passed them on. The public they reached was in many cases impressive: for example, the annual agricultural show of England's *Royal Agricultural Society* drew, between 1860 and 1900, on average around 110,000 visitors, with a maximum of nearly 220,000 in 1897; the Dutch *Katholieke Boeren- en Tuindersbond* (Dutch Catholic Farmers' and Horticulturalists' Union) had a good 150,000 members by 1939; and in that same year the *Belgische Boerenbond* (Belgian Farmers' Union), which operated throughout Flanders and Walloon Brabant, boasted more than 100,000 male and as many female members, or easily half the Flemish agricultural population. A further feature of the *Boerenbond* was its demand that the monthly meetings of its farmers' and farmers' wives' guilds always be accompanied by a lecture; in addition, it also organised lectures, short courses and conferences itself (Van Molle, 1990: 194–197).

Table 8.1 Lectures organised annually by the *Belgische Boerenbond*, 1905–1939

Period	Average number/year
1905–1908	215
1912–1913	570
1922–1924	3,043
1927–1929	5,587
1930–1934	6,299
1935–1939	5,988

Source: Blomme, 1992: 264.

III.6. The press and advertising

Little has hitherto been said about the agricultural press, which, from the late eighteenth century onwards, nevertheless found itself with a central part to play in the educational offensive. Books, periodicals, almanacs, brochures and pamphlets served as fixed values in the formation of the desired 'modern' farming opinion. Indeed, the context lent itself to this: the decline in illiteracy, the fall in printing costs, the rise in the number of skilled agronomists seeking a public and the general regard for the printed word as a means of uplifting the population. The young science of agriculture became a successful component of the expanding scientific journalism, with both professionals and amateurs feeding a swelling flood of publications (Béguet, 1990; Daum, 1998, 458–471). In 1837, France offered prizes for the editing of *Manuels d'agriculture*, which had to be adapted for the various agricultural regions of the country. Belgium began in 1848 with a *Bibliothèque rurale*, which published a good thirty titles in the following ten years, but which a hamstrung cost/benefit ratio then caused to fold. Between 1888 and 1910, 146 agriculture engineer graduates from Louvain produced 850 publications, or an average of nearly six per year. Certain authors displayed a degree of missionary zeal, with one Flemish agricultural engineer, Hector Miserez, penning before the First World War forty or so brochures devoted to the improvement of hop cultivation (Association des Ingénieurs, 1910: 6–7, 255–287). In order to make the outpouring of writings available, municipalities, schools and farmers' associations set up agricultural libraries (Trigaut, 1899).

In every country, parallel with the increase in the number of agricultural institutions, there was a rise in the number of periodical titles: general and specialised scientific journals, newspapers and magazines for the broad rural public, information sheets, advertising papers, commercial periodicals and all types of members' news-sheet. Many periodicals offered their readers a mixture of technical information about agriculture, market news, political opinion, regional and association news, relaxation and advertising. On the eve of the Second World War, the *Belgische Boerenbond* alone was publishing nine different titles, three of which in French and six in Dutch: there were titles for male and female members, the youth and board members, one for agriculture and horticulture, and one for keepers of poultry and small stock. The agricultural press created a separate agricultural world that stood apart from the world reflected in the general press, and, in creating it, helped to bring into being and maintain the rift between town and countryside.

Lastly, mention has to be made of the manufacturers and merchants of fertilisers, cattle feed, machinery, etc. as players on the educational market, whose role has hitherto been least examined in research. The fact that they did not shrink from aggressive sales techniques is evident. In Belgium, swindlers with adulterated fertilisers and cattle feed were reported in 1895, going from door to door with the visiting card of the Minister of Agriculture in their hands (Van Molle, 1989: 205). Using printed matter both implicitly and explicitly for their commercial ends, producers and traders directed cleverly pitched advertising at the farming community. By frequently advertising in the magazines of farmers' associations, they indirectly acquired those associations' seal of approval for their products. Due to the lack of itinerant professional advisers, farmers were easily persuaded to continue right through to the inter-war years to turn to the representatives of feed and fertiliser companies for advice (Duby and Wallon, 1976, IV: 65).

III.7. Uniformity and variation

What was the perceived significance of this wide-ranging system of educating? For contemporaries, it was clear, as, for them, it was a question of doing what was necessary ultimately to bring every farmer to the desired level of professional expertise. From the late nineteenth century on, agricultural education rapidly came to be seen as an achievement, the sense of which was in fact no longer seriously discussed. Pride rose in proportion to the number of initiatives and the number of farmers reached. The World

Exhibition at Paris in 1900 gave a prominent place to agricultural education, featuring France, Germany, The Netherlands and Belgium as countries with a developed form of it, as well as Denmark, Sweden, Italy, Austria, Hungary and Russia. At the prompting of the competent ministries, international congresses were held in 1900 and 1905 on the subject and in 1908 and 1913 on home economics, including agricultural housekeeping. A later initiative was the publication during the late 1930s of three volumes on the state of agricultural education in Europe and North America by the *International Institute of Agriculture*, which had its headquarters at Rome. International comparison of this sort encouraged competition and thus further development. By the time the Second World War broke out, Western Europe and North America had virtually uniform systems of agricultural education (*Rapports et comptes rendus,* 1900; *2e Congrès international,* 1905; *Rapports et comptes rendus,* 1908; *Rapports et comptes rendus du 2e congrès,* 1913; *Agricultural Education,* 1935–1936; *Agricultural Education,* 1938).

Reassessment of developments since 1800 throws up three points. In the first place, there was never a clear dividing-line between the science and the various forms of instruction. As members of agrarian associations, teachers and publicists, agronomists themselves took care of the dissemination of their scientific findings. Indeed, the question has to be asked of whether strict lines can be drawn between pure science, applied science and knowledge acquired empirically. Laboratory research, *in situ* experiments and the many forms of agricultural education formed a complex network of professional and personal relationships within which there was constant interaction. In this respect, mention can be made of Leopold Frateur (1877–1946), veterinary surgeon and renowned specialist in livestock improvement: he was professor at the Louvain Faculty of Agriculture, headed his own zootechnical research institute, was a member of the Central Board of the *Belgische Boerenbond*, published numerous articles in the members' news-sheet and gave hundreds of lectures to local cattle-breeding syndicates.

Secondly, it is often difficult to determine the difference between State and private initiative, at least as regards the continent, where private educational institutions and instruction were subsidised, legally recognised and sometimes taken over by the State. State institutions such as veterinary schools and research stations trained highly skilled staff, conducted analyses and published, but did so partly for the use of the private initiative.

Thirdly, matters were ordered differently in England, at least until the end of the nineteenth century. In relative terms, agricultural research lagged behind (the work done at Rothamsted aside), association life was much more disparate until 1914 and State intervention in agricultural education was a half-century later in coming than on the continent. Whereas the German *Länder* led the way in laboratory research, agricultural education and the co-operative organisation of the farming classes, the English public stuck with exhibitions, competitions and field trials, besides giving particular attention to the mechanisation of agriculture. This all became apparent at the World Exhibition at Paris, where England exhibited not items about agricultural education, but agricultural machinery and experiments of John B. Lawes and Joseph H. Gilbert, who together headed Rothamsted (Vander Vaeren, 1902: 1). It is no easy matter to explain the difference in the development of research and education, although – as Paul Brassley has already pointed out – certainly a key factor was the absence of any sustained guidance and financing of either by the British government. It was a gap that the less evolved British agricultural associations were unable to close before the First World War; contrary to their continental counterparts, moreover, they did not receive any government subsidy for specific services. Brassley also mentions the reluctance, until late in the nineteenth century, of British universities as ivory towers of pure science to open their portals to applied science and to respond to the interests and needs of a wider public. These few points of explanation are in turn part of a range of socio-economic and political differences, and indeed differences of mentality, that is much broader and more difficult to grasp. During the Agricultural Invasion, France and Germany turned to protectionism and Belgium and The Netherlands invested heavily in reorienting agriculture towards the intensive cultivation of vegetables, fruit and ornamental plants, and intensive cattle, small-stock and dairy farming. The typical, British, large estates, on the other hand, entered a period of decline during the 1880s; while Great Britain, confident in the face of the world, was engaged in further expansion of her industrial and colonial empire, she appeared oblivious to the state of her agriculture. It was only after the turn of the century, and feeling the hot breath of American and German competition on her neck, that her agricultural science and education, with State support, began to make up leeway. By 1914, again according to Brassley, the necessary framework had taken shape (Brassley, 2000: 618–622, 647–649).

IV. The reception of the message

In order to impart the science successfully to farmers, appropriate educational methods had to be found. To introduce book-learning was no obvious step. To make this point clear, a brief examination of the method of instruction pertaining in the agrarian world before the educational offensive is necessary. Knowledge was passed from father to son, from mother to daughter, from master to farm-hand, from old to young. There was no specific age at which one learnt; learning began as soon as the child was ready for it and took place on the job, chiefly through illustration and imitation, meaning that no prior book knowledge needed to be acquired and even that few words in fact had to be used. For centuries, farming people had learnt from the great book of experience. Moreover, learning did not take place in a separate bookish atmosphere, but at the place of work; learning and work were not set apart from each other, but were organically interwoven with the process of growing to adulthood. Nor was the teacher an external educator from an alien background, but a person getting on with his occupation in a familiar environment. Learning was thus just a part of general upbringing and social intercourse in the local community.

It was typical of the nineteenth-century academic that he had little respect for this traditional way of learning. Around 1850, Charles Morren wrote indignantly that the farmer was the only Belgian citizen 'die zijne kunst niet leeren kan, noch zich op de hoogte houden van den vooruitgang der wetenschap. Hij ontvangt slechts overleveringen' ('who is unable to learn his craft or keep abreast of advances in science. All he takes in is tradition') (quoted in De Bruycker, 1911: 71). Farmers were unschooled, whereas it was precisely with schooling that nineteenth-century respectable society equated knowledge and ability.

It is nevertheless noticeable that the traditional rural method of learning was not discarded blindly. The *fermes-écoles* applied a method that was very close to it in two ways: in the first place, education was provided in a functioning farm – less alienating than in a school – and by instructors who were at the same time cultivators; secondly, it was based on observation, demonstration and imitation in practice. Other recognisable forms of 'illustrative pedagogy' that permitted farming people to learn with their eyes were competitions, agricultural shows, demonstrations of agricultural machines, and experimental or demonstration fields. Furthermore, and something newer in the equation, such things provided the occasion for informal gatherings and dialogue among the various characters in the agricultural sector, whether it be at village or a broader level.

Besides these methods of illustrative pedagogy, however, the directing elite, including government and farmers' associations, turned increasingly towards more academic and classical forms of agricultural education for both children and adults. Their approach was interventionist. If necessary, the desired modernity had to be hammered in, and an example of this was the Dutch educational manual for primary schools *Landbouwkundig schoolboek*, which won an award in 1783 and which was conceived in terms of a catechism, a form of highly drilled learning through questions and answers designed to be learnt by heart, thereby obviating the necessity of repeating the reading phase. Indeed, the catechism format was very popular throughout the nineteenth century as a means of imparting knowledge to and forming opinion among the working classes and it was made frequent use of in the socialist and catholic election propaganda directed at workers and farmers alike.

It was only at the end of the nineteenth century that a rapidly growing body of writing appeared that explicitly set out the function and methodology of agricultural education (Proost, 1897; Strauch, 1903; De Vuyst, 1913). Before World War I, there were two German reviews devoted entirely to didactics: *Land- und forstwirtschaftliche Unterrichtszeitung* (Vienna) en *Landwirtschaftliche Schulzeitung* (Leipzig). It is noticeable that those writings used a very broad definition of the subject, which included various forms of vulgarisation and self-education. However, most attention was focused on education in the school. Paul De Vuyst (1863–1950), Director-general of the Belgian Ministry of Agriculture and author of a handbook on agricultural education that ran into two editions before the First World War, made a distinction among the aims of elementary vocational education (for boys between the ages of twelve and fourteen) of secondary vocational education and of tertiary (academic) agricultural education. The first level was to suffice for the sons of smallholders and the second for those who were later to run medium-sized farms; the last, he deemed necessary for the sons of gentlemen farmers and big landowners. What he proposed for the first level was no longer simply learning by doing, but to submit pupils to a methodical, empirical approach. For the second level and above, his proposal was for a strictly academic approach: 'Il (l'agriculteur) lui faut une préparation intellectuelle très étendue, il doit se rendre compte du pourquoi des choses, il doit posséder des notions exactes sur la chimie et sur la physiologie végétale et animale, sur les propriétés des terres et des engrais; il doit connaître les parasites et les maladies qui ravagent les cultures. Les jeunes agriculteurs doivent être mis à même d'aborder la lecture de revues ou d'ouvrages spéciaux, et d'entendre avec fruit les conférences agricoles en vue de perfectionner leurs opérations journalières' ('He, the farmer, requires very broad intellectual preparation, he must verify the 'why' of things, he needs to have a good grasp of chemistry and of plant and animal physiology, of the properties of soils and fertilizers; he must know about the parasites and sicknesses that destroy crops. Young farmers must also be trained to read magazines and specialist reading-matter, and to draw benefit from what they hear at agricultural conferences, with a view to perfecting their daily operations') (De Vuyst, 1913: 97–98, 164–165).

These aims were certainly ambitious. They inevitably implied that the traditional system of education would largely be replaced by a transfer of knowledge through external agents, in the classroom, in the meeting-room of the local farmers' association, or through reading-matter. Farming people were treated to a discourse on physics, chemistry, botany, zoology and zoo-technics, soil science, agricultural economics, farm book-keeping and commercialisation. Discourse was also used to make clear to them that they should adapt their farms to the economic imperatives of the world market. It was also used systematically to detach them from their local setting and integrate them into the educated world – *das Bildungsbürgertum* – that employed the universal language of science. For the farming community itself, this must have been a radical and even alienating change. It was being made subject to a sort of intellectual 'colonisation', a process of *Entbäuerlichung* (being weaned away from the old farmer's identity)[15] and retraining to become what De Vuyst called 'le cultivateur du progrès' ('the modern farmer'). Over and above the traditional differences among small, medium and large farms, a new dichotomy was created, namely that between the traditional and the modern, between the technically inferior and the technically superior.

The methodological writings leave no room for doubt about the so-called 'modernisation of agriculture' being a process orchestrated from the top through many avenues of instruction, but what effect did the instruction have? The tangible effect is fairly easy to measure; it can be done on the basis of the number of seed-drills, threshers, milking machines, tractors, etc., and the expanding use of chemical fertilisers and other aids. We can also calculate the increasing number of agricultural colleges and students attending them, the number of periodicals and sometimes their circulation, too. Additionally, we can calculate the amount of money devoted to agricultural research and education – certainly insofar as public funds are concerned – and, to a certain extent, the financial input of farmers' associations. In this way, it is perhaps possible to estimate how many farmers were touched by some form of agricultural education. But what was the relationship between the various forms of instruction and the process of innovation? The impact of the agricultural press ought not to be rated high at all: until the mid-nineteenth century, periodicals were often very luxurious and expensive publications. In 1889, fewer than one in ten English farmers was subscribing to an agricultural periodical (Goddard, 2000: 673). Prior to 1940, the level of participation in classical forms of agricultural education was nowhere high. For the nineteenth century as a whole, the number of pupils at agricultural schools was minimal and it was only during the twentieth century that the intake of those schools shifted (albeit very slowly) from the rural elite to the actual farming population. A survey conducted in France in 1955 revealed that just 5% of those currently engaged in agriculture at that time had been at an agricultural school, and that was more than 100 years after the first piece of French legislation concerning agricultural education (Boulet, 2003: 39). The Belgian agricultural census of 1979 indicated that barely 22.4% of persons running farms or similar enterprises had followed secondary agricultural education and a fifth of those had not got beyond lower secondary level (Everaert, 1986).

Figures, of course, do not tell the whole story. The impact of agricultural school education cannot be rigorously assessed, because representative samples of farmers with and farmers without that education, together with statistics on their farm's productivity,

[15] The term *Entbäuerlichung* was launched in this context by the German agricultural historian W. Achilles in 1989, see Trossbach and Zimmerman, 1998: 147–149.

do not exist (Brassley, 2000: 646). Indeed, when one considers that farmers also learn from each other, where is the dividing line between the skilled and the unskilled? Where is the dividing line between those giving instruction in agriculture and the farmers who are the object of that instruction? Who is influencing whom? Goudswaard cites ten or so examples of big farmers and landowners from the first half of the nineteenth century who, on their own farms, provided their tenants or poor children from the vicinity with free agricultural education (Goudswaard, 1986: 92–93). Some of those who gained a diploma took on the role – each in his own family and immediate circle – of intermediary between agricultural science and practice. It could well have been the case that, in turn, farmers' sons with diplomas and who had gone into agricultural research or public administration exercised a bottom-up influence. On the selling side, reference has also to be made to the numerous personal and financial links between (bigger) farmers and manufacturing: more particularly, sugar refineries, dairies (co-operative or otherwise) and preserved food factories. Farmers also innovated in response to guaranteed sales. However, more archival and prosopographic research is required if the links between the various players are to be made apparent. Who, in fact, were the members of the *sociétés savantes* or learned societies and the farmers' associations? Who studied at agricultural schools and what employment did those with diplomas go into?

A more difficult task is to throw light on the change in mentality among farmers and their preparedness to put what they had learnt into practice. To classify them as 'conservative' would be to use a cliché and be too simple. In an article, Haushofer stated that farmers had never displayed any *generelle Bildungsfeindlichkeit* (general hostility towards education) and that the most that could be said was that there was a difference between the progressive- and the conservative-minded (Haushofer, 1970). There are interesting examples that confirm that farmers did not adopt innovation regardless, including the published, elaborate letter written by Ollivier Le Diouron (he called himself 'a simple farmer') in 1843 to the sub-prefect of his district in France, in which he describes how he was prompted to abandon the old routines in favour of the improved instruments of which he had for years been a notorious opponent (Le Diouron, 1843 [2002]). There are, of course, explanations for the proverbial resistance of farmers to change. Farmers perhaps try harder to maintain their income than to raise it, try to minimise risk rather than maximise profit. Their so-called 'conservatism' may spring from a lack of capital turnover or from the habit of systematically investing profit in the purchase of land, rather than in capital goods. Duby, for example, calls French farmers 'trop terrienne' (Duby and Wallon, 1976, III: 465–467) ('as paying too much attention to the factor of land'). Furthermore, the small farm was little suited to mechanisation, and the tenancy system, including *métayage* (whereby a tenant farmer paid a proportion of his rent in kind), served to smother innovation. A further curb, lastly, was agricultural protectionism (as in France), which served to safeguard the interests of the big landowners.

A number of other factors could have forced or prompted farmers to innovate. These include the flight from the countryside, which encouraged the replacement of manual labour by mechanisation; tax pressure, the free trade policy of the late nineteenth century (as in Belgium and The Netherlands); and expanding credit facilities. Blomme states (1992: 264) 'the enormous growth of professional agricultural associations' in Belgium as 'the most plausible explanation' for the acceleration in the growth of productivity during the inter-war period. The compelling appeal of the farmers' associations to their members certainly appears to have been very convincing, but interpretations in recent

German writings seem to be less one-sidedly top-down. Micro-research into daily life indicates that individuals and groups in the past could be self-willed in their strategic decision-making. Those writings emphasise the bottom-up process: more particularly, the active part played by farmers themselves in looking for new techniques, broader markets, the defence of interests vis-à-vis the government, etc. They had recourse to innovation only when innovation was viable in terms of finance and organisation of their households, and when it would manifestly provide benefit. It was thus not pressure from external experts, but the *Eigensinn* or self-will of farmers that was the ultimate deciding factor in the process of innovation (Gudermann, 2001: 432–433, 448–449).

Rather than a vertical, top-down diffusion of knowledge, therefore, knowledge circulated and did so on the basis of negotiation between farmers and the entire network of experts that began increasingly to surround them from early in the nineteenth century. Supply and demand led to an accumulation of knowledge and also to selection and compromise. The transmission of new farming techniques appears to have been intensely varied, diffuse and very informal. Over and above the institutions (government, schools, associations, industry, etc.) and their methods of communication (periodicals, exhibitions, advertising, etc.), individual persons and informal social networks have probably played a much greater role than one would originally have thought. What, for example, was the role of the veterinary surgeon nominated by the Frenchman André Sanson in 1858 as the missionary of agricultural progress (Denis, 2000: 223–227)? The changes in farming practices are the result of a very complex process of interaction among government, scientists and publicists, teachers and professional advisers, industrialists, traders and bankers, and farmers (see Figure 8.3).

Figure 8.3 The knowledge network

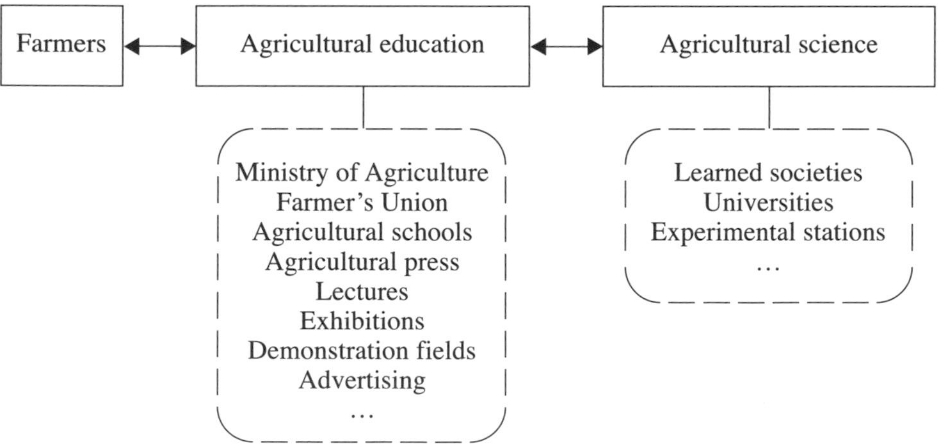

V. The hidden curriculum

At this point, it is useful to return to the central question, namely that of the aim of the educational offensive in agriculture during the 1800–1940 period. Ostensibly, of course, it was to raise agricultural productivity, improve the methods of production (including the use of labour-saving techniques) and maintain the level of farmers' incomes. However, close reading of the primary sources teaches that there was much more. The hierarchical system of rural education was also created to maintain and reinforce class and gender differences. It was a paternalistic system with a much wider agenda than simply the dissemination of science and techniques into the sphere of the rural public: in the minds of the agricultural reformers, the idea of social control and disciplining was never far away. Education was used as a political instrument, whereby the transfer of knowledge was not neutral, but intentional in the sense of combating the exodus from the countryside and maintaining the social, political and religious *status quo*.[16] According to De Vuyst, the aim of agricultural education was chiefly to mould farmers and their womenfolk into the 'desired' type of citizen, highly professional and worthy: 'Les professeurs d'agriculture doivent, en outre, contribuer à l'éducation morale et sociale des élèves, leur inspirer une haute et ferme conviction de l'importance sociale de la profession de cultivateur, et les rendre aptes à la remplir dignement' ('Those teaching agriculture, moreover, have to contribute to the moral and social education of their pupils, to inspire them to an elevated and firm conviction of the social importance of the profession of farmer, and to provide them with the ability to follow that profession with dignity') (De Vuyst, 1913: 264 and 384–385).

An excellent example of this is provided by the teaching of home economics. Until recently, the womenfolk of farmers have been excluded on three fronts as a subject of research: agricultural history has been researched wholly from the male angle; women's history has given central place to proletarianised urban female workers or high-profile feminists; and educational history has ignored female vocational training. This situation is now changing.[17] With the technological innovations in arable farming, dairying, poultry-keeping and horticulture, it has become increasingly clear to what degree the nature of women's work has changed: from being productive to being much more reproductive and centred on the family. An inevitable consequence of the technological and economic changes in agriculture, moreover, has been the change in the relationship of man to woman: on the one hand, farming adopted innovation and gained respectability; on the other, domestic tasks continued to be done in traditional fashion and lost status. A gulf thus arose between farming and housekeeping. The tasks of women became restricted to the assigned roles of the female as cook (making excellent marmalades and cakes), seamstress, cleaning woman, and mother for the upbringing of the next generation of farmers, and thus increasingly excluded that of small-scale producer on the farm (vegetables, flowers, poultry and dairy products). How was it, then, that women could continue to be tied to the countryside? During the late nineteenth century, agricultural housekeeping instruction was introduced and was presented as a major weapon in the struggle to stem the exodus from the countryside (De Vuyst, 1913: 121–126). Via that

[16] See, among others, Digby, 2000.

[17] See, for example, 'Women and rural history', special issue of *Rural History,* 5, 1994: 2; Morgan, 1996; Thompson, 1999; Van der Burg, 2002.

instruction, the womenfolk of farmers were separated not only from urban womenfolk, but also from the typical male labour on the farm. The fact of giving young rural girls their own education made them – in the same way as young boys – respectable. In turn, the Women's Institutes in Great Britain and the *Boerinnenbonden* (Associations of Farming Women) in The Netherlands and Belgium contributed to the idea that housewives were 'skilled workers' within the narrow confines of the private sphere. In 1919, France launched *semaines agricoles féminines* (agricultural weeks for females) with the prime purpose of 'garder des jeunes filles à la campagne qui soient heureuses d'être des femmes de paysan' ('keeping young girls in the countryside who would happily become farmers' wives') (Couffin, 2000: 63–66).

In the same way that they promoted their rural housewives to 'skilled workers', the Nation States attempted to transform their peasants into modern farmers. The agricultural transformation process had to contribute towards national prestige on the back of a process of economic upgrading. However, farmers were never permitted to renounce their background, their *Heimat*, and particular socio-cultural nature. In the discourse, they were presented as the most authentic of people, the guardians of national morals and culture. For their part, the churches lauded farming people as being the most devout of believers. In order to maintain the link between belief and the farming classes, village priests assumed the role of agricultural advisers and played a leading part in farmers' associations (e.g. Baratay, 2003). Political parties, especially on the conservative wing, turned to the farming community as their 'natural' electorate. State, religious and political interests converged in the programmes for public and private agricultural education, in which farming people, male and female, were made into the desired type of citizen, the eminent representatives of the authentic folk and race, the most stable guardians of political order and the faith. The reduction of rural life to the level of folklore was not far away. England measured her true 'Englishness' by the country's typical landscapes and cottages. The Netherlands promoted herself with windmills and laughing peasant girls in clogs. Despite the urbanisation and industrialisation, Flanders was labelled as being authentically rustic and religious. In Germany, *Heimat*-mindedness triumphed in literature, music and folklore, and ultimately in the *Blut und Boden* (Blood and Soil) theory.[18] The result of the entire offensive to introduce science and technical modernity through education was a painful paradox. Despite the praising of the authenticity of the rural way of life and virtues, the educational offensive in respect of the agricultural sector led directly to an accelerated decline in agrarian employment, the disappearance of picturesque landscapes and the rise of bourgeois and urban lifestyles.

[18] See, among others, Matless, 1998; Burchardt, 2002; De Jong, 2001; Van Molle, 1989; Pil, 1990; Corni, 1990; Eidenbenz, 1993.

Bibliography

2e Congrès international de l'enseignement agricole. Rapports et documents préliminaires (1905) Liège.

50 jaar Nederlandstalig diergeneeskundig onderwijs aan de R.U.G. (1984) Ghent.

125 jaar tuinbouwonderwijs te Vilvoorde. 1848–1973 (1973) Vilvoorde.

Agricultural education in the world (1935–1936) vols 1 and 2: *Europe*, Rome.

Agricultural education in the world (1938) vol. 3: *North America*, Rome.

Aelbroeck, J.L. van (1830) *L'agriculture pratique de la Flandre*, Paris.

Agulhon, M., Désert, G. and Specklin, R. (1976) 'Apogée et crise de la civilisation paysanne 1789–1914', in G. Duby and A. Wallon (eds), *Histoire de la France rurale*, vol. III, Paris.

Antoine, R. and Hennebert, G. (1985) 'La Faculté des sciences agronomiques de l'Université catholique de Louvain 1778–1985', *Revue trimestrielle des amis de l'Université catholique de Louvain,* 2, pp. 3–66.

'Association des Ingénieurs sortis de l'Institut Agronomique de L'Université de Louvain. Liste des publications des membres' (1910), *Revue générale agronomique*, 19, 6–7, pp. 255–287.

Baratay, E. (2003) 'Les soutanes à la ferme. Le clergé et l'essor de l'élevage XVIIIe–XXe siècles', in *Chrétiens et sociétés XVIe–XXe siècles. Bulletin du Centre André Latreille et de l'Institut d'Histoire du Christianisme*, Lyon, vol. 10, pp. 109–131.

Barral, P. (1966) 'Note historique sur l'emploi du terme "paysan"', *Etudes rurales*, pp. 72–80.

Béguet, B. (1990) *La science pour tous. Sur la vulgarisation scientifique en France de 1850 à 1914*, Paris.

Blomme, J. (1992) *The economic development of Belgian agriculture 1880–1980. A quantitative and qualitative analysis*, Brussels.

Boulaine, J. (1992) *Histoire de l'agronomie en France*, Paris.

Boulet, M. (ed.) (2000) *Les enjeux de la formation des acteurs de l'agriculture, 1760–1945*, Dijon.

Boulet, M. (ed.) (2003) *La formation des acteurs de l'agriculture en France. Continuités et ruptures, 1945–1985*, Dijon.

Bourrigaud, R. (2000) 'Aux origines des organisations professionnelles agricoles: les comices agricoles du siècle dernier', in M. Boulet (ed.), *Les enjeux de la formation des acteurs de l'agriculture, 1760–1945*, Dijon, pp. 173–177.

Brassley, P. (2000) 'Agricultural science', in E.T.J. Collins (ed.), *The agrarian history of England and Wales, vol. VII/1: 1850–1914*, Cambridge, pp. 594–649.

Bruycker, C. De (1911) 'Het land- en tuinbouwonderwijs', in *Vlaamsch België sedert 1830. Studiën en schetsen bijeengebracht door het algemeen bestuur van het Willemsfonds ter gelegenheid van het jubeljaar 1905*, vol. V, Ghent, pp. 62–134.

Burchardt, J. (2002) *Paradise lost. Rural idyll and social change in England since 1800*, London.

Burg, M. van der (2002) *'Geen tweede boer'. Gender, landbouwmodernisering en onderwijs aan plattelandsvrouwen in Nederland, 1863–1968*, Hilversum.

Buttress, F.A. (1950) *Agricultural periodicals of the British Isles, 1681–1900, and their location*, Cambridge.

Buyst, E., Smits J.-P. and Zanden, J.L. van (1995) 'National accounts for the Low Countries: the Netherlands and Belgium 1800–1990', *Scandinavian Economic History Review,* XLIII: 1, pp. 53–76.

Campbell, B. and Overton, M. (1998) 'L'histoire agraire de l'Angleterre avant 1850. Bilan historiographique et état actuel de la recherche', *Histoire et Sociétés Rurales,* 9, pp. 77–105.

Cocaud, M. (2000) 'Des cadres pour la rénovation agricole: les sociétés d'agriculture', in M. Boulet, (ed.), *Les enjeux de la formation des acteurs de l'agriculture, 1760–1945*, Dijon, pp. 199–206.

Code van het landbouwonderwijs (1935) Brussels.

Collins, E.T.J. (ed.) (2000) *The agrarian history of England and Wales,* vol. VII/1 and VII/2: *1850–1914*, Cambridge.

Collins, E.T.J. (2000) 'Rural and agricultural change', in E.T.J. Collins (ed.), *The agrarian history of England and Wales*, vol. VII/1, pp. 138–140.

Corni, G. (1990) *Hitler and the peasants. Agrarian policy of the Third Reich*, New York.

Couffin, C. (2000) 'Les semaines agricoles féminines', in M. Boulet (ed.), *Les enjeux de la formation des acteurs de l'agriculture, 1760–1945*, Dijon, pp. 63–66.

Daum, A. (1998) *Wissenschaftspopularisierung im 19. Jahrhundert. Bürgerliche Kultur, naturwissenschaftliche Bildung und die deutsche Öffentlichkeit, 1848–1914*, Munich.

David, J. (1975) *Boeken betreffende de Belgische landbouw*, Centre belge d'histoire rurale, Publication No. 47, Louvain.

Dejongh, G. (1999) 'New estimates of land productivity in Belgium, 1750–1850', *The Agricultural History Review* 47: 1, pp. 7–28.

Dejongh, G. (2000) 'De Belgische landbouw in een periode van transitie, 1750–1850. Een nieuwe bijdrage tot een onvoltooid debat', *Belgisch Tijdschrift voor Filologie en Geschiedenis,* 78, pp. 470–496.

Denis, B. (2000) 'André Sanson et les vétérinaires "missionnaires du progrès agricoles"', in M. Boulet (ed.), *Les enjeux de la formation des acteurs de l'agriculture, 1760–1945*, Dijon, pp. 223–228.

Dictionnaire de l'Académie Française (1798), Paris.

Digby, A. (2000) 'Social institutions', in E.T.J. Collins (ed.), *The Agrarian History of England and Wales, vol. VII/1: 1850–1914*, Cambridge, pp. 1465–1500.

Duby, G. and Wallon, A. (eds) (1976) *Histoire de la France rurale*, vols III and IV, Paris.

Eidenbenz, M. (1993) *"Blut und Boden". Zu Funktion und Genese der Metaphern des Agrarismus und Biologismus in der nationalsozialistischen Bauernpropaganda*, Bern.

Encyclopaedia Britannica (1999) CD-Rom.

Encyclopédie ou dictionnaire universel raisonné des connoissances humaines (1770), Yverdon, vol. I.

Everaert, H. (1986) *Boer en landbouwonderwijs*, Brussels. (LEI-nota's; 103).

Finlay, M. (1988) 'The German agricultural experiment stations and the beginnings of American research', *Agricultural History*, LXXII: 2, pp. 41–50.

Gadisseur, J. (1990) *Histoire quantitative et développement de la Belgique au XIXe siècle*, vol. IV, 1a*: Le produit physique de la Belgique 1830–1913, Agriculture*, Brussels.

Gadisseur, J. (1993) 'Nouvelles perspectives pour l'histoire de l'agriculture belge', *Académie royale de Belgique. Bulletin de la Classe des Lettres et des Sciences Morales et Politique,* 6e série, vol. IV, pp. 41–63.

Gervais, M., Jollivet, M. and Tavernier, Y. (1976) 'La fin de la France paysanne de 1914 à nos jours', in G. Duby and A. Wallon (eds), *Histoire de la France rurale*, vol. IV, Paris.

Goddard, N. (1989) 'Agricultural literature and societies', in G.E. Mingay (ed.), *The agrarian history of England and Wales,* vol. VI: *1750–1850*, Cambridge, pp. 361–383.

Goddard, N. (2000) 'Agricultural institutions: societies, associations and the press', in E.T.J. Collins (ed.), *The agrarian history of England and Wales, vol. VII/1: 1850–1914*, Cambridge, pp. 650–690.

Goossens, M. (1993) *The economic development of Belgian agriculture: a regional perspective, 1812–1846*, Brussels.

Goudswaard, N.B. (1986) *Agrarisch onderwijs in Nederland 1783–1983. Hoe het wor(s)telde en groeide*, Culemborg.

Gudermann, R. (2001) 'Neue Forschungen zur Agrargeschichte', *Archiv für Sozialgeschichte*, 41, pp. 432–449.

Halleux, R. [et al.] (eds) (1998–2001) *Geschiedenis van de wetenschappen in België 1815–2000*, 2 vols, Brussels.

Haushofer, H. (1963) *Die deutsche Landwirtschaft im technischen Zeitalter, Deutsche Agrargeschichte*, vol. V, Stuttgart.

Haushofer, H. (1970) 'Bauer und Schule in der Geschichte', *Zeitschrift für Agrargeschichte und Agrarsoziologie*, pp. 1–15.

L'Institut agricole de l'Etat à Gembloux. 1860–1910 (1910) Brussels.

Jas, N. (2000) *Au carrefour de la chimie et de l'agriculture. Les sciences agronomiques en France et en Allemagne, 1850–1914,* Paris.

Jong, A. de (2001) *De dirigenten van de herinnering. Musealisering en nationalisering van de volkscultuur in Nederland 1815–1940*, Nijmegen.

Klemm, V. (1992) *Agrarwissenschaften in Deutschland. Geschichte – Tradition, von den Anfängen bis 1945*, St. Katharinen.

Le Diouron, O. (2002) *Progrès agricoles d'un simple fermier* (Guingamp, 1843) (Archives départementales Côtes d'Armor); reprint in *Histoire et Sociétés Rurales,* 17, pp. 209–217.

Lindemans, P. (1952) *Geschiedenis van de landbouw in België*, 2 vols, Antwerp.

Mammerickx, M. (1967) *Histoire de la médecine vétérinaire belge, Verhandelingen van de Koninklijke Academie voor Geneeskunde van België*, 2nd series, Book V, No. 4, Brussels.

Matless, D. (1998) *Landscape and Englishness*, London and New York.

Mayaud, J.-L. (2000) 'Les comices agricoles et la pédagogie de l'exemple dans la France du XIXe siècle', in M. Boulet (ed.), *Les enjeux de la formation des acteurs de l'agriculture, 1760–1945*, Dijon, pp. 253–257.

Moerman, J. (ed.) (1970) *Liber memorialis 1920–1970. Faculteit van de Landbouwwetenschappen*, Ghent.

Molle, L. van (1984) 'Le centenaire du Ministère de l'Agriculture. La politique agricole belge dans son contexte économique, social, politique et administratif au cours de la période 1884–1984', *Agricontact*, No. 154, pp. 1–147.

Molle, L. van (1989) *Katholieken en landbouw. Landbouwpolitiek in België 1884–1914*, Louvain.

Molle, L. van (1990) *Chacun pour tous. Le Boerenbond belge 1890–1990*, Louvain. (KADOC-Studies; 9).

Morgan, M. (1996) 'Jam making, cuthbert rabbit and cakes: redefining domestic labour in the Women's Institute, 1915–1960', *Rural History,* 7, pp. 207–219.

Muth, H. (1968) '"Bauer" und "Bauernstand" im Lexicon des 19. und 20. Jahrhunderts', *Zeitschrift für Agrargeschichte und Agrarsoziologie,* 16, pp. 72–98.

Overton, M. (1996) *Agricultural revolution in England: the transformation of the agrarian economy 1500–1850*, Cambridge.

Paarlberg, D. and Paarlberg, Ph. (2000) *The agricultural revolution of the twentieth century*, Ames (Iowa).

Pastoret, P.P., Mees, G. and Mammerickx, M. (1986) *De l'art à la science ou 150 ans de médecine vétérinaire à Cureghem*, Brussels.

Pil, L. (1990) (ed.) *Boeren, burgers en buitenlui. Voorstellingen van het landelijk leven in België vanaf 1850*, Louvain.

Proost, A. (1897) *La pédagogie moderne et la pédagogie empirique*, Brussels.

Rapports et comptes rendus du congrès international de l'enseignement agricole (1900) Paris.

Rapports et comptes rendus du congrès international de l'enseignement ménager de Fribourg (1908) Fribourg.

Rapports et comptes rendus du 2e congrès international de l'enseignement ménager tenu à Gand (1913) Brussels.

Russell, E.J. (1966) *A history of agricultural science in Great Britain 1620–1954*, London.

Schling-Brodersen, U. (1989) *Entwicklung und Institutionalisierung der Agrikulturchemie im 19. Jahrhundert: Liebig und die Landwirtschaftlichen Versuchsstationen*, Braunschweig.

Sigaut, F. (2000) 'Les origines européennes de l'enseignement agricole', in M. Boulet (ed.), *Les enjeux de la formation des acteurs de l'agriculture, 1760–1945*, Dijon, pp. 489–496.

Slicher van Bath, B.H. (1960) *De agrarische geschiedenis van West-Europa 500–1850*, Utrecht-Antwerp.

Strauch, K. (1903) *Didactik und Methodik der Unterrichts im landwirtschaflichen Schulen*, Leipzig.

Sykes, J.D. (1981) 'Agriculture and Science', in G.E. Mingay (ed.), *The Victorian Countryside*, vol. 1, London, pp. 260–272.

La Terre récompensée. Primes d'honneur, prix et médailles agricoles en France de 1857 à 1895 (2002) CD-Rom, Dijon.

Thompson, L. (1999) 'The promotion of agricultural education for adults: the Lancashire Federation of Women's Institutes, 1919–45', *Rural History*, 10: 2, pp. 217–234

Trigaut, J. (1899) *Les bibliothèques agricoles en Belgiques et dans les pays étrangers*, Brussels.

Trossbach, W. and Zimmerman, C. (eds) (1998) *Agrargeschichte, Positionen und Perspectiven, Quellen und Forschungen zur Agrargeschichte*, vol. 44, Stuttgart.

Vaeren, J. Vander (1902) *L'enseignement agricole à l'étranger*, Brussels.

Vaeren, J. Vander (1913) 'L'organisation et la situation actuelle de l'enseignement agricole en Belgique, *Revue des questions scientifiques*, pp. 5–39.

Vuyst, P. De (1913[2]) *L'enseignement agricole et ses méthodes*, Brussels.

Wetham, E.H. (1978) *The agrarian history of England and Wales*, vol. VIII: *1914–1939*, Cambridge.

Zanden, J.L. van (1991) 'The first green revolution: the growth of production and productivity in European agriculture, 1870–1914', *Economic History Review*, 44, pp. 214–239.

Zimmerman, C. (1999) 'La modernisation des campagnes allemandes (XIXe–XXe siècle). Les apports de l'historiographie récente en Allemagne', *Histoire et Sociétés Rurales*, 11, pp. 87–108.

9 Changing tastes. The role of scientific and medical discoveries in changing the modern diet

Rayna GAVRILOVA, University of Sofia

The changes in behaviour of individuals and social groups over time has elicited strong interest not only among historians (for whom it was natural), but also among philosophers, sociologists, economists and anthropologists, to name just a few. The choice among various possibilities that humans make in their everyday life is especially attractive as it brings together considerations from opposite spheres: private and public, individual and collective, economic and cultural. The recurring and stable patterns of liking and disliking are generally referred to as *taste*. The first association that comes to mind in this respect, is food.

This chapter is an attempt to tie together the loose ends of a few specific factors affecting the changes in human nutrition: the influence of scientific and medical knowledge on food and eating. This subject has been the centrepiece of a research effort only sporadically.[1] Many related disciplines such as the history of science, history of medicine, nutritional science, biology or sociology have touched on problems such as the birth of scientific nutrition, the discovery of the chemistry of food elements, the development of public health, the interest and redefinition of 'good' nutrition, etc.[2] But so far, only Kamminga and Cunningham's collection (1995) has addressed the relationship between scientific knowledge and nutritional culture in a systematic way.[3]

In the following pages I shall try to summarise and present the major contributions in our understanding of the role of science in shaping food preferences. The chapter has no ambition to add new knowledge, but rather to place the existing one in its social and cultural setting. My background of a historian and my experience as an anthropologist clearly define the paradigm within which I conceive of my task, i.e. the methods and the questions asked in the field of cultural studies. Anthropological tastes account for my holistic approach and interpretative position, while my sympathies with the *nouvelle histoire* bring in the interdisciplinarity and the interest in the *longue durée*.

---

[1] As Kamminga and Cunningham (1995: 1) wrote in their Introduction: 'Very little work has hitherto been done within the history of nutrition which firmly links the scientific domain with the social, which looks at the relation between the production of scientific knowledge about nutrition and the social and political variations that have entered into the promotion and application of such knowledge. The aim of this book is to draw attention to these cultural dimensions to the history of nutrition science'.

[2] Teuteberg (1992a: 4) wrote 'The modern nutrition-related sciences... have until now maintained an exclusively scientific focus... there is some brief information on the genesis of nutritional science but one generally doesn't know how far the inventions of the natural sciences actually changed people's food habits'.

[3] Interesting although passing observations are provided by Oddy and Burnett, 1992; Barker, McKenzie and Yudkin, 1966. The general influence of science on culture is discussed by Cooter and Pumfrey, 1994. A general survey of the chemical composition of food, nutrition, deficiency diseases, food disorders, diets and major debates between scientists is provided in Kiple and Coneè Ornelas, 2000.

I. What is taste?

In order to find our way in the maze of publications on taste I propose to use a definition phrased (albeit not invented) by Heinz-Dieter Meyer (2000: 36) of taste being 'collective interpretative activity in specific institutional context and through culturally defined rhetorics'. It is immediately clear that when we talk about taste we confuse two distinctive phenomena – aesthetic taste and gustatory taste. The mutual referrals between the two concepts are evident and important, even more so in my culturalist approach to changing food habits. When Roland Barthes (1961) writes that between the real product and the consumer 'there is a considerable production of false perceptions and values', he is considering precisely this socially shared interpretative activity. Some of the most influential sociological texts have used taste, displayed by certain social groups as important analytical tool, moving freely between the taste as consumer behaviour and as social behaviour (the 'taste-as-refinement') (Simmel, 1950; Veblen, 1953; Bourdieu, 1979).

However, this correlation is not unproblematic. Senses play a significant role in the very foundation of the Western civilisation as the 'metaphorisation of the senses was widely used to found philosophical ideas' (Synnott, 1991: 174).[4] 'Even if there is a commonality between aesthetic taste and gustatory taste [...] it does not follow that there is an isomorphism', claims Shiner (1979: 163) and locates the difference in the presence or absence of 'objective' (my term) criteria that could be indicated and defined, not only demonstrated. In the eighteenth century taste lost its sensuousness, and became the model for forming judgements. This ambiguity of the taste as a measure of judgement and a measure of sensual experience, once again, roots it simultaneously in two fields that impose a parallel examination of its biological and cultural aspects.

Another important distinction when we speak about gustatory taste, is the difference between taste as attribute of food itself (good, bad, salty, sour, etc. – the sensation of a dissolved object penetrating the taste buds on the tongue and surfaces of the mouth), and the taste of the consumer as attitude toward the food. The second meaning is the one used in this chapter, as I am not interested in the contribution of chemical science to the quality of food per se (i.e. improving taste), but rather in the social impact of scientific knowledge on collective nutritional behaviour.

The cultural phenomenon of the taste is strongly correlated but not identical to the social phenomenon of food habits (food choice). The taste is an interpretative activity objectivised in food choice, and as a cultural phenomenon it displays the characteristics of a 'pattern'.[5] It is shaped mainly by cultural and economic considerations (though

---

[4] As Hume noted in the middle of the eighteenth century: 'Where the organs are so fine as to allow nothing to escape them, and at the same time so exact as to perceive every ingredient in the composition, this we call delicacy of taste, whether we employ the term in the literal or metaphorical sense' (Hume, 1965: 11). In Kant's classification based in subjectivity or objectivity the objective senses, hearing and sight and, with qualifications, touch, give the subject knowledge of the external object without arousing consciousness of the affected organ. The subjective senses, taste and smell, give the subject 'more [of] an idea of our enjoyment of the object than knowledge of the external object' (Kant, 1974: 32–33). Taste is slightly more complicated since it has been metaphorised into the sense of judgement.

[5] In the sense of Benedict, 1959.

sometimes by political ones, for example food boycotts). The relative weight of the two groups of factors is assessed differently by different authors, depending of their methodological preferences.[6] This pattern in the nutritional culture of a specific group or area as affected by the production of new scientific knowledge deserves further investigation. Further in this chapter the economic factors (and the problem of food choice altogether) are omitted on purpose from the analysis in order to devote my full attention to the influence of 'ideas' on 'ideas', the impact of knowledge on culture.[7]

The best research on the evolution of gustatory taste as social behaviour and the factors, which influenced it, is Stephen Mennell's *All Manners of Food* (1985). Mennell brings together the ideas of several authors and discusses the most influential schools of thought on taste, answering essentially the question 'how social groups develop standards of taste' (p. 2). Cecilia Novero (2000) has studied this evolution in Germany and reached a similar conclusion that 'the use of the discourse of taste (good versus bad taste) as a socially discriminating factor has been the fundamental characteristic of bourgeois cuisine since its inception'.

The question 'What influences gustatory taste?' has elicited a number of interesting studies focusing on ethnicity and age, standards of living and nutrition (e.g., Devine et al., 1998; Devine et al., 1999; Fogel, 1986). Claude Fischler devotes one chapter to the formation of taste in his book *L'Homnivor*e (1990), and he explores the elements and the mechanisms of this formation – the genetic transmission, the cultural transmission and the mechanisms of acquiring new tastes. Most authors agree in general on the necessity of an open approach (Claudian et al., 1969; Corbeau, 1988). Which comes first – culture, nature or economics? – is, as usual, a misleading question, but I am inclined to sustain the idea that culture should be considered a primary factor as all other factors necessarily operate through the culture. Traditional food habits – sometimes called foodways – are part of the traditional culture, the influence of which lingers on long after the radical social transformation of modernity. The choice of certain kinds of generic foods over others, the ways and means of preparation, and especially the festive and ritual foods form a substratum of food preferences that not only persists as specific tastes but also influences the selective choice of innovation. The culinary culture of the social group is shaped by ecological settings and the natural availability or possibility to grow certain foods and ecology, as Marvin Harris (1985) believes, influences not only tastes but deeper nutritional structures. Economic structures and technology influence availability and distribution, but economic value is often translated in cultural and social terms (Aymard, 1973; Hammond, 1976). Pre-modern societies possess a comprehensive system of assessing good and bad food, based on experience, prescriptions and proscription of cultural and social order and availability. The modern era articulates this knowledge in the form of science and popular science. Although it would be very difficult to distinguish between the impact of the modern way of life in urbanised and industrialised environment, and the impact of the scientifically informed choices, there is no doubt that advance in the science of nutrition has radically reshaped the attitude

[6] Perrot (1961: 292), for instance claims that economic factors play a lesser role in changing food habits than changing tastes, which, on their turn, are transformed by 'the greater awareness of the rules of nutrition'.

[7] A general conceptual model of the influence of economic factors could be found in Furst et al. (1996). See also Golton (1986).

toward food and eating. Knowledge about biology assumes an important place in the existential thinking (Booth, 1982; Leviz, 1991). According to Adel den Hartog (1992), after the disappearance of malnutrition as a social danger, the problem was essentially transformed into a cultural one: how to make the best possible choice from a large range of different foods, how to use our sense of judgement.[8]

II. Nutrition in the modern era

Any attempt to sum up the fundamental cultural transition to modernity in a few introductory sentences is bound to be reductionist. Among the several parallel developments that changed the European scene it would be helpful to list these selected few, which affected more directly the way people perceived and ate food. It was already mentioned above that the concept of taste emerged as a general cultural category precisely at the threshold of the early modern era, and developed intensely during the nineteenth century. Two central concepts of the modern and enlightened era are especially relevant to this development: the trust in rational knowledge and the valorisation of the autonomy of the individual. The first concept changed dramatically the ways in which humans related to the external world and to their own bodies (Lupton, 1994 and 1996). The reinvention of the sciences and the humanities set out the sequence of discoveries that changed the structures of everyday life, including tastes and eating. The establishment of the concept of the autonomy of the individual on the other hand had two special consequences: the legitimisation of the right to and the widening of the practice of free choice, and the acceptance of the physical body as an integral part of the human essence. To these cultural dimensions we should add two social and historical ones: the emergence of the socially and culturally powerful bourgeoisie/middle class/urban class, and the establishment of the nation and the nation-state as form of the social contract, intended to secure the autonomy of the individual.

This enormous cultural shift affected every practice and created specialised discourses. The feeding of the humans left the premises of the private sphere and assumed central place in public economics, science, culture and politics.[9] The emergence of gastronomy as 'discourse and practice of culinary excellence', as retraced by Pricilla Parkhurst Ferguson, is one telling example. The culinary refinement spread among the urban classes and especially among the middle classes, and acquired the status of mass social phenomenon. The *good eating* assumed the traits of an art. The word 'taste' was used

---

[8] Of course we should not neglect the other major factors such as raise in income (and availability of more expensive and not readily available foods); the transformation of the domestic life, the education and advertising. Researchers of modern food trends (such as Variyam and Golan, 2002) put in first place the change in food prices and income levels, but believe that the growing scientific evidence linking health to diet has significant and growing importance. Income, urbanization and health concerns are cited as primary factors by Regmi and Gehlhar, 2001: 3.

[9] For Michel Foucault (1970) the dominant *epistème* of the nineteenth and twentieth centuries centers around the presumably fictional creation of the individual body or 'man', clearly visible in the development of the three basic models of the human sciences – economics, biology, and philology.

already to denote not only specific gustatory sensation, but attitude (as exemplified by Brillat-Savarin, 1841). Simultaneously (and not incidentally), the interest in food entered a different realm – the scientific one – and gradually, as we shall see further, nutrition was raised to the status of science. Good nutrition became a norm both for individuals and for states following the concept of the Enlightenment that strong and autonomous individuals have an enlightened mind and a strong body. A strong body was achieved through rational feeding, and rational feeding depended on scientific ideas. The nation-state on its turn was perceived as an organic entity of individuals each one contributing to the national body; strong individuals bred healthy and strong nations. Nutrition became also politics (Ellerbock, 1987; Hildreth, 1987; Tanner, 1997).

By the end of the nineteenth and the beginning of the twentieth century, nutrition displayed all the elements of a subculture with its institutions (committees, scholarly societies, research centres, health centres, and spas), its discourse; its practices and ideologies. All of them claimed to be founded into science: the link between good nutrition and rationally/scientifically defined nutrition was established once forever as a dominant and universal principle (Mani, 1976). Ethical considerations (such as some vegetarian positions), even when in contradiction with the scholarly knowledge of the time, never attacked it directly and sought its bases in yet other scholarly arguments (Jabs et al., 1998).

World War I changed dramatically the entire mindset of the European world. The war tragedies brought a new sense of mortality, vulnerability, awareness of the fragility of the human body. Attitudes to food and nutrition were affected not only by the shortages and the famines, but also by the new and unprecedented involvement of governments in the previously mostly private sphere of eating. Rationing, procurement, norms and legislation entered forcefully everyday life (Bonzon and Davis, 1997). Their shaping was once again based in first place on scholarly findings and recommendations (Hull, 2002).

The inter-war period was marked with reinforced belief in rationality and efficiency. Efficiency meant a healthy and fit body, i.e. rational nutrition and diet. Modern cooking was supposed to be clean, organised and easy, appealing to women of all classes. This was a powerful impetus to change attitudes and taste (Novero, 2000). Another important development of the inter-war period was the mass qualitative change in diet – the conclusion of what Hans Teuteberg (1992b: 119) calls the 'diet revolution', which he locates between 1880 and 1930. The malnutrition disappeared almost everywhere even in less developed European nations to the south and the north (Toivonen, 1992). Good eating thereafter was not defined by abundance but by selection.

The Second World War was another dividing point. The exhaustion of the resources and the prolonged rationing system (in Britain this lasted until 1954) among many other changes led to the ascription of an even higher status and raised desirability of meats, tea, sugar, cheese, bacon, cooking fats, butter and margarine, chocolate, sweets, and eggs. At the same time the war was a major factor in producing mass changes and affecting collective behaviour – the quality of food was strictly monitored; the restaurant prices were fixed (Tims, 1976: 99).

The return to normality was marked by a sharp raise in the consumption of protein products, refined products (such as white bread), and fatty and sugary foods. I should note, however, the appearance of one significant divide. While the West transformed itself into a consumer society, the East followed the trend of the pre-war totalitarian regimes with heavy ideological overtones and strong involvement of the state in everyday life. Consumerism was delayed, as nutrition was not supposed to bring satisfaction but 'satisfaction of necessities' (in the new speak of the communist regimes). Choice stopped to be an issue in the Eastern part of the continent as it was eliminated altogether as a distracting factor. Shortages of specific foods, due to bad management or used as a carefully manipulated tool of control became a norm in the socialist 'camp'.

In Western Europe the indulgence of the 1950s was followed by a new shift in tastes and choices, starting in the 1960s. It became evident that new health problems, related to nutrition were arising with an alarming speed: coronary heart disease, hypertension, diabetes, dental caries, peptic ulcer, diverticular disease, gallstones, obesity, cancer in certain sites including colon, pancreas and breast, constipation and haemorrhoids (Jones, 1994). The response was the emergence of the 'new cuisine'. In France the move toward light meals, fast cooking techniques, lighters sauces, and the use of colours was already visible in the 1960s. In Germany preoccupation with the health aspects of food started in the early 1960s with the promotion of new kinds of breakfast, attention to the contents of vitamins and minerals, etc. The cult of slimness equalling smartness reached a new peak, becoming the overwhelming cultural norm (Wildt, 2001: 76–77), and leading to the mass spread of lipophobia (Baudrillard, 1970; Tanner, 1997).

III. The contribution of the sciences and medicine

The modern era established the science as leading authority in all matters human and social and the rational argumentation as leading factor of individual behaviour and social policy. The reinvention of chemistry, biology and medicine was to affect considerably the change of tastes and food choice. On the onset of this scientific revolution it would have been difficult to make a clear division of labour between their respective fields. Most of the significant advances were produced by individuals who worked simultaneously in more than one field and through co-operation and complementarity of efforts coming from different countries and professions. According to Mark Weatherall (1995: 189ff), until the beginning of the twentieth century the experts of good nutrition in Britain were almost exclusively physicians, while 'pure' knowledge came from scientists in Germany and the US (Liebig, Voit, Rubner, Atwater). This heterotopia – both geographic and scholastic – and the accumulated knowledge made possible the constitution of two new fields: nutrition science, and public health in the second half of the nineteenth century (Todhunter, 1976). The development of the nutritional science in general is well presented, as it has evolved as a meeting point of biochemical research and public health issues (McCollum, 1957; Mani, 1976; Todhunter, 1976).

Adel den Hartog (1992) claims that the first scientific study of nutrition was produced by the Dutch physician and chemist Gerrit Jan Mulder (1803–1880) in 1847. Another early attempt to bring together all the aspects of nutrition was the book of the German medical instructor Eduard Reich (1860), which was a survey of the existing literature of the time (Teuteberg, 1992b: 110). Dietrich Milles (1995) suggests that food became a point of public and official interest in the nineteenth century thanks to the growing concerns about its adulteration and the necessity to provide more adequate nutrition. The first and earliest preoccupation focused on what is good food, and roughly coincided with what Teuteberg called the food revolution. The second major direction of development looked into what is bad food and how eating can endanger the humans.

A good, short presentation of the history of studying food is provided in Kamminga's and Cunningham's Introduction (1995: 4) that retraces the accumulation of discoveries, which led to the articulation and argumentation of the relationship between food and health. From mid-eighteenth century, they observe, chemists such as Antoine Lavoisier, Joseph-Louis Gay-Lussac or Michel Eugène Chevreul have been exploring the content and the 'working' of the food elements. In 1827 William Prout identified the three major components of food – sugars, oils and albumines, which received their modern names as carbohydrates, fats and proteins during the 1840s. Justus Liebig and his laboratory marked the beginning of modern scientific research of the 'animal' (organic) chemistry – the relation between ingested and excreted foodstuff, and the way its chemical transformation within the body produces energy. As at this time one of the assumptions, held not only by the growing number of biochemists but also public figures, cookbook authors, politicians and the general public, was that meat is the most important nutrient. Liebig developed probably the first widely marketed food supplement and substitute, the extract of meat (Finlay, 1995: 48–49). The next major contribution was Max Rubner's research of the comparative energy value of the nutrients, measured in calories (linked to the law of conservation of energy) and the possibility to substitute food items for one another, provided that they carry the same calorie value. By the end of the nineteenth century the food was demystified and described into consistent and measurable scholarly categories (Bekaert, 1991). The next step was the establishment of recommended food norms that were supposed to eliminate the unbalance of the diet of millions of Europeans (Milles, 1995: 85; Weatherall, 1995: 189ff).

At the beginning of the twentieth century, a new 'qualitative' period did start: further research established that besides the caloric value, the composition of food was vital: the necessity to intake vital amino acids that cannot be synthesised by the body, the importance of microelement. The first among them became known under the name of vitamins. In 1912 Casimir Funk suggested that beriberi, pellagra, and scurvy were due to deficiency of substances, which he called 'vitamines'. At the same time McCollum (1957) and Davis identified a fat-soluble factor (vitamin A) that was necessary for the growth of rats. A significant contribution was also made by Gowland Hopkins (Jones, 1994). The Warsaw Institute of Hygiene and Funk made big advance in the research of vitamins in the 1920s, discovering new kinds of vitamins and establishing the quantities necessary for good nutrition, the factors influencing their content in food, et cetera. By 1930 the structure of vitamin A was established, and during the next ten years scientists discovered three new fat-soluble and seven water-soluble vitamins (Todhunter, 1976; Wilson, 1975; Ziegler and Filer, 1996). The work of many scholars from different countries led to the isolation of the thiamin and the other group B vitamins; in 1935 the vitamin C and its effects were identified (Carpenter, 1986). The newer science of nutrition came into being. (McCollum and Simmons, 1929).

The second line of scientific investigation was the research on the hazardousness of food (McKeown, 1983; Apfelbaum, 1989 and 1998). In the eighteenth century took shape the certainty that 'sudden deaths' (hearth attacks) and angina pectoris were linked to obesity, hearth degeneration and hardening of arteries (Schwartz, 1986: 214). The adulteration of foods was evidenced as a serious problem as early as 1820 (Accum, 1820). The discovery of the germs revolutionised the understanding of the disease and substantiated the suspicions that food could be dangerous. Bruno Latour (1988) excellently researched the social effect, the public opinion and institutional reactions. The main problem with medicine before the twentieth century, however, was that the germ theory of disease was not generally accepted. People understood that certain diseases tended to be infectious and were spread from the sick to the healthy, but the exact mechanism of such contagion was a mystery. Not until after the 1890s was the role of germs and other micro-organisms generally appreciated among doctors (Anderson, 1990). The food poisoning by botulinum toxin was diagnosed in 1890s by the Belgian E. Van Ermengem who isolated the anaerobic bacterium bacillus botulinus (Devriese, 1999). The use of the microscope and the study of the body at the end of the nineteenth century, the devices to measure blood pressure and electrocardiograph machine in the first decades of the twentieth, established the correlation between high blood pressure, arrhythmia, hearth disease and fat (Schwartz, 1986: 215). Cholesterol, discovered in 1823 and monitored since the 1880s, was experimentally proven to cause plaques by the 1930s.

IV. Changing diets, changing tastes

As pure science moved into levels of understanding, which challenged the general level of comprehension, nutritionists of all sorts assumed the responsibility to design and promote norms and recommendations of good nutrition. They used extensively (and not infrequently arbitrarily) the latest in chemistry, biology and medicine to craft new 'diets'. The process started back in the late nineteenth century when the word 'diet' stopped meaning the composition of everyday food, and came to denote the radical departure from normal eating to achieve a medical or, usually, cosmetic result.

The amazing popularity of the varied nutritional regimes was one of the glowing illustrations of the modern culture, where the unquestionable faith into the scientific discoveries espoused the new awareness of the body as important part of the public self. Being negligent of your body, i.e. health and eating habits, was a telling evidence of internal misbalance: either a negligence to the opinion of the others or act of moral and intellectual laxity (Sobal, 1999). This new mindset spread primarily among the urban classes, which were not preoccupied with subsistence anymore, but by mid-twentieth century this group already formed the majority in Europe.

'The popular preoccupation with diet, a sociological feature of the twentieth century, emerged in the 1890s, accompanying a surge in discovery in the nutritional laboratory', writes Margaret Barnett (1995: 156), 'hundreds of books and articles, cartoons, jokes and jingles, catchy advertisements and even plays and novels... lectures, food reform societies... sport fixtures... sanatoria and spas. Campaign for pure food swept Western Europe and the US in 1906'. Vegetarianism appeared in Germany, Britain and the United States by mid-nineteenth century, followed by the milk diet, the Salisbury diet, the fruitarianism and the nutarianism, and the raw food diet. The 'fasting fad' appeared in 1902, together with the Fletcherism – the recommendation to intake less food through very long mastication (Barnett, 1995; Blix, 1970). The Kempner diet consisted of boiled or steamed rice (with no salt) and fruit, but – against its claims – it had very little effect on reducing blood pressure (Hamdy, 2001). In a more comprehensive way the concept of what is a good diet was studied by Major-General Sir Robert McCarrison between 1919 and 1935, based on extensive observations and research: the importance of whole foods and natural foods, coinciding with the whole bread campaign (Weatherall, 1995: 183–184, 195). Cleave, Burkitt, and Trowell made another major contribution with their research on the role of dietary fibre, removed during the refining process. The importance of polyunsaturated fatty acids was discovered in 1956 by Dr Hugh Sinclair in Oxford. In 1965 Ansel Keys brought again the issue of the fats and cholesterol in relation to heart disease and since then, evidence on the association of specific foods with certain health conditions has grown dramatically. A database of medical literature shows 13 English-language articles linking fats and cholesterol to heart disease in 1965 and 82 in 1996 (Variyam and Golan, 2002).

The long exposure of European public to the discourse of 'healthy' food led to easily observable change in the very perception and attitude to food and different food items (Caplan, 1997; Girois et al., 2001). Health came to be associated with unrefined foods by large sectors of the middle class, especially as a result of the natural food movement pioneered by Graham and Kellog (Gusfield, 1992: 82). Simple foods existing among traditional peoples gained marked support as being natural healers. The yoghurt, strongly supported by Elie Metchnikoff, deputy director of the Pasteur Institute of Paris and Nobel Prize winner in 1908, became widely liked food all over Europe (Barnett, 1995: 157). The establishment of the taste for fresh, natural and unrefined foods was attested by the famous study of Charles and Kerr (1988). The participants in this study emphasised the importance of 'proper meals'. These were described as meals involving 'good foods', which were defined as 'fresh' and 'natural' rather than convenience foods, and which were cooked rather than cold or heated up. But the preference of fresh and natural foods over artificial and modified ones is not the only change. J. Variyam and E. Golan (2002: 13) state that 'Over the past half century, consumption patterns of many food commodities have shifted dramatically in the face of changing consumer demand', and namely the consumption of eggs, whole milk, red meats, butter, lard had dropped markedly, while the consumption of reduced-fat milk, poultry, salad and cooking oils, etc. has risen dramatically. Processed high-calorie products, such as meat, beverages, bakery products, and snack foods still account for about 34 percent of global food trade, up from 18 percent in 1980 (Regmi and Gehlhar, 2001), but the distribution is quite uneven with Europe being on a different trend from the developing world. The rise of the organic foods movement is another example of changing consumer behaviour and tastes (Lohr, 2001). World-wide markets for organic foods are expanding, with annual growth rates of 15 to 30 percent in Europe, the United States, and Japan for more than 5 years. As many as 20 to 30 percent of consumers surveyed in Europe, North America, and Japan claim to purchase organic foods regularly. While there is interest in organic foods among higher income, better educated population segments in nearly every country, consumers in the United States, Europe, and Japan are driving the growing demand for these goods. Another aspect is the cultural interpretation of diets and the new stereotyping, based on it (Saddala and Burroughs, 1981).

The list of the sequence of nutritional regimes and recommendations is impressive. Some of them appealed to the common sense and expectations; others tended to create distaste and even fears.[10] The food scares are powerful factor affecting choices and, over time, tastes (Buzby, 2001; Apfelbaum, 1998). The unhealthy image of the animal fats, especially lard, led not only to its replacement with vegetable cooking fats but to

[10] Claude Fishler mentions the famous absinthe scare, 'les trois aliments meurtriers: l'alcool, la viande et le sucre' (Fischler, 1989: 37).

gradual collective dislike for foods, cooked in lard. Even more dramatic was the 1996 announcement in Great Britain of a possible link between bovine spongiform encephalopathy (BSE), or mad cow disease, in cattle and a new strain of Creutzfeldt-Jakob disease in humans, which led to dramatic declines in beef consumption in Europe, particularly in the United Kingdom. Louise Crewe, exploring the policies of consumption, the food scares (campylobacter, salmonela, e-coli) in UK and public anxieties, mentions the opinions of others that 'consumption has increasingly become more reflexive and more risky; (...) increasing concerns over food safety and quality, along with growing consumer mistrust of scientific knowledge and government agencies' assessment of risk, has led to a consumer culture of suspicion' (Crewe, 2001: 630–631; see also Sulkunen et al., 1997: 15; Peterson and Lupron, 1996).

V. How changes did take place

One of the most important questions related to the impact of scientific advance on food choice and eating habits is how the scholarly discourse reached the public. As most of the other major social and cultural innovations of the modern epoch it spread through printed text.[11] P. Ferguson (1998: 599) emphasises that 'gastronomy constructed its modernity through an expansive culinary discourse and, more specifically, through texts. Gastronomic texts were key agents in the socialisation of individual desire and the redefinition of appetite in collective terms'. The mass production and availability of texts on food, cooking and eating extended the previously limited circle of connoisseurs and professionals to include everyone who possessed the literacy level and modest resources to access the last and the latest in advise on good nutrition. The major sources of 'knowledge' on cooking and eating were the advice books and guides, the media and the different form of education (Claudian et al., 1969). Many of the texts referred also to culture, tradition, taste, effectiveness and so on – but health was the strongest argument since the onset, and became in the 1970s and 1980s a primary target of this literature. Dorfman (1992) suggests that food industries, women's magazines, cookbooks and advertisers have all contributed to reproducing the message that providing a 'proper' meal is the key for women to a successful marriage and home life.

The role of the media in promoting new nutritional trends was a key factor (Fieldhouse, 1995; Krondl and Lau, 1982; Lupton and Chapman, 1995). Frederick Gowland Hopkins, who won a Nobel Prize for his research on vitamins, published his first discussion on their role not in a medical journal but in the *Daily Mail* (Weatherall, 1995: 180). Sociological research has shown (Luxton, 1980; Becker et al., 1999; DeVault, 1992; and particularly Murcott, 1983), that it were usually women who shouldered the responsibility for choosing, shopping, preparing and cooking 'proper meals' for the 'proper family', and that they privileged their partners' and children's preferences over their own tastes. Therefore women's magazines played primary a role in promoting new nutritional information as they reached at the heart of eating habits and preferences. The role of women's

[11] On the role of the printed texts as medium of modernization see Andersen, 1983. The standard theory of consumer choice, at least for many decades following Marshall, held that the chief or only constraint facing consumers was income.

and fashion magazines was very important in yet another aspect: they were instrumental in strengthening collective consciousness about the association between taste in dress, decoration and in life style in general, and the proper nutrition (Ferguson, 1982; Tims, 1976). In Jean Corbeau's (Corbeau and Poulain, 2002: 33ff) research on nutrition patterns the strong distaste for 'traditional' fatty cuisine is clearly cultivated by the lifestyle industry. In addition, the magazines had the advantage to be frequently renewable knowledge (unlike cookbooks, which were usually bough for longer periods of time). A good case study is Catherine Salzman's (1985) who studied Dutch culinary history between 1945–1975 through women's magazines and cookbooks (see also White, 1970). The advice literature, which emerged in Europe gradually during the course of the nineteenth century, was another key medium. The famous cookbook of Henriette Davidis, published in 1844 had sixty-three editions (Reagen, 2001). By the late nineteenth century, manufacturers had begun to distribute free cookbooks and pamphlets in order to promote their products, based on the newest scientific discoveries.

Education, especially of young girls was another and very important channel of spreading nutritional knowledge, especially if we keep in mind that the very nature of education called for a rational and argumentative presentation with heavy reference to science and knowledge. Women's charitable organisations in Berlin, for instance, established housewives' association and published a widely read magazine, *Die Deutsche Hausfrauen-Zeitung*, for over thirty years (Reagen, 2001). The evolution of the left-right divide (more money vs. more education) in Britain in the 1940s, saw a 'remarkable conversion' of the left-wing nutrition pressure groups toward more education and the establishment of 'household schools' (Smith and Nicolson, 1995: 290). 'Home economics' was extensively researched and taught in the Netherlands (Den Hartog, 1992). Domestic science started to be taught in schools first in Sweden, 1865, in Germany in the 1870s, in France in 1882, and in Britain in the 1880s (Mennell, 1985: 230).

The next major vehicle in bringing scientific change in everyday consideration was the political one. The role of medical knowledge in social change in general became of interest to scholars after the 1970s, and a few good researches lay the ground for understanding the social role of medicine (McKeown, 1976 and 1979; Imhof, 1976, 1981, 1984; Imhof and Larsen, 1976). The evolution of medical goals from the healing of the individual to securing the health of the nation and good living conditions is researched by Rosen (1958, 1974), Hildreth (1987), and Stevens (1966).

The century between 1840–1940 saw the rise of the modern nation-state, of modern laboratory science, and state intervention in science and people's diet. The 'reciprocal relation between state and citizen in terms of health and wellbeing' (Kamminga and Cunningham, 1995: 1) was translated in the elaboration and the acceptance of the role of the state as factor in decision making in everyday choices. The rationing, the food norms, the legislation on hygiene, prevention of diseases and adulteration of food, the school feeding, the standards of food quality were all elements of a giant educational wave affecting the life of every citizen through the mechanisms of the normative base and the institutions (especially schools, social establishments and the army). After World War II, and especially after the 1970s when the quality of food became a major public issue, the governments of most European countries took even harder position in regulating nutritional issues (James and Ralph, 1991 and 1997; Helsing, 1997; for the USA: Barkan, 1985). Germany, for instance, has at present some of the strictest consumer protection laws in the world. But it should be noted that this situation reflects conscious public choices and opinions – 81 percent of respondents put in first place the quality, not the price of food items, when stating their preferences (Walsh et al., 2001: 76). The establishment of the European Union was another step in elaborating and enforcing food standards. Western European countries have adopted certification systems that guarantee the origin of fresh and processed meats and attest the authenticity of organic and natural production (Regmi and Gehlhar, 2001). A system of vertical and horizontal directives provides standardisation of hygiene requirements.

In conclusion, we could state with some certainty that rational knowledge, provided by science and medicine and the informed choices, made by individuals and social groups, have concurrently helped the emergence of a different, pointedly modern attitude to food and eating. This attitude deeply affected taste as the sense of judging food and led to yet another controversy that we have to deal with in our everyday lives: the tension between the rational taste that we are taught to trust and the gustatory taste, which leads us, oh so often, to crave things that we know for certain we should not.

Bibliography

Accum, F. (1820) *Treatise on the adulteration of food and culinary poisons*, London.

Andersen, B. (1983) *Imagined communities*, London.

Anderson, G.M. (1990) 'Parasites, profits, and politicians: Public Health and Public Choice', *CATO Journal*, 9, 3, pp. 557–579.

Apfelbaum, M. (1989) 'La recherche face aux peurs du siècle', *Autrement Nourriture*, 108, pp. 180–183.

Apfelbaum, M. (ed.) (1998) *Risques et peurs alimentaires*, Paris.

Aymard, M. (1973) 'The history of nutrition and economic history', *Journal of European Economic History*, 2, 1, pp. 207–219.

Barkan, I.D. (1985) 'Industry invites regulation: The passage of the Pure Food and Drug Act of 1906', *American Journal of Public Health*, 75, pp. 18–26.

Barker, T.C., McKenzie, J.C. and Yudkin, Y. (eds) (1966) *Our changing fare: Two hundred years of British food habits*, London.

Barnett, L.M. (1995) '"Every Man His Own Physician": Dietetic fads, 1890–1914', in H. Kamminga and A. Cunningham (eds), *The Science and Culture of Nutrition, 1840–1940*, Amsterdam, pp. 155–178.

Barthes, R. (1961) 'Vers une psycho-sociologie de l'alimentation moderne', *Annales: Economies, Sociétés, Civilisations*, pp. 977–986.

Baudrillard, J. (1970) *La société de consommation*, Paris.

Becker, P.E. and Moen, Ph. (1999) 'Scaling Back: Dual-Earner Couples' Work-Family Strategies', *Journal of Marriage and the Family*, 61, 4, pp. 995–1007.

Bekaert, G. (1991) 'Caloric consumption in industrializing Belgium', *Journal of Economic History,* 51, pp. 633–655.

Benedict, R. (1959) *Patterns of Culture*, Boston.

Blix, G. (ed.) (1970) *Symposium on food cultism and nutrition quackery*, Stockholm.

Bonzon, T. and Davis, B. (1997) 'Feeding the cities', in J. Winter and J.-L. Robert (eds), *Capital cities at war. Paris, London, Berlin 1914–1919*, New York, pp. 306–342.

Booth, D.A. (1982) 'How nutritional effects of food can influence people's dietary choices', in L.M. Barker (ed.), *The psychobiology of human food selection*, Westport, Co, pp. 67–84.

Bourdieu, P. (1979) *La distinction*, Paris.

Brillat-Savarin, A. (1841) *Physiologie du goût*, Paris.

Buzby, J. (1999) "Food safety and international trade in the twenty-first century', *Choices: The Magazine of Food, Farm & Resource Issues*, Vol. 14, Issue 4, pp. 23–28.

Buzby, J. (2001) 'Effects of food-safety perceptions on food demand and global trade', in A. Regmi (ed.), *Changing structure of global food consumption and trade*, Washington, pp. 55–66. (Agriculture and Trade Report WRS-01-1).

Capatti, A. (1989) 'L'avenir archaïque de la gastronomie', *Autrement* (Nourritures), 168, pp. 18–23.

Caplan, P. (ed.) (1997) *Food, health, and identity*, London.

Carpenter, K. J. (1986) *The history of scurvy and vitamin C*, New York.

Charles, N. and Kerr, M. (1988) *Women, food and families*, Manchester.

Chasis, H. (1950) 'Salt and protein restriction: effects on blood pressure', *JAMA*, 142, pp. 326–331.

Claudian, J., Serville, Y. and Tremolières, F. (1969) 'Enquête sur les facteurs de choix des aliments', *Bulletin de l'INSERM*, 24: 5, pp. 1277–1390.

Cooter, R. and Pumfrey, S. (1994) 'Separate spheres and public places: Reflections on the history of science popularization and science in popular culture', *History of Science*, 32, pp. 237–267.

Corbeau, J.-P. (1988) 'Trois scénarios de mutation des goûts alimentaires', in *Actes du colloque transfrontalier sur le Goût*, Dijon, pp. 321–328.

Corbeau, J.-P. and Poulain, J.P. (2002) *Penser l'alimentation. Entre imaginaire et rationalité*, Toulouse.

Crewe, L. (2001) 'The besieged body: geographies of retailing and consumption', *Progress in Human Geography*, 25, 4, pp. 630–631.

Davidson, S. [et al.] (1979) *Human nutrition and dietetics*, Edinburgh.

DeVault, M. (1992) *Feeding the family: The social organization of caring as gendered work*, Chicago.

Devine, C., Sobal, J., Bisogai, C. and Connors, M. (1999) 'Food choices in three ethnic groups: interactions of ideals, identities and roles', *Journal of Nutritional Education*, 32, pp. 86–93.

Devine, C., Connors, M., Bisogni, C. and Sobal, J. (1998) 'Life course influences on fruit and vegetable trajectories: a qualitative analysis of food choices', *Journal of Nutritional Education*, 31, pp. 361–370.

Devriese, P. (1999) 'On the discovery of clostridium botulinum', *Journal of the History of the Neurosciences*, 8, 1, pp. 43–50.

Dietrich, M. (1995) 'Working capacity and calorie consumption: The history of rational physical economy', in H. Kamminga and A. Cunningham (eds), *The Science and Culture of Nutrition, 1840–1940*, Amsterdam, pp. 76–89.

Dorfmann, C. (1992) 'The garden of eating: the carnal kitchen in contemporary American culture', *Feminist Issues*, 12, quoted in G. Valentine (1999) 'Eating in: Home consumption and identity', *Sociological Review*, 47, p. 493.

Ellerbock, K.-P. (1987) 'Lebensmittelqualitat vor dem ersten Weltkrieg: Industrielle Produktion und staatliche Gesundheitspolitik', in H.-J. Teuteberg (ed.), *Durchbruch zum modernen Massenkonsum. Lebensmittelmarkte und Lebensmittelqualitat im Stadtewachstum des Industriezeitalters*, Muenster, pp. 127–189.

Ferguson, M. (1982) *Forever feminine: Women's magazines and the cult of femininity*, London.

Ferguson, P. Parkhurst (1998) 'A cultural field in the making: Gastronomy in 19th-century France', *The American Journal of Sociology*, 104, 3, pp. 597–620.

Fieldhouse, P. (1995) *Food and nutrition, customs and culture*, London.

Finlay, M. (1995) 'Early marketing of the theory of nutrition: The science and culture of Liebig's extract of meat', in H. Kamminga and A. Cunningham (eds), *The science and culture of nutrition, 1840–1940*, Amsterdam, pp. 48–75.

Fischler, C. (1990) *L'Homnivore*, Paris.

Fischler, C. (1989) 'Les aventures de la douceur', *Autrement*, 168, pp. 32–39.

Fogel, R.W. (ed.) (1986) *Long-term changes in nutrition and the standard of living*, Research topics for Section B7 of the 9th International Economic History Congress, Berne.

Foucault, M. (1970) *The order of things: An archaeology of the human sciences*, New York.

Furst, T., Connors, M., Bisogni, C., Sobal J. and Falk, L. (1996) 'Food choice: a conceptual model of the process', *Appetite*, 26, pp. 247–265.

Giles, J. (1997) *Women, identity and private life in Britain, 1900–50*, London.

Girois, S.B., Kumanyika, S.K., Morabia, A. and Mauger, E. (2001) 'A comparison of knowledge and attitudes about diet and health among 35- to 75-year-old adults in the United States and Geneva, Switzerland', *American Journal of Public Health*, 91, pp. 418–24.

Golton, L. (1986) 'The rules of the table: Sociological factors influencing food choice', in C. Ritson, L. Golton and I. McKenzie (eds), *The food consciousness*, Chichester, pp. 127–153.

Gusfield, J.R. (1992) 'Nature's body and the metaphors of food', in M. Lamont and M. Fournier (eds), *Cultivating differences. Symbolic boundaries and the making of inequality*, Chicago, pp. 81–90.

Hamdy, R.C. (2001) 'Hypertension: A turning point in the history of medicine… and mankind', *Southern Medical Journal*, 94, 11, pp. 1045–1047.

Hammond, P. (1976) 'Changing tastes and coherent dynamic choice', *Review of Economic Studies*, 43, pp. 159–173.

Harris, M. (1985) *Good to eat: riddles of food and culture*, New York.

Hartog, Adel P. den (1992) 'Modern nutritional problems and historical nutrition research, with special reference to the Netherlands', in H.-J. Teuteberg (ed.), *European Food History. A research review*, Leicester, pp. 56–70.

Helsing, E. (1997) 'The history of nutrition policy', *Nutrition Reviews*, 55, pp. 1–4.

Hildreth, M. L. (1987) *Doctors, bureaucrats, and public health in France, 1888–1902*, New York–London.

Hull, A., (2002) 'Food for thought?: The relations between the Royal Society Food Committees and Government, 1915–19', *Annals of Science*, 59, 3, pp. 263–285.

Hume, D. (1965) *On the standard of taste and other essays*, Indianapolis.

Imhof, A. (1984) *Die verlorenen Welten. Alltagsbewältigung durch unsere Vorfahren und warum wir uns heute so schwer damit tun*, Munich.

Imhof, A. (1981) *Die Gewonnenen Jahre. Von der Zunahme unserer Lebenspanne seit 300 Jahren oder von der Notwendigkeit einer neuen Einstellung zu Leben und Sterben*, Munich.

Imhof, A. (ed.) (1980) *Mensch und Gesundheit in der Geschichte*, Husum.

Imhof, A. and Larsen, O. (1976) *Sozialgeschichte und Medizin*, Stuttgart.

Jabs, J.A., Devine, C.M. and Sobal, J. (1998) 'A model of the process of adopting vegetarian diets: health vegetarians and ethical vegetarians', *Journal of Nutritional Education*, 30, pp. 196–202.

James, W.P.T. and Ralph A. (1991) 'National food policies, in M. Estwood, C. Edwards and D. Parry (eds), *Human nutrition*, London, pp. 4–20.

James, W.P.T. and Ralph, A. (1997) 'Nutrition policy in Western Europe: National policies in Belgium, the Netherlands, France', *Nutrition Reviews*, 55, 11, pp. 4–20.

Jones, F.A. (1994) 'New concepts in human nutrition in the twentieth century: the special role of micro-nutrients', *Journal of Nutritional Medicine*, 4, pp. 99–114.

Kamminga, H. and Cunningham, A. (1995) 'Introduction', in H. Kamminga and A. Cunningham (eds), *The science and culture of nutrition, 1840–1940*, Amsterdam, pp. 1–14.

Kant, I. (1974) *Anthropology from a pragmatic point of view*, The Hague. (Translated by M.J. Gregor).

Kiple, K.F. and Coneè Ornelas, K. (eds) (2000) *The Cambridge world history of food*, 2 vols, Cambridge.

Krondl, M. and Lau, D. (1982) 'Social determinants in human food selection', in L.M. Barker (ed.), *The Psychology of Human Food selection*, Chichester, pp. 139–151.

Latour, B. (1988) *The Pasteurization of France*, Cambridge Mass.

Leviz, C. (1991) 'Changing food habits: An introduction', *Food and Foodways*, I, pp. 1–13.

Lohr, L. (2001) 'Factors affecting international demand and trade in organic food products', in A. Regmi (ed.), *Changing structure of global food consumption and trade*, Washington, pp. 222–231. (Agriculture and Trade Report WRS-01-1).

Lupton, D. (1994) *Medicine as culture: Illness, disease, and the body in Western societies*, London.

Lupton, D. (1996) *Food, the body and the self*, London.

Lupton, D. and Chapman, S. (1995) '"A healthy lifestyle might be the death of you": discourses on diet, cholesterol control and heart disease in the press and among the lay public', *Sociology of Health and Illness,* 17, pp. 477–494.

Luxton, M. (1980) *More than a labour of love: Three generations of women's work in the home*, Toronto.

Mani, N. (1976) 'Die wissenschaftliche Ernährungslehre im 19. Jahrhundert', in Heischkel-Artelt, E. (ed.), *Ernährung und Ernährungslehre im 19. Jahrhundert*, Göttingen, pp. 438–452.

McCollum, E.V. (1957) *A history of nutrition: The sequence of ideas in nutrition investigations*, Boston.

McCollum, E.V. and Simmons, N. (1929) *The newer knowledge of nutrition. The use of food for the preservation of vitality and health*, New York.

McKeown, T. (1976) *The modern rise of population,* New York.

McKeown, T. (1979) *The role of medicine: Dream, mirage, or nemesis?* Princeton, N.J.

McKeown, T. (1983) 'Food, infection, and population', *Journal of Interdisciplinary History,* 14, pp. 227–248.

Mennell, S. (1985) *All manners of food: Eating and taste in England and France from the Middle Ages to the Present*, Oxford.

Mennell, S. and Murcott, A. (1992) 'Patterns of food consumption', *Current Sociology*, 40, 2, pp. 35–41.

Mennell, S., Murcott, A. and Otterloo, A. van (1992) *The sociology of food. Eating, diet and culture*, London.

Meyer, H.-D. (2000) 'Taste formation in pluralist societies: The role of rhetorics and institutions', *International Sociology*, 15, 1, pp. 33–57.

Milles, D. (1995) 'Working capacity and calorie consumption: The history of rational physical economy', in H. Kamminga and A. Cunningham (eds), *The science and culture of nutrition, 1840–1940*, Amsterdam, pp. 75–96.

Murcott, A. (ed.) (1983) *The sociology of food and eating: Essays on the sociological significance of food*, Aldershot.

Novero, C. (2000) 'Stories of food: Recipes of modernity, recipes of tradition in Weimar Germany', *Journal of Popular Culture*, 34, 3, pp. 163–181.

Oddy, D.J. and Burnett, J. (1992) 'British diet since industrialization: a bibliographic study', in H.-J. Teuteberg (ed.), *European food history. A research review*, Leicester, pp. 19–44.

Payer, L. (1988) *Medicine and culture*, New York.

Perrot, M. (1961) *Le Mode de vie des familles bourgeoises, 1873–1953*, Paris.

Peterson, A. and Lupron, D. (1996) *The new public health: Health and self in the age of risk*, London.

Reagen, N. (2001) 'The imagined Hausfrau: National identity, domesticity, and colonialism in Imperial Germany', *The Journal of Modern History*, 73, 1, pp. 54–86.

Regmi, A. and Gehlhar, M. (2001) 'Consumer preferences and concerns shape global food trade', *Food Review*, 24, pp. 111–143.

Reich, E. (1860) *Die Nahrungs- und Genussmittelkunde, historisch, naturwissenschaftlich und hygienisch begründet*, 2 vols., Göttingen.

Rosen, G. (1958) *A history of public health*, New York.

Rosen, G. (1974) *From medical police to social medicine*, New York.

Rozin, P., Fischler, C., Imada, S., Sarubin, A. and Wrzesniewski, A. (1999) 'Attitudes to food and the role of food in life: Comparisons of Flemish Belgian, France, Japan and the United States', *Appetite*, quoted in J. Friedlander, P. Rozin and R. Sokolov (1999) 'Everyday life: Ordinary pleasures, rituals and taboos', *Social research*, vol. 66, 1, p. 78.

Saddala, E. and Burroughs, J. (1981) 'Profiles in eating: sexy vegetarians and other diet-based social stereotypes', *Psychology Today*, 15, pp. 51–57.

Salzman, C. (1985) 'Margriet's advies aan de Nederlandse huisvrouwen. Continuïteit en verandering in de culinaire geschiedenis van Nederland 1945–1975', *Volkskundig Bulletin*, 11, 1, pp. 1–27.

Schwartz, H. (1986) *Never satisfied. A cultural history of diets, Fantasies and fat*, New York.

Shiner, R. A. (1979) 'Sense-experience, colors and tastes', *Mind*, 88, pp. 161–178.

Simmel, G. (1950) *The sociology of Georg Simmel*, New York.

Smith, D. and Nicolson, M. (1995) 'Nutrition, education, ignorance and income: A twentieth-century debate', in H. Kamminga and A. Cunningham (eds), *The science and culture of nutrition, 1840–1940*, Amsterdam, pp. 288–318.

Sobal, J. (1999) 'Sociological analysis of the stigmatisation of obesity', in J. Germov and L. Williams (eds), *Sociology of food and nutrition: The social appetite*, Oxford, pp. 187–204.

Spree, R. (1981) 'Zu den Veränderungen der Volksgesundheit zwischen 1870 und 1913 und ihren Determinanten in Deutschland, vor allem in Preussen', in *Arbeiterexistenz im 19. Jahrhundert. Lebensstandard und Lebensgestaltung deutscher Arbeiter und Handwerker*, Stuttgart, pp. 235–292.

Stevens, R.-M. (1966) *Medical practice in Modern England*, New Haven.

Sulkunen, P., Holmwood, J., Radner, H. and Schilze, G. (1997) *Constructing the new consumer society*, Basingstoke.

Synnott, A. (1991) 'Puzzling over the senses: from Plato to Marx', in D. Howes (ed.), *The Varieties of Sensory Experience*, Toronto, pp. 78–97.

Tanner, J. (1997) 'Industrialisierung, Rationalisierung und Wandel des Konsum und Geschmacksverhaltens im europäisch-amerikanischen Vergleich', in H. Siegrist, H. Kaelble and J. Kocka (eds), *Europäische Konsumsgeschichte. Zur Gesellschafts- und Kulturgeschichte des Konsums (18.–20. Jahrhundert)*, Frankfurt–New York, pp. 606–616.

Teuteberg, H.-J. (1992a) 'Agenda for a comparative European history of diet', in H.-J. Teuteberg (ed.), *European Food History. A research review*, Leicester, pp. 1–16.

Teuteberg, H.-J. (1992b) 'The diet as an object of historical analysis in Germany', in H.-J. Teuteberg (ed.) *European Food History. A research review*, Leicester, pp. 109–128.

Teuteberg, H.-J. and Edema, J.P. (eds) (1983) *Nutritional behaviour as a topic of social sciences*, Frankfurt am Main.

Tims, B. (1976) *Food in 'Vogue'; six decades of cooking and entertaining*, London.

Todhunter, E.N. (1976) 'Chronology of some events in the development and application of the science of nutrition', *Nutrition Reviews*, 34, pp. 353–365.

Toivonen, T. (1992) 'Class, countries and consumption between the World Wars. A comparison of the structure of expenditure in Estonia, Finland and Sweden in the 1920s and 1930s', *Acta Sociologica*, 35, pp. 219–233.

Turner, B.S. (1982a) 'The discourse of diet', *Theory, Culture and Society*, 1, pp. 23–32.

Turner, B.S. (1982b) 'The government of the body: Medical regimens and the rationalization of the diet', *British Journal of Sociology*, 33, pp. 254–269.

Variyam, J. and Golan, E. (2002) 'New health information is reshaping food choices', *Food Review*, 25, pp. 13–18.

Veblen, Th. (1953) *The theory of the leisure Class*, New York.

Walsh, G., Mitchell, V. and Hennig-Thurau, T. (2001) 'German consumer decision-making styles', *Journal of Consumer Affairs*, 5, 1, pp. 73–95.

Weatherall, M. (1995) 'Bread and newspapers: The making of "A revolution in the science of food"', in H. Kamminga and A. Cunningham (eds), *The science and culture of nutrition, 1840–1940*, Amsterdam, pp. 179–212.

White, C. (1970) *Women's magazines 1693–1968*, London.

Wildt, M. (2001) 'Promise of more. The rhetoric of (food) consumption in a society searching for itself: West Germany in the 1950s', in P. Scholliers (ed.), *Food, drink and identity*, Oxford, pp. 63–80.

Wilson, L.G. (1975) 'The clinical definition of scurvy and the discovery of vitamin C', *Journal of Hist Med Allied Sciences*, 30, pp. 40–60.

Ziegler, E.E. and Filer, L.J. (eds) (1996) *Present knowledge in nutrition*, Washington.

10 The rise of supermarkets in twentieth-century Britain and France

Isabelle LESCENT-GILES, University of Paris – Sorbonne

This chapter deals with the rise of large-scale food retailing in twentieth-century Britain and France. By the end of the twentieth century, these two countries were competing with the US at the leading edge of world retailing. Since the 1960s, they pioneered many of today's trends in technology, strategy and business organisation, such as scanning technology at point of sale or the automation of central warehouses. This leading edge is partly historical: in the second half of the nineteenth century, the rise of the middle classes and gradual improvements to average standards of living in both countries provided opportunities for the rise of department stores (Bon Marché, Selfridges) and retailing chains (Sainsbury's, Boots). But France and Britain, like most European countries, fell behind the US in the first half of the twentieth century. America became the land *par excellence* of mass markets and large-scale retailing. It is remarkable that France's and Britain's food retailers were able to catch up from the 1950s onwards, to the point where, in the 1980s and 1990s, they were introducing technological innovations, such as electronic points of sale and integrated logistics, at the same pace as their US counterparts. They also pioneered new ways of selling food. In the United Kingdom (UK), Marks & Spencer was the first to introduce imports of fruit and vegetables by plane so as to offer ripe tomatoes and green beans in winter. It is also credited with developing industrially produced ready meals, such as Indian curry, Italian pasta and French chicken *chasseur*. As for France, its leading retailers pioneered the hypermarket format (larger units selling both food and textiles) in the 1960s, then exported it to the rest of Europe, and, with mixed success, to the US in the 1970s.

One may therefore argue that food retailing is one area where old Europe, led by France and Britain, did not trail behind the US in terms of technology and size. This chapter looks at the evolution of leading French and British retailers' strategy and organisation in order to understand why they achieved world-class status in the late twentieth century. It starts by an overview of the historical factors that shaped these two countries' retailers since the nineteenth century. It then focuses on the impact of new technology in three key areas (product innovation, sales and marketing, and business organisation) since the 1960s.

I. The rise of large-scale retailing in France and Britain from the 1860s to the 1990s

I.1. The emergence of mass markets in the nineteenth century

Some of the patterns of today's food distribution can be traced back to nineteenth-century Britain and France. With the Industrial Revolution came the rise of the urban middle classes. In the second half of the nineteenth century, they started to integrate the upper echelons of the skilled working class to form a mass market for food, textile and household goods. Several entrepreneurs jumped on the bandwagon and created the first multi-unit retailers. This period saw the rise of Boots the chemist, Sainsbury the grocer and Marks & Spencer's penny bazaars. They catered for the urban middle classes and an aspiring working class, driving down prices. This in turn helped diversify eating habits, with increased consumption of goods that had been luxury items until then, such as tea, sugar, chocolate and beef.

Food manufacturers were the other beneficiaries, and strove to find new, cheaper ways to produce food industrially, in order to create a mass market for their goods. This trend was particularly marked in confectionary, with the rise of a spate of new firms such as Cadbury's, Fry or Rowntree. They pioneered both new food technology and new marketing methods, including branding and advertising. Indeed, the end of the nineteenth century saw the balance of power between food retailers and manufacturers tip in favour of manufacturers. Consumers gave their trust to brands rather than relying on shopkeepers for advice and information about provenance. From the 1890s to the 1970s, food manufacturers were the driving force behind technological change and product innovation.

Innovation in manufacturing did not stifle innovation in retailing though. The second half of the nineteenth century saw the rise of the department store, such as France's Bon Marché and Britain's Selfridges. Although mainly focused on textiles and household goods, they created a shop format and marketing ethos that most multi-unit food retailers followed. The family-owned small and medium size shop, with most goods in drawers, was replaced by wider outlets with several counters specialising in different types of goods. Each counter had its own specialised staff. Most of the goods were on display. How best to display foodstuffs, household goods or clothing became the key to success, and was central to the rise of the new science of marketing. Pricing was the other key to success: in the old days, pricing was something you discussed with the shop owner, and could prove very intimidating for the lower middle classes. The new generation of multi-unit chains promoted price 'transparency': goods on display had price tags, so as not to force modest customers to ask for it. Prices were fixed, and bargaining was on the way out. Marks & Spencer pushed this to the extreme with its penny bazaar. Founded by an immigrant with little knowledge of English, its formula borne out of necessity, 'don't ask the price, it's a penny', soon proved central to its extraordinary success. Bargaining was replaced in non-food retailing by the new concept of 'seasonal sales'. In food retailing, 'promotions' on goods reaching the end of their shelf life or seasonal foodstuffs, became a tool for attracting new customers.

The third characteristic of food retailing to have emerged in the nineteenth century is the domination of family firms and, thanks to a low-entry level, the rise of new major players at each generation. Food retailing was in the nineteenth century a land of many opportunities for young aspiring men with little capital and little or no education. It remained so until the late twentieth century.

Last but not least, nineteenth-century Britain pioneered a new form of retailing aimed specifically at the struggling working classes: the co-operative movement. It was in part a reaction to the hated practice of truck-shops, outlawed in the early nineteenth century but still practised in many parts of the country until the 1850s. Lack of existing shops in many new industrial districts gave manufacturers the possibility of paying their staff partly in nature, with foodstuff that was often overpriced and of poor quality. The 'coop' provided an alternative for these isolated communities. It aimed to pool the buying power of the local working class in order to negotiate lower prices with wholesalers and manufacturers. Some of these co-ops even became manufacturers themselves to push prices even lower. The 'coops' became a driving force in British retailing until the 1960s. They then started to decline through the rise of supermarkets that strove to make food shopping enjoyable and seduced the younger, more affluent housewives. It was never as popular in France, although it accounted for a reasonable share of the market. Ironically enough, the concept of 'co-operatives', exported to the US, came back to Europe in the 1970s under the new form of discounters and 'buying clubs' that are strangely reminiscent of the Rochdale pioneers.

I.2. 'Back to basics': food consumption in the age of two world wars and a major depression

From the 1890s onwards, manufacturers took the upper hand on retailers. Thanks to growing consumer loyalty towards well-known brands, manufacturers influenced prices and decided on discounts for bulk buying. From the 1930s onwards, marketing science gave them the new power of influencing the way their goods were displayed in the shops: shelf allocation (place in shop, height, and space devoted to its brand) became central to the power game between manufacturers and retailers in the larger units (department stores and multi-unit retailers).

In this period, Europe fell behind the US in term of innovation in food retailing. Two world wars and the 1930s depression hit Europe even harder than the US. The rise of mass consumption that can be seen in the US from the early 1920s onwards, exemplified by the rise of Ford and GM motorcars, did not happen in Europe until the late 1950s. The first half of the twentieth century was for European retailers a time for returning to basics: hard-pressed consumers wanted fewer trimmings and focused on lower prices. This helped the rise of the multi-unit retailers in Britain, such as Sainsbury and Tesco in food retailing. Consumers moved increasingly towards industrially produced foodstuffs, from bread to canned vegetables and meat. The 1930s–1950s were a golden age for corned beef, tinned salmon and tinned peas. France stood apart from this trend, though: unlike Britain, France remained a largely rural country, attached to traditional home cooking and local specialised shops (bakers, butchers and so on). Retailing, therefore, remained more traditional in France, and multi-unit food retailers did not really take off in France until the 1950s. In both countries, change was slowed by the two world wars: rationing froze rationalisation in retailing, both in food and non-foodstuff. In France, many small shopkeepers' profits rose and gave them a new lease of life. In the UK, the black market was less developed, but government fixed prices and quantities, which discouraged concentration of the food market. Change was delayed until after the end of rationing and the return to affluence.

I.3. The age of affluence: adapting the American model of mass consumption to European tastes

Food retailing entered a period of far-reaching change from the end of the Second World War, with the rise of major supermarket chains in all European countries. Pioneered in the United States, the concept of self-service and large retail units was not adopted in Europe before the late 1950s. But from then on, both countries started to catch up on the

US. Change was driven by some owners of existing multiple food chains, such as Tesco and Sainsbury in the UK, as well as new entrants, such as Associated Dairies. In France, several entrepreneurs built major food retailing empires, such as Leclerc and Carrefour, in less than two decades. In the 1970s, they were given a new boost by the introduction of information technology. Electronic points of sale and computer data analysis gave retailers a privileged access to a mass of information about consumer habits and behaviour. This tipped the balance of power back in favour of retailers. Manufacturers increasingly relied on retailers to give them information about the market that marketing firms could not produce. But this change is only the tip of the iceberg.

Changes in post-war services mirrored changes in industry, in scale and nature. The business of retailing was completely transformed by the introduction of information technology, from the electronic tills of the 1960s and the Electronic Points of Sales (EPOS) and scanning in the late 1970s and 1980s, to the integrated management software systems of the 1990s. Automation was not confined to car, steel, or chemical plant, but rapidly spread in services too. In retail, this push for automation focused on points of sales until the late 1970s, and then spread to stock management, deliveries, and shelf stocking. It had the same effect as in industry: an increase of repetitive work and pressures for increased speed at check-out in the 1960s and 1970s, with a move back to flexibility, diversity and 'softer' notions of consumer service from the late 1980s onwards. Many historians consider this change to amount to a 'third industrial revolution', heralding the move to the 'Age of Information Technology' (IT).

Along with new processes came new products, with a constant tension between ongoing standardisation and pressures for differentiation of products along geographical and sociological lines. For retailing, that meant constant arbitrage between a reduction in the number of product lines and global sourcing, and a customer yearning for differentiation that increased in the 1980s and 1990s. From the same sources (customers, consumer associations, and governments) came conflicting pressure for better quality, better service, but reduced or at least stable prices. Product innovation was one major answer. In came convenience food, such as frozen foods and chilled ready meals, so-called 'healthy food', ranging from fresh fruit and vegetables in the 1960s to organic food in the 1990s, and more diversified foods including foreign dishes and regional produce. With increased sophistication in sales and marketing techniques, these new products quickly became part of every day life for the European consumer.

These changes in processes and products were matched by a new business organisation. Everywhere in Europe, retailers copied the Americans in a search for growth to achieve economies of scale and enhanced market share. All European countries saw the consolidation of food retailing into a handful of major players and the concentration of operations into bigger units. In the 1990s, the introduction of integrated management slowly translated into an increased centralisation of operations, with many decisions removed from the shop floor. That meant a redefinition of the role of supermarket manager, transformed into a 'team leader', and the development of support personnel in the form of experts, in IT, marketing and sales, financial management (in particular in property management) and public relations (PR).

These trends were not specific to retailing. The issue of Americanisation is as pervasive here as in other industrial or service sectors, but probably more controversial and more 'visible' in retail. And retail does indeed provide an ideal vantage point from which to assess the impact of Americanisation in Europe. As a consumer industry, it experienced at first hand the impact of socio-economic change. Spiralling urbanisation, rising average incomes, and increases in female employment led to new trends in consumers' behaviour. Although the percentage of household expenditure devoted to food fell, net food expenses rose with the rise in living standards. Food expenditure moved away from basic items, such as bread, tea and sugar, and increased on formerly 'luxury' items, in particular on alcohol, fruit and vegetables.[1] But the single most important new trend in retailing since the 1960s is the rise of convenience food in the form of prepared meals.

Retailing has been at the front line of negative reactions to Americanisation from governments, producers and consumers. It is also at the heart of the battle between modernising and conservative forces in European societies. Supermarkets have come to embody all that is good or bad in Americanisation: on the one hand, cheaper baskets of goods, increased access to fresh fruit and meat, ability to do a 'one stop shop' and extra leisure time; on the other hand, increased standardisation of food and decreased personal contact with retailers, not to mention the hotly-debated topics of choice, access, and quality.

But the experiments of the supermarket chains to find suitable answers reflect the ambiguity built into the 'Americanisation debate'. In the 1990s, one saw a move away from standardisation, large units and out-of-town shopping, towards store segmentation according to the sociological composition of the area, and, within the supermarket giants themselves, the reintroduction of smaller centre of town stores. In the same way, one sees a return to 'traditional values' in the design and outlay of stores, that include an attempt at 'fitting architecturally' into the environment, the reintroduction of fish, vegetables or deli counters, and the return of packers at point of sales. This, however, must not be interpreted as a move away from the American model, as one sees the same trends sweeping through American market chains in the same 1990s. Besides, both European and American food retailers are now focusing on surviving a new and more worrying challenge, that of the discount stores, in particular that of Wal-Mart.

I.4. The benefits of comparing the development paths of French and British retailing since the 1850s

In this context, what is the specificity of European supermarkets in the last 50 years? The first answer is the rapid catch up of Europe on America in the field of retailing. From the mid-1970s onwards, technological leadership is shared between Europe and America. EPOS and scanning were introduced in Europe only a few years after America, and today, Tesco and Carrefour are amongst the leading retailers in the field of fleet management, stock management, and systems integration. European retailers are amongst world-class players in terms of size, profitability and multi-nationalisation, with the 1970s and 1980s seeing a reverse flow of European investment into the US retailing market (often with mixed results, given the difficult competitive market there). The story of European retailing since the 1950s is therefore not one of traditional 'catch up' on America, but rather one of crossing influences between two continents. It is also one

[1] See for example the Economist Intelligence Unit's publication *Marketing in Europe*, for regular updates on the spread of household expenses and shares of particular foodstuff.

of success (in terms of innovation and company profits) matched with constant controversy: supermarkets in Europe have been at the heart of ideological debates on the future of European societies. The debate on Americanisation touches on many issues, including the wider debate on globalisation, the benefits and dangers of capitalism (the battle of the giants versus the small producers and retailers), the quality of food (most of all genetic engineering) and ethical trading, to cite but a few.

These issues are common to all supermarket chains across the world. This chapter focuses on France and Britain, though. These two countries were chosen because they have consistently been leaders in the field of European food retailing since the 1850s. The comparative approach is necessary, as they have followed radically different approaches in terms of business organisation and strategies. It highlights some of the pitfalls of purely 'national histories' of a specific branch. It also helps in identifying common challenges and answers, and assessing how far strategies and structures are shaped by national histories and development paths dating back to the middle of the nineteenth century.

France and Britain differ in their responses to the emergence of mass markets and global retailing in four key areas. The first one is in store format: the impact of hypermarkets has been limited in Britain by the importance of strong existing multiples in clothing (Marks & Spencer), drugs (Boots), furniture/DIY (BHS) and electrical goods (Currys). France, by contrast, has pushed the hypermarket format to a form of art. The second difference lies in ownership structure: licensing is more developed in France. There, most major players (such as Leclerc) are no more than a centralised purchasing organisation for independently held shops. Carrefour is the most notable exception. Britain, on the other hand, has been dominated by four or five major supermarket chains that own all their stores. Again, this has been shaped by history. British supermarkets have been developed by retailers that built their store capacity over several generations. France did not have major multi-unit retailers in food until the 1950s, and the need to build store capacity quickly meant that the licensing form was probably the easiest way forward.

The third and fourth differences are in marketing strategy. British supermarkets have been a major force behind the internationalisation of customer tastes, in the field of food and wine, whilst French ones have often played the 'terroir' card, actively promoting regional produce, as a PR gesture proving their 'good food practice'. This must be replaced in the wider context of consumer tastes: the British middle class increasingly turned outward in the last 40 years, and travelled the world for business and pleasure (the dreaded 'package holiday'). This trend is less marked in France. One of the possible explanations is that French society went through a faster sociological change since the 1950s, with a rapid change from a rural to an urban society. Its response was to turn to its cultural roots and reinvent them (a process also seen today in fast changing Asian countries, such as Japan and Korea). The fourth major difference lies in geographical markets. British retailers concentrated on their home markets. French retailers moved across borders aggressively. This reflected the continuing lobbying strength of France's small retailers, who successfully managed through their MPs to block the expansion of major supermarket chains through tough planning laws. French supermarket chains therefore sought expansion abroad. British retailers had no such need until the mid-1970s, when toughened planning legislation produced the same effects: they in turn looked at other countries (America first, and later on Central Europe) to boost their profits.

These differences must not be exaggerated though, and one can see a slow convergence of European supermarket practices in the 1990s. This chapter will now look in more detail at three areas where the nature and spread of innovation was shaped by the conflict between international convergence of international information technology since the 1950s and national historical factors: product innovation, sales and marketing, and business organisation.

II. Product innovation since the 1960s

The two key words of the last 40 years in product innovation are convenience and quality, mitigated by a constant pressure to keep costs low. Lifestyle changes have favoured food that is both easier and quicker to prepare. Hence the rise of two new major markets, those for frozen food and ready chilled meals. From the 1970s onwards, though, a sizeable segment of the market has succumbed to a new fashion for 'healthy' food. Most of the new products launched in the 1980s and 1990s, from yoghurts to cereals through to mineral water, stress the 'healthy diet' element. Marketing campaigns started by stressing the low-fat (margarine) and low sugar (particularly in soft drinks) quality of the products and has now turned towards added vitamins and nutrients (iron, magnesium...).

II.1. The search for convenience: frozen foods and ready chilled meal

New techniques for freezing food were developed as early as the late nineteenth century and first used to transport food over long distances, such as Argentinean meat to the UK on specially equipped boats. By the 1890s, refrigerated and frozen beef already dominated the Smithfield food market in London. The well-loved ice-cream van appeared on the streets of London.[2] But fridges and freezers were tools of the trade and did not enter most households before the last third of the twentieth century. Churn freezers for households were invented in the USA in the mid-1860s and became available in Britain soon afterwards. But they remained a luxury item until the late 1960s (Oddy and Oddy, 1998: 298).

Change came in the 1970s, driven by supply. Prices of domestic freezers dropped dramatically. But the key element in the development of the frozen food market was the creation of the combined fridge/freezer. In 1974, 88 percent of French households had a fridge but only 12 percent had freezers. But in the 1970s, prices fell and many households acquired a combined fridge/freezer. For once, rural households had a higher rate of penetration than urban households: in 1974, 40 percent of French farmers owned a freezer (*Marketing in Europe*, March 1975: 39). At the same time, frozen foods became widely available in Europe and started to compete successfully with canned and dried foods. French consumption of frozen foods rose to 36,000 tons in 1971, creating a market worth 331 million francs (Table 10.1). This was a remarkable increase of 32 percent in volume and 38.5 percent in value over the previous year, but still well behind the US. Sea fish was the fastest growing item, accounting for 52 percent of all sales of frozen

[2] See for example Thomson, 1994.

foods, with vegetables coming next with about 1/3 of the frozen food market. On the other hand, sales of frozen meat, poultry and rabbit remained low.

Table 10.1 Household consumption of canned and frozen food in France in the early 1970s (million French francs)

	1969	1970	1971	% Δ 1971/1970
Frozen foods	172	239	331	+38.5
Of which fish	129	174	266	+52.9

Sources: *Marketing in Europe*, 25 April 1973: 10.

In the UK, frozen foods were distributed both by specialist freezer centres, such as Iceland, and by supermarket chains, who promoted frozen food as a quality alternative for canned and dried meals in the semi-prepared and prepared meals sectors. By the 1980s, frozen foods were a common feature in European weekly purchases of food. Most of the major food manufacturers entered the frozen food market in the 1960s. From 1957, for example, Unilever re-entered the European frozen food market under the Iglo brand (Oddy and Oddy, 1998: 296–297). Supermarkets also developed their own brands.

The story of frozen foods is a classic 'catching up' story of Europe on the US. In the chilled meal sector, however, European food retailers were amongst the pioneers. In the 1970s, Marks & Spencer were the first to introduce a chilled ready meal range in their shops. Marcus Sieff believed that the convergence of European lifestyles towards the American model, in particular rising rates of female employment and increasing numbers of single households, created new market opportunities. This was a stroke of genius. Marks & Spencer had identified a relatively upmarket niche that was occupied in the States by small family deli shops but did not exist in Europe. They invested heavily in food quality in order to replicate by industrial means, in large quantities, the taste of family cooking. Marks & Spencer was in that sense the Henry Ford of family cooking. Food was manufactured in their own factories, or subcontracted to major food manufacturers on the basis of detailed specification of sourcing, manufacturing methods, quality control, and price. Thus started a tradition of close partnership that mirrored the relationship Marks & Spencer had with its suppliers in the clothing sector.

Product innovation in chilled ready meals included technological breakthroughs in food manufacturing, in order to replicate industrially the essence of home cooking. The first dishes on offer included the well-loved shepherd's pie. But innovation also laid in the introduction of dishes imported from abroad. Marks & Spencer pioneered the introduction of Italian, Indian, French, and Chinese dishes, at a time when most supermarkets stuck to basics like bangers and mash or steak and chips. Marcus Sieff, then leader of Marks & Spencer, recognises the increasing internationalisation of Britain's middle classes, thanks to the explosion in package holidays abroad. Foreign dishes became Marks & Spencer's top selling lines, as well as sandwiches. Standard food supermarkets soon realised the potential of the ready meal, and produced their own version at a lower cost, usually through own-brands. By the 1990s, a survey of British cooking habits revealed that curry had become the national dish of Britain, thanks to the multiplication of Indian take-away, but also to supermarkets' 'ready to heat' versions. Unfortunately, research into the importance of chilled ready meals to food retailing and its impact on other forms of ready meals (canned, dehydrated, frozen) remains largely confined to market research and needs to be reassessed in an historical perspective.

II.2. The explosion of the 'healthy food' segment

The first advertising campaigns promoting 'healthy foods' started in Europe in the 1960s. Governments teamed up with producers' association, such as the UK's Milk Board, and launched advertising campaigns for milk, fresh fruit and vegetables. They had little impact. In the 1970s, attention focused on labelling. British and French consumers' associations lobbied Parliament for legislation imposing the disclosure of the nature and quantity of all ingredients. This was argued on grounds of consumer choice, on medical grounds (allergies, in particular to peanuts), and on more general health grounds (to spot fatty, sugary foods or those with undesirable additives). That battle was won by the introduction of new legislation and successful consumer pressure on manufacturers.

Supermarkets in France and the UK played a leading role in relaying consumer pressures in favour of food labelling in the late 1970s. Their motives are clear: at a time of tightening planning regulations and rising consumer militancy on prices and quality of food in supermarkets, this was a golden opportunity to foster a more 'consumer-friendly image' amongst consumers and their organisations, and government departments. Supermarkets also seized the opportunity to create own-brand 'healthier ranges' in order to compete with established brands. But this new, winning strategy was partly the consequence of new technology: the recently introduced electronic points of sales (EPOS, see below) gave the supermarkets' marketing teams early warnings of rising consumer demands for 'healthy products'. This gave them a leading edge on manufacturers. More research needs to be done on this issue, though.

The trend in favour of healthier foods then moved on to a new issue in the 1990s: that of 'local' versus 'global' sourcing. Supermarkets have been instrumental in changing European eating habits. They popularised international brands, such as America's Heinz ketchup sauce and Coca Cola drinks, Britain's Cadbury chocolates and France's Perrier water. They tempted middle class consumers into trying new 'exotic' foods, such as bananas in the 1960s, mangoes in the 1980s. Overall, food-buying habits converged in Europe and the US. For many consumers, major international brands offered the best guarantees in terms of information, traceability and enhanced hygiene. Only large-scale firms could offer the food hygiene standards inspired by the pharmaceutical industry. This obsession with hygiene started in the US in the 1930s and slowly spread to Europe from the 1960s onwards. One example is cellophane packing for fruit and vegetables, popular in Northern European countries and shunned in Mediterranean ones, where consumers like to poke their fingers in search of ripeness and freshness.

This debate resurfaced in the 1980s and '90s, with a succession of food scares, from salmonella in eggs to BSE and genetically modified foods. They revealed a clear split in European population between partisans of maximum hygiene and partisans of 'authenticity'. The 'salmonella in eggs' rumpus in the UK was typical of consumer attitudes to food safety. The former argued that low standards of hygiene in small units were responsible for the infection, while the other half argued that the standard use of antibiotics in large-units had contributed to the spread of the disease. Cunningly, UK supermarkets sought to capitalise on both attitudes by offering standard '*fresh farm* eggs' and 'free range' eggs. This is but one example of a wider trend. In France, and to a lesser extent in Britain, supermarkets, best known for offering a wide range of international brands since the 1950s, started to promote 'regional products' from the 1970s onwards. France led the way in the 1970s. Many supermarket chains installed colourful temporary counters of *produits du terroir*, such as foie gras, sauerkraut and dried sausage. Consumers were attracted by free tasting and 'special deals'. British supermarkets followed these trends in the 1990s, in particular Tesco, which seems to have its ear closer to the ground than its once big rival Sainsbury, and installed permanent 'deli' counters selling fresh produce such as fish, unpacked meat or cheese. This policy is now widespread in Europe and has now reached the US itself.

Temporary counters are rented out to local producers of semi-industrial size for a short period of time, so as to create novelty for the consumer. Permanent 'deli' counters are usually staffed in house. But occasionally, they are contracted out to small operators with experience in a particular trade, to attract new customers or more frequent trips to the supermarket. This is often the case of fishmongers. The Deauville small 'Champion' supermarket, on the French Normandy coast, has eaten into the larger edge of town clientele of the Leclerc and SuperU by farming out a fish counter to an old-established local fishmonger. It now attracts a wide clientele from permanent residents and week-enders. They come for the high quality fish and the local accent, and come out with the weekly shopping. In the UK, Morrison has adopted a 'Market Street' format for the layout of its food halls (Competition Commission, 2000, II, 89). And this strategy has long been that of small supermarket chain Waitrose. The development of organic food and of regional 'quality labels', such as France's *Appellations d'Origine Contrôlée* (AOC) and 'label rouge' are part of the same search for healthier 'authentic' produce. It stems from the conviction that 'traditional' food is healthier than its modern version.

But the search for healthier food is only part of the explanation for the success of 'local produce'. According to some sociologists[3], consumers react to rapid socio-economic change by clinging to brands that remind them of bygone days. This lies at the heart of the 'Bonne Maman' (i.e. 'granny' in Belgium) jam's success. The label pictures a granny hand-stirring jam in an old fashioned copper cauldron. Industrial groups have been quick to exploit this nostalgia. For example, most battery eggs sold in France and Britain are packed into boxes adorned with pictures of old-fashioned farms that remind the buyer of farming as pictured in children's books. The appeal of food safety increased with the various food scares of the 1990s. Many consumers reacted by joining the ranks of the 'nostalgic' in search of yesterday's farming. Britain saw in the 1990s a spectacular revival of old-style cheeses and breeds of cows and pigs. France brought some old animal breeds, such as the blacklegged pig of the Pyrenees, back to life. Organic farming also falls within this trend.

[3] Such as those of *Cofremca*, a European research institute on socio-economic trends.

III. Innovation in sales and marketing since the Second World War

III.1. The parallel introduction of self-service and supermarkets in the 1960s

Undoubtedly, self-service has been the most dramatic innovation in European grocery shopping since the Second World War. Pioneered in the US before the War, it was introduced in Europe in the mid-1950s. By 1973, it had achieved dominant market share. Britain led the way. By the mid-1960s, 7,000 UK self-service shops controlled 20 percent of national grocery sales. But only 400 of these shops were supermarkets. The remaining 6,600 were convenience stores (*Marketing in Europe,* 1960, General notes: 690). 'Supermarkets', defined as large outlets selling food and basic household goods, were the second major retail innovation in post-war Europe. In the 1950s, the multiple food retailers that had appeared in Europe from the late nineteenth century started to convert their shops into larger self-service outlets. Their aim was to provide a one-stop shop for both food and basic household goods. Between 1961 and 1971, the number of outlets owned by British multiples (food and non-food) shrank by 4.7 percent whilst turnover increased by 130.5 percent (Tucker, 1978: 15). The table below charts the development of supermarkets in the UK:

Table 10.2 Numbers of supermarket outlets in the UK

1958	1959	1960	1961	1962	1963	1972
175	286	367	572	996	1,366	2,110

Source: *Retail Business*, September 1964: 3, derived from *Self-service and Supermarket Directory,* and *Marketing in Europe*, 1973.

Supermarkets were a huge success in France too. But since multiples were not as well established, newcomers such as Leclerc and Carrefour led the way. Carrefour, created in 1959 by the merger of two medium-sized family retailers, opened its first supermarket in 1960 in the alpine town of Annecy ('How Carrefour', 1978: 331). By 1973, it had 21 hypermarkets (giant supermarkets selling a wider range of non-food goods). But the Royer legislation, which toughened planning regulations, slowed its expansion, so that by 1977, they opened only 6 new stores, bringing its hypermarkets to a total of 27.

Differences in Europe remained high. For example, supermarkets' share of fruit and vegetable sales in 1972 ranged from 3–4 percent in Italy to 90 percent in Sweden. Britain lagged behind Sweden but remained in the top segment, with about a third. France, with 17 percent, was close to the bottom segment (*Marketing Review*, 1973). Overall, small food retailers fared better in Southern Europe. Marketing Office attributed these to the more demanding consumers in terms of meat, fruit, and vegetable quality, and to more protectionist legislation, particularly in France and Italy. Supermarket share of food sales also varied according to product type. In France, fresh produce such as fruit and vegetable, meat and bread were bought by a majority of customers in specialist shops, but supermarkets controlled a third of the processed foods segment, with a dominant position in prepared meals and baby foods (*Marketing Review*, 1973). In other words, the supermarket share was highest in new products such as convenience food. On the contrary, supermarkets in Britain soon came to dominate sales of bread and, but to a lower extent, meat and fruit.

The rise of supermarkets was driven by market change rather than technological change. A 1969 survey of French shoppers conducted by Marketing Office, a French market research organisation, concluded that speed of service was the main motivation behind supermarket shopping. Next came the fact that one could buy both food and non-food lines in the same shop. However, by 1973, cheaper prices had become the customers' first motivation in visiting superstores, unsurprisingly when one considers the then rate of inflation and the looming economic crisis (*Marketing in Europe*, 25 April 1973: 2). The motivations were roughly similar all over Europe. Speed of service and an attractive environment were two other key features in taking customers away from traditional shops (*Marketing in Europe*, 1973: 691). British consumer surveys of the 1970s suggest that part of the decline of the British co-ops can be attributed to the feeling amongst the younger middle and lower middle class housewives that the co-ops were downmarket, cheap, and frumpy.

By 1972, Germany had the highest number of supermarkets (2,802), followed by France (2,060) and the UK (2,110). However, this is a reflection of the size of these countries. When one considers the number of supermarkets per head, Denmark and Belgium took the lead. Britain and Switzerland followed. France belonged to a third group, together with the Netherlands and Germany, but well ahead of Southern Europe.

Table 10.3 Supermarket penetration in Europe in January 1972

	Total numbers	Million of square meters	Outlet per million inhabitants
West Germany	2,802	2.026	46
Great Britain	2,110	1.435	58
France	2,060	1.462	40
Netherlands	622	0.7	47
Belgium	606	0.473	62
Italy	600	0.41	11
Switzerland	335	n.a.	55
Denmark	325	0.11	65

Source: Marketing in Europe, March 1974: 3.

In most of Europe, supermarkets sold only food and basic household goods. France, however, imagined a different format in the 1960s. Hypermarkets were characterised by larger sales space and a wider range of products, most notably textiles (clothes, bedding, towels etc). By 1972, France had 209 hypermarkets with between 25,000 and 200,000 square feet of selling space each, as against 2,334 supermarkets (officially defined, as most things in France, as shops of between 4,000 and 25,000 square feet. According to *Libre Service Actualités*, both formats controlled 21.9 percent of total food sales; a big increase compared with the 1960s, but still only one fifth of the market.[4]

[4] The hypermarkets accounted for 7.4 percent of food sales and 4.75 percent of all retail sales, with the supermarkets controlling a further 14.5 percent of food sales and 6.5 percent of all retail sales (LSA, *Atlas des supermarchés en France*, quoted by *Marketing in Europe*, 23 February 1973: 2).

Table 10.4 Hypermarket penetration in Europe at 1/1/1972

	Total numbers	Total square mile (million)	Outlet per million inhabitants
West Germany	370	2.07	6
Belgium	46	0.309	5
France	147	0.83	3
Great Britain	22		0.4
Italy	1	0.08	0.006

Source: Marketing in Europe, March 1974: 3.

In the 1970s, hypermarkets spread to Germany and Belgium, and by 1972, France, who had pioneered the concept, was trailing behind Germany and Belgium in terms of number of outlets per head (Table 10.4).

Tables 10.3 and 10.4 prove that France's more stringent planning regulations did not succeed in stemming the rise of supermarket shopping. By 1972, France had as many supermarkets per head as Germany and the Netherlands. In the long term, customer pressures proved more powerful than the vested interests of small retailers (Ogenyi, 1999: 43), in spite of the poujadist movement and the lobbying of small retailers. Consumers voted with their feet. French consumers surveyed in 1973 by the *Centre d'Information Civique*, said they preferred to shop in supermarkets, followed, in descending order, by chain stores, co-operatives and independent retailers (*Marketing in Europe*, May 1974: 1).

In Britain, planning regulations made the opening of hypermarkets difficult, so that supermarkets remained the multiples' favoured format (Ogenyi, 1999: 141). Sainsbury, Tesco and Safeway were the biggest players, but Marks & Spencer fared well in some areas, in particular fruit, and preserved the position gained in 1964 as Britain's largest fruit retailer (*Retail Business,* September 1964). Supermarket development in the UK exhibited marked regional variations, though. Greater London, the South, the Southeast, and the Northwest had the highest levels of supermarkets. The lowest levels were found in the East and West Riding, the North, Scotland and Wales. This roughly reflected patterns of urbanisation and wealth, with the exception of the Northwest. Initially situated near city centres, supermarkets soon targeted the suburbs, where they could open larger outlets with increased parking space.

The 1980s and '90s saw a continuation of some of the trends that emerged in the 1950s. The share of the small independent food shop (grocers, specialised butchers, fishmongers etc) continued to decline. By the end of the 1980s, independent food retailers accounted for only 37 percent of the French food market, against 56 percent for all supermarkets, 6 percent for department stores and less than 1 percent for the coops. This supermarket 'revolution' was driven by changes in consumers' aspiration rather than by new technology. But the concentration of food retailing into a handful of large supermarket groups has had profound repercussions on both marketing techniques and business organisation, through the use of new technology. The most significant is Electronic Points of Sales (EPOS), introduced in the 1970s. This allowed supermarkets to cut down queues, reduce the number of empty shelves through better information about real sales, and, up to a point, cut prices.

III.2. The EPOS revolution from the 1970s to the 1990s

The major complaint of supermarkets' shoppers from the 1960s to the 1990s was queues at the tills. Supermarkets looked at new technology to automate points of sales. In 1973, *The Economist* estimated that only about 100 (mostly non-food) retail organisations in the UK had electronic registers (usually Sweda 700 terminals), and that out of the 700,000 cash registers in operation, only about 10,000–15,000 had a throughput sufficient to justify the switch to electronic registers. But their adoption was accelerated by the introduction of VAT following Britain's entry into the European Economic Community (EEC) in 1973 (*Marketing in Europe*, May 1973, 3).

The major innovation in till automation, though, was the import, in the late 1970s, of American scanning technology. French and British supermarkets led the way, ahead of non-food retailers and manufacturers. In 1978, a mere 6 years after scanning was first introduced in US retailing, Tesco introduced a pilot scheme in its Wellingborough store. Scanners at the till were linked to the central stock control computer and to the minicomputers in the group's warehouses ('Tesco searches', 1978: 957; Walsh, 1993: 94). Sainsbury followed in the same year. Continental Europe quickly caught up: by 1981, Germany had 38 stores with optical scanners, France and Sweden 11, Italy 10, Switzerland 9, the UK 8, Austria and the Netherlands 3, Norway 2 and Belgium 1. By the mid-1980s, Europe still lagged behind the US for EPOS density, but by the late 1980s, all major European supermarket chains had converted to EPOS. The primary aim was to cut down the time it took the operator to tally the purchases. Scanning relied on the co-operation of food manufacturers, as goods had to be coded first, but this was completed in the 1980s. It worked so well that goods moved through the till faster than the customer's ability to pack them. This in turn brought experiments in European supermarkets in the 1990s to introduce partly automated packing or to bring back the packers of yesteryear at peak times.

By the mid-1980s users began to realise the potential of EPOS beyond increased speed at checkout. In this race for best use of computer technology, European retailers were neck to neck with American retailers, accrediting the hypothesis of technological convergence between Europe and the US. By then, microcomputers had appeared at store level. One application was in staff management. New computer software was developed to analyse the flow of customers at the till throughout the day and allowed store managers to calculate the number of tills needed at different times and on different days (Maggart, 1981). Tills and stores were also redesigned to speed up operations. Operators were often checked, whether officially or unofficially, for speed. This amounted to a new form of taylorism, at the very moment taylorism was dying in industry.

But the impact of EPOS was not limited to points of sale. By bringing together information provided by individual checkouts at store or head office level, sales staff accessed new information on customer purchases that helped with marketing and supply management. In fact, EPOS only improved on the possibilities offered by the first electronic tills of the late 1960s. As early as 1968, the UK's National Computing Centre stressed the possible applications of computers in retailing (National Computing Centre 1968). Its survey of UK trade (food and non food, wholesale and retail) showed that 178 retailers already used electronic applications for sales analysis, 177 for invoicing, 163 for stock recording, 157 for sales ledgers. 39 companies used critical path analysis, a higher rate than in other industries. 33 companies experimented with electronic applications for choosing the location of warehouses, in spite of the teething problems of this very new technology (National Computing Centre 1968). Still, it was food and household goods manufacturers, such as Procter and Gamble, Kimberly-Clarke, Littlewoods, Unigate and Petrofina, that led the field, well ahead of supermarket chains. The Lewis Partnership was the only supermarket chain to invest in this technology by 1968 (National Computing Centre 1968, appendix 4). By 1973, however, all major French and UK supermarket chains used information provided by electronic tills to optimise shelf-allocation. That year, a UK working group headed by the Institute of Grocery Distribution and supported by 11 food retailers co-operated with equipment and systems manufacturers to produce a report on existing and future methods to optimise shelf allocation in food stores (Marketing Review, 31 October 1973: 3). By the 1980s, EPOS data was helping in the process.

Paradoxically, whilst computerisation is often associated with decentralisation in heavy industry, it increased centralisation in retailing. Up to the 1980s, store managers had a high degree of control over contracts for fresh foods, deliveries, and shelf allocation. In the 1990s, the search for a balance between increased flexibility and cost cutting led all major supermarket chains to centralise information produced at point of sales and analyse it at head office. There, experts in marketing redefined store strategy and decided on shelf allocation and product mix (Spilsbury et al., 1993: 4). The first software packages analysing stores' customer base went on sale in the early 1970s. Produced by Los Angeles's firm Urban Decision Systems, they studied local demographic patterns to help firms select the best sites for new stores (Reynolds, 1992: 270). The 1980s saw the development of new software analysing the customer profile of existing stores, under the name 'category management' (Competition Commission, 2000, II: 91). They allowed top management to see which products sell in which shop and which don't. They also produced figures of sales per employee, per square metre and per shelf, enabling comparisons within each store as well as between all stores. British and American food retailers were the first to develop and adopt this new type of software.

It has had two major consequences. First, store managers came under increased pressure to match best practice within the group, making managers more accountable to head office than ever before. They also lost the authority to decide on shelf allocation, product mix and stock levels. The managers' empirical experience of what was sold in the shop, who were its clients, and what they wanted has superseded by a more scientific and efficient tool. But this tool is too complex and too costly to be left at store level, at least for the time being. The job of store manager has therefore changed beyond all recognition. His job is not to make decisions on stock, shelf allocation, product mix, and stock any more, but has been refocused on customer relations and team leadership. Its main task is now to keep a well-motivated workforce that understands the importance of keeping consumers happy. A third consequence of customer profiling has been an overhaul of supermarket formats to reflect local differences. Tesco now offers six different supermarket formats, from the petrol station cum convenience store through to the hypermarket.

Still, category management has had its drawbacks. Data interpretation proved more difficult than expected. And front line staff has not always accepted change easily. It proved particularly difficult in the UK, where qualifications are lower than on the continent and a greater proportion of the workforce is made of pensioners returning to work for a few hours a week to top up their pensions. While perfectly happy at stocking up shelves and putting up prices, they were more reluctant to adopt technological innovations such as hand held scanners for stock control on the shelves. They also quietly resisted moves towards increased customer service and somehow don't quite fit into the new face of supermarket assistants dreamed of by top management and marketing experts (Spilsbury et al., 1993: 4). Still, since many UK supermarkets are unwilling to increase levels of permanent staffing and increase pay, they will have to work round this problem. Maybe training opportunity directed towards early school leavers will provide an answer. It is already experimented by French and, to a lesser extent, UK supermarkets.

III.3. The search for differentiation in the 1990s

Americanisation was supposed to create a standardised mass market in Europe. European integration was to lead to the emergence of the 'euro consumer'. This, at least, was the received wisdom of the 1970s. In fact, marketing gurus and, before them, supermarkets realised that the European market, in spite of some convergence, remains for the time being a myth. For a start, the size of the total food market and per capita revenues still varied widely in European countries. In 1988, West Germany and France were by far the biggest markets, with 96 and 89 billion dollars respectively; Britain was in the 'medium-sized market' group, just behind Italy, with 50 billion dollars. Followed a group of small markets (less than 16 billion dollars), comprising all the countries of Northern Europe. There are also major differences in the rate of female employment. In 1988, 51 percent of Danish women worked, 40 percent of British women and nearly as many in Portugal, but less than a third in Italy and just over a quarter in Spain. Female working patterns affect household wealth and consumption of convenience foods, so that supermarkets offered different product mix in each country. Finally, tastes still vary widely across borders: the British consume more canned pasta than the French, and still favour one brand (Heinz) above own-label brands.

But national differences do not paint the whole picture. Sociologists and retailers increasingly segment societies beyond national differences, according to life style. The London upper-middle class is not so different from the New York or Paris ones. Supermarkets have always adapted product mix to the sociological composition of their catchment areas: olive oil, avocado pears and melons remain a quaint luxury in the industrial heart of the Midlands, whilst in London it has become part of the weekly shopping list for the affluent upper middle classes. In the same way, the 'bobos' (*bourgeois bohèmes*) of Paris echo the 'Islington set' of London and the Heights-Ashbury crowd of San Francisco in their choices of food and shopping venues.

These national and sociological differentiations have been reinforced in the 1990s by a new taste for differentiation. In the affluent Western societies, the consumer of the 21st century wants to be made to feel 'special' and yearns for the village of yesteryear. Without relinquishing the lower prices, car parks and long opening hours. Retailers have tried to turn this trend to their advantage. Locally sourced or products bearing quality labels are bought at a premium by consumers eager to differentiate themselves and buy a sense of 'authenticity'. The success of French *foie gras* on the East Coast of America or the world-wide boom in demand for premium olive oil from Tuscany and Provence are testimonies to these trends. Some will be sold on the Internet, but most will be bought in deli counters of the local supermarkets. Globalisation in the food industry will see two coexisting trends: one for cheaper standardised goods, the other for top quality 'speciality' products that sell world-wide thanks to a local label.

This search for so-called 'authenticity' goes beyond products, into supermarkets' buildings and layout. Retailers have now moved away from the 'utilitarian shed', left to hard discounters, to seek inspiration in traditional buildings. Tesco has adopted a 'mock-Tudor style', complete with black beams. In France, retailers are experimenting with regional styles. The 'Mistral 7' edge-of-town shopping centre in Avignon replicates a market town environment, with half the surface occupied by an Auchan hypermarket and the other half let to small non-food multiples and two-storey 'boutiques' made to look like a typical Provencal town. Shop windows on the ground floor are shaped in half-circles like medieval shops and the first floor boasts blind windows overlooking a small Mediterranean balcony complete with iron railings and potted olive trees. In Normandy, black beams are painting on the outside to fit in with local style. Inside, the layout of supermarkets increasingly seeks to create a 'market-like' feel. Counters, once considered as a dangerous waste of space and labour, are now appearing everywhere. This trend first appeared in France as early as the late 1970s, but is now spreading to Europe and some US towns. One feels a move away from uniformity and size towards product customisation and increased service levels, in retailing as in most other consumer industries (cars, kitchen equipment etc.).

IV. The impact of new technology on organisations: reinventing the business of distribution

Product innovation and changes in sales and marketing strategies are a reflection of the changing business structure of retailing. It is led by changes in both technology and the business environment. The relationship with suppliers has seen some of the most radical changes.

IV.1. The manufacturer/retailer relationship: who won the power struggle?

In the late 1960s, British supermarkets followed America's lead in centralising food buying at either regional (Tesco) or head office level (Sainsbury). Smaller convenience stores themselves formed joint buying groups. In 1973, for example, the 11 Spar voluntary chain organisations, led by the British, German and Dutch Spar groups, formed Intergroup Trading, with a buying power estimated at $34 million (Marketing in Europe July 1973: 2). This eliminated the middlemen and gave them an advantage in price and quality, as goods arrived faster on the shelves. Some went further: Marks and Spencer, having pioneered central buying techniques in the UK, became famous for their exacting food standards: buying direct from the farms meant they could impose their standards and quality requirements. European retailers such as Carrefour, Sainsbury, or Ahold pushed central buying further in the 1990s with the reinforcement of foreign investment. This went hand in hand with another trend that started in the 1950s and accelerated in the 1990s: that of the globalisation of food sourcing. Increasingly, brands are sold under the same name all over the world and local brands have to go global to survive. French and British food manufacturers, such as Danone and Cadbury's, have recognised this and invested aggressively abroad over the last 15 years.

Still, the supermarket chains have had the upper hand over manufacturers since the late 1970s. Centralisation means that the majority of food manufacturers' products now sell through only a handful of negotiators, who can – up to the point that consumers won't miss the brand – go elsewhere. This has been shown by all inquiries instigated by governments worried about competition, one of the most comprehensive being the 2000 inquiry of Britain's Competition Commission into supermarkets (2000 II, chpt.2: 229–259). It showed, for example, that Britain's top four retailers made 70 percent of the UK sales of the two leading washing powder brands, Ariel and Persil (Competition Commission, 2000: 231). But the power of supermarkets must not be exaggerated: individually, most UK and French retailers still represent only a fraction of the sales of leading food multinationals. For example, Tesco accounts for Procter and Gamble's biggest UK supermarket sales but that represents just 0.85 percent of the group's total sales worldwide (Competition Commission, 2000: 231). But this is changing. Today, all major retailers are moving towards international sourcing deals negotiated at head office level with the major multinational food manufacturers. They also try to take advantage of food manufacturers' differentiated price policy across countries to buy supplies for all their supermarkets in the cheapest country. For example, Asda, bought in 2000 by the wonder child of US discounting Wal-Mart, now sources some of its branded goods in Germany to take advantage of cheaper prices there. And although Asda by itself only takes 0.47 percent of Procter and Gamble's total sales, the Wal-Mart group, owner of Asda, now represents 13 percent of Procter and Gamble's global sales (Competition Commission, 2000: 231).

To increase their negotiating power, major retailers are also joining forces with other retailers in international consortium to take advantage of B2B (Business to Business) Internet dealing and online auctions. Britain's Tesco, John Lewis, Marks & Spencer and Safeway and France's Auchan, Casino and Cora are now all members of the World Wide Retail Exchange (WWRE). Created in March 2000 to allow Web-based transactions between retailers and suppliers, it kicked off by organising several online auctions for canned goods.[5] WWRE adds to B2B auctions software for collaborative planning, forecasting, and replenishment (CPFR). But again, this does not represent the whole picture of supermarkets' supply strategies. Increasing local sourcing for some produce mitigates this increasing globalisation. In fact, what is happening is a split in the market between 'commodity'-type goods, such as golden apples, sugar and canned vegetables, bought through these modern channels, and upmarket 'niche products', such as old-fashioned species of fruit, vegetables or cattle, bought in a more traditional way.

But two new factors have reinforced the power of supermarket chains in the last decade. Internet sourcing via auctions allows retailers to compare prices internationally and push them down as long as there is no shortage of that particular produce. Last but foremost, the introduction of EPOS has given retailers a wealth of free information on consumer behaviour that food manufacturers don't have. The latter depend on expensive market research bought in, that is usually less comprehensive. Retailers can therefore anticipate future market trends before manufacturers. The latter are seeking to react by striking partnerships with supermarkets, but this is being watched carefully by European regulators worried about price fixing.

[5] Competition Commission 2000, May–June 2001: 5 and 23. Two other such exchanges are *GlobalNetXChange* and *Transora*.

IV.2. The redesign of supply chains since the 1970s

It is not only the balance of power that has changed since the 1960s between food manufacturers and retailers. The whole supply chain has been redesigned in order to reduce stocks. In retailing, a key component of store profitability is the sales space as a proportion of total space. Reducing stocks allows more shelf space to be installed within any given store. But reduction of stocks may increase the risk of having these empty shelves that so infuriate customers. Again, EPOS data electronically transferred to head office has provided some answer to this long-term conundrum for all shopkeepers. It allows head office to judge in real time what is being bought and send targeted orders to suppliers. Some retailers are even moving to a direct transfer of data from store to suppliers, in a wide ranging review called 'Efficient customer response' (ECR).

But as French industrialists and retailers found to their cost when transporters went on a national strike a few years ago, passing on information to suppliers is not enough. The real key to reducing stocks and coincidentally to offering fresher goods is faster delivery of goods from suppliers to retailers. This has been a priority for all retailers, with French and UK leading supermarket chains at the forefront of innovation. Since the early 1990s, logistics systems and transport fleets have been radically transformed. Sainsbury was one of the pioneers of such moves when it created its first central distribution systems in the late 1980s. Suppliers delivered to a central warehouse from which goods were dispatched to individual stores.

Centralisation went hand to hand with increased automation. The major innovation of the 1980s was the use of pallets. Goods thus packaged moved from the suppliers' lorries to the warehouse, back into the company's lorries, and then directly onto the shelves, thus cutting substantially the need for manhandling. These trends were pioneered in the States, but quickly introduced in the UK and then in continental Europe. The 1990s brought the extension of bar coding from individual items to pallets. Wrapped in cellophane and bar-coded, pallets are moved in warehouses by automated forklifts with scanning equipment. The latest innovation in the field of warehouse management is the use of hand scanners, carried by forklift truck operators. They enable head office to assess stocks in warehouses in real time and track their every move from factory to store.

These central distribution systems, once confined to dry groceries, were extended in the 1990s to fresh goods. This not only reduces costs but also allows retailers to provide fresher goods than most small shopkeepers. Most leading supermarket chains now have three deliveries a day from central warehouses. They take into account changing shopping patterns during the day. In the morning, old pensioners and housewives with children are in a majority, so that items such as baby food, milk and nappies are essential. In the afternoon, shopping patterns are more varied, but the evening sees the dash of young professionals in search of top ups and ready meals. The leading edge retailers of today are now experimenting with 'mixed deliveries' in consolidated warehouses, where several suppliers combine pallets for store deliveries. They are also investing heavily into fleet management, whereby lorries are fitting with satellite positioning systems, that enables head office to redirect them to respond to changing store needs as the day progresses. Warehouses and stores are given advance notice of the precise arrival times of lorries, cutting down of waiting times.

Tesco has been one of the most enthusiastic defender of these new trends in logistics. Experts now admit that this has been a major driver of its increased market share, as customers came to value the freshness of its goods and the reduction in empty shelves. On the other hand, Sainsbury suffered from being first mover. Warehouses set up in the 1980s were inadequate for the storage of fresh foods and automated handling. Still, one cannot but be surprised by its lack of reaction as Tesco and Asda reaped major competitive advantage from their rapid adoption of the new technologies. Recently though, Sainsbury has been, with Tesco, at the forefront of in-house collaborative systems with suppliers to improve information flows and reduce lead times, with, respectively, the TIE and SID systems (Distribution and Logistics, 2001: 9).

IV.3. Technology and economies of scale

Constant pressure from customers for lower prices, particularly in times of economic crisis, such as the 1970s, proved a potent stimulus for mergers in Europe. Consolidation occurred in all Western European markets in the 1960s, and even more in the 1970s. According to A.C. Nielsen, the number of grocery outlets dropped between 1965 and 1972 by 38 percent in Belgium, 35 percent in Sweden, 29 percent in Germany, 27 percent in the Netherlands, 25 percent in the UK and 24 percent in France. This led to high levels of concentration on the American model: the top 2 percent of grocery stores in France and Belgium accounted for half of total turnover.[6] But Nielsen stressed that the 'rationalisation' of shop numbers in relation to population still lagged in Europe in comparison to the US. Northern Europe, and in particular Sweden, was closer to the American model than Southern Europe. Sweden's average turnover for grocery outlets was £373,000, very close to the US figure of £389,000, whilst the figures for Spain and Portugal were £17,000 and £14,000 respectively (*Marketing in Europe*, January 1975: 1–2). Concentration accelerated in the 1980s and 1990s. Table 10.5 suggests that by 1999, Germany had the highest levels of concentration, followed by Austria, Belgium, the UK, and France.

Table 10.5 Share of top five food retailers in Europe in 2000

European country	% of food retailers' sales
Germany	80.3
Austria	74.3
Belgium	70.8
UK	64.8
France	58.1
Ireland	57.1
Netherlands	51.8
Switzerland	49.6

Source: Competition Commission, 2000, II: 94.

[6] Still according to Nielsen, quoted in *Marketing in Europe*, January 1975: 2.

At the same time, companies started to look beyond national borders. The 1970s saw the rise of the first food retailing multinationals, with varying degrees of success. The Americans were the first to cross the Atlantic, driven by ferocious competition at home. They invested first in the UK, then in Germany and France. The Southland Corporation, for example, linked up with Cavenham Foods in the UK, which gave it a 50 percent share in several food multiples in the UK, before looking for potential investment in France and Germany (Marketing in Europe, 22 January 1973: 3). European retailers soon followed suit. In 1973, Marks & Spencer, one of Britain's most successful retailers and definitely a trend setter and a pioneer at that time, decided to open its first continental store in Brussels, planning for a three storey building selling textiles and food, complete with a restaurant and a hairdresser. But in the end it was the Paris store that opened first, followed by Brussels and Lyons (Marketing in Europe, 23 February 1973: 3; February 1974; May 1975: 2). Carrefour was the first French retailer to internationalise. It looked to Southern Europe first, then to neighbouring countries such as Belgium, usually through 50/50 joint ventures with local partners. For example, it bought 50 percent in Italmare, which opened its first hypermarket near Milan in September 1972, offering 35,000 lines, as well as banking services and a travel agency (the latter a joint venture with Club Méditerranée). Carrefour's first hypermarket in the UK, near Cardiff was so successful its manager had to go on television and ask customers to defer their next visit as access roads became saturated (Marketing in Europe, December 1972: 1). By 1977, Carrefour had 26 hypermarkets abroad, including 10 in Spain and 3 in Brazil (How Carrefour, 1978: 331). It then tried to import the 'hypermarket' format into the US, with mixed fortunes (European retailers, 1985: 33). The US market had become a priority for European retailers. Their acquisitions from 1973 to 1984 are summarised in Table 10.6.

Table 10.6 American acquisitions by European food retailers 1973–1984

Year	European buyer	Nationality	Acquisition	percent held	Initial investment ($ million)
1975	Delhaize Frères	Belgian	Food Town Stores	52	27
1977	Franz Haniel & Cie	German	Scrivner	100	27.7
1977	Ahold NV	Dutch	Bi-Lo	100	60
1978	Docks de France	French	Lil-Champ	35	3
1979	Tengelmann	German	A&P	50	80
1980	Promodes	French	Red Food	100	36
1980	Delhaize Frères	Belgian	Food Giant	100	36
1980	Albrecht	German	Albertson's	6.6	18
1981	Ahold NV	Dutch	Giant Foods	100	35
1983	Sainsbury	British	Shaw's Supermarkets	21	20.1
1984	Casino	French	Thriftmart	>50	n.a.

Source: Compiled from Kacker, 1985: 30–31.

By 1990, European (food and non-food) retailers' investment in the US was valued at $13 billion (Sternquist and Kacker, 1994: 4). The top ranking international operators in the US by value of grocery turnover were, in descending order, Tengelmann (Germany), Delhaize Le Lion (Belgium), Ahold (Netherlands), Carrefour (France), Albrecht (Germany), Promodès (Halley, 1985) (France), and J. Sainsbury (UK) (Burt, 1991). By then, France had 24 retailers with 2,336 outlets in 11 other Western European countries, 320 outlets in the US, 35 in Japan and 102 elsewhere, mainly East Asia and South America (see Table 10.7). Carrefour was the most international. UK supermarkets, which had lagged behind in the internationalisation process of the 1970s and '80s and kept their investment to one market, the US for Sainsbury, Ireland for Tesco had caught up by 1990 (Investors Chronicle, 22 December 1978: 957). The top 30 British retailers owned 1,335 retail outlets in Western Europe, 350 in the US, none elsewhere.

Table 10.7 International network of major European retailers in 1990 (food and non-food)

	Number of retail firms involved	Number of Western European countries covered	Number of retail outlets in Western Europe	Number of Eastern European countries covered	Number of retail outlets in Eastern Europe	Number of retail outlets in the USA	Number of retail outlets in Japan	Number of retail outlets in other regions
Germany	14	9	670	5	64	2,050	2	
France	24	11	2,336			320	35	102
UK	30	11	1,335			350		
Sweden	16	11	337	3		4		21
Netherlands	8	8	423			644		
Belgium	5	7	2,911	1		810	520	1

Source: Sternquist and Kacker, 1994: 9.

Consolidation and internationalisation were only two of the tools used by supermarket chains vying for survival in the increasingly competitive environment of Europe since the 1960s. Their most important tool remained price cuts. This was particularly true in the 1970s, with consumers pressed with inflation and recession. Tesco's price cutting initiative, 'Operation Check-Out', in the summer of 1977, was an astounding success that went beyond its promoters' expectations and established it as a dangerous competitor for the then market leader Sainsbury (Investors Chronicle, 23 November 1977: 645 and 5 May 1978: 393). Sainsbury was forced to follow a few months later with its 'Discount 78' operation, which raised its market share from 7.3 to 7.8 percent and more than offset its cost (Investors Chronicle, 9 June 1978: 968).

Food retailers have also diversified into petrol, medicine, and books, forcing the hand of government and manufacturers in the process. Edouard Leclerc in France, and Tesco in the UK, have been at the forefront of this battle. Sainsbury developed own-label over the counter medicine such as paracetamol and plasters, claiming they were on average 40 percent cheaper (quoted in Ogenyi, 1999: 247). Increasingly, these retailers have used the non-official 'grey market' to buy designer jeans, electrical products, and designer sportswear and flog them in their shops at a fraction of the price (Ogenyi, 1999: 247).[7] European retailers also followed the US lead in own label, which aimed to provide customers with equivalent or better quality goods at lower prices. Tesco tried to erase the memory of its low price, low quality own label goods of the 1960s. It relaunched in the 1980s its Tesco label, then declined it in the 1990s into a low price, medium price and higher priced upmarket range of own label (Tesco Value, Tesco, Tesco Finest) (Competition Commission, 2000, II, 87). Following figures testify to this, with the UK closest to the US, with 23.8 percent of total retail sales in 1990, versus 11.9 percent, 11.6 percent and 10.6 percent in Germany, France and the Netherlands, and only 5.1, 2.9 and 2.6 percent in Spain, Sweden and Italy (Hakansson, 2000: 7). By 1999, own labels represented 35.3 percent of all food retail sales, ranging from 28 percent at Somerfield to 41.8 percent at Asda and 42.4 percent at Sainsbury (Competition Commission, 2000, II: 88).

These strategies came increasingly under attack in the 1990s by the rise of hard discounters. Discounting first originated in the UK and Germany after the Second World War, although one might consider that the co-operative movement was in many ways a precursor of discounting techniques (in particular buying clubs). Discounting was more developed in these two countries by the 1970s than elsewhere in Europe. Britain's leading discounter of the 1970s, Kwik Save, was typical. But the threat was contained by the fact that discounters sold only a limited range of goods in small town centre stores (Investors Chronicle, 16 December 1977: 889). Shopping centres in the 1970s explicitly excluded discounters from becoming tenants on the grounds that it took these centres downmarket, with their largely working class customer base, whilst suburban shopping centres targeted the more affluent middle classes. It was the case in Birmingham New Street, which refused access to interested discounters. France was the exception, maybe because most French discounters belonged to 'normal' supermarket chains such as Carrefour. In 1971, Carrefour opened a discount store within the Caen-Mondeville shopping centre (Retail business, February 1971: 2). But the real threat of discounting came in the 1990s with the arrival of American hard discounters such as America's giant Wal-Mart. Its take-over of Britain's Asda sent shivers down the spine of all European supermarket chains, although it is still too early to assess its impact on the structure of European food distribution.

[7] Tesco sold Levi 501 jeans, Calvin Klein underwear, Adidas sportswear.

Bibliography

Burt, S.L. (1991) 'Trends in the internationalisation of grocery retailing: The European experience', *International Review of Retail, Distribution and Consumer Research*, 1, pp. 487–515.

Competition Commission (2000) *Supermarkets: A report on the supply of groceries from multiple stores in the UK*, London, 2 vols.

'Distribution & Logistics' (2001) special issue of *French Chamber of Commerce in Britain, Info*, May–June.

'European retailers try to transfer their unique styles to US Market' (1985) *Wall Street Journal*, 23 April, p. 33.

Halley, P. (1985) 'The internationalization of the promodès Group', in *International trends in Retailing*, Chicago.

Hakansson, P. (2000) *Beyond private label: The strategic view on distributor own brand*, Stockholm.

'How Carrefour makes its mark in the retailing war' (1978) *Investor's Chronicle*, 28 April, p. 331.

Investors Chronicle (1977–1978), London.

Kacker, M. (1985) *Transatlantic trends in retailing*, Westport.

Maggart, M. (1981) 'Determining electronic point-of-sale cash register requirement', *Journal of retailing*, 57, 2.

Marketing in Europe (1970–1974).

Marketing Review (1973) 'Europroduce', *Marketing in Europe*, 22 January, p. 2.

National Computing Centre (1968) *Computers in distribution: A brief survey of computer application in the distributive trades*, Manchester.

Oddy, D.J. and Oddy, J.R. (1998) 'The iceman cometh: The effect of low temperature technology on the British diet', in M. Scharer and A. Fenton (eds), *Food and material culture*, East Linton, pp. 287–309.

Ogenyi, O. (1999) *Retail marketing*, London.

Retail Business (1964–1971).

Reynolds, J. (1992) 'Managing the local market: Information technology applications in retailing', *Journal of Information Technology*, 7, pp. 267–277.

Spilsbury, M., Toyes, J. and Davies, C. (1993) *Occupation and skill change in the European retail industry* (report 247), London.

Sternquist, B. and Kacker, M. (1994) *European retailing's vanishing borders*, London.

'Tesco searches for new growth areas' (1978) *Investor's Chronicle*, 22 December, p. 957.

Thomson, J. (1994) *Victorian London street life in historic photographs*, New York.

Tucker, K.A. (1978) *Concentrations and costs in retailing*, Farnborough.

Walsh, J. (1993) *Supermarkets transformed. Understanding organizational and technological innovations*, New Brunswick.

11 Milky ways. Dairy, landscape and nation building until 1930

Barbara ORLAND, Swiss Federal Institute of Technology, Zurich

I. Perceptions of dairy farming landscapes

In 1823, August Niemann remarked that dairy farming in Holstein was one of the most unusual branches of *'our fatherland's'* economy (1823: 1). 'Fatherland' as it was used in this phrase, meant only the entire region of Holstein, a province in the northern part of Germany. Only here, Niemann went on, this branch of agriculture had played a central role over the course of centuries, a role, which was of similar importance in only three other countries – the mountainous regions of Switzerland, Holland (more precisely, Frisia, a northern province of the Netherlands), and Ireland. In no other German state and in no other region in Europe 'is the dairy farming a national manufacture to the extent that it is in Holstein' (p. 3).

The cause of this local farming specialisation to Niemann was self-evident. From his point of view, dairy farming could only develop as a lucrative enterprise under special natural conditions and was thus bound to precisely defined regions. Butter from Holstein was of high quality thanks to 'the nature of Holstein's soil and its hardy herbage' as well as the local dairy breeds and 'their peaceful, comfortable pastures on green-fenced paddocks' (p. 3). Although the importance of milkmaids' virtuosity and the careful, cleanly, and well-planned order adhered to in skimming and buttering were also noted, these aspects took second place.

About fifty years later, reports on the development of dairy farming again referred to the landscape of entire regions. In contrast to their forerunners, however, authors now attributed the quality of local products to dairying techniques, rather than to local agricultural conditions. 'The refinement and taste of the goods are due less to the quality of the milk, than to the exact handling, the manipulation in preparing cheese and butter' (Kleinpeter, 1879: 18). Competition was no longer on the level of landscape, animals, and fodder conditions. The various Dutch cheeses 'are just as good when produced along the Lower Rhine as in Holland itself' (p. 19). What determined the quality of the goods were instead the objects and methods of science and technology. A regional agriculture that was unable to utilise the latest knowledge from science and technology had no hope of asserting itself in the marketplace.

At the same time, however, experts sought to uncover specific regional dairy practices, and they found them in history. 'Among the domestic branches of industry, in Bavaria, dairying occupies an outstanding position'. These were the opening words of the special issue on the history of the Bavarian dairy published by the *Süddeutsche Molkerei-Zeitung* on 4 June 1929 (Zum Geleit, 1929: 1). In commemoration of its fiftieth anniversary, the journal organised an exhibition devoted to underlining the economic power of the Bavarian dairies within a national context. While the founding narratives of agricultural co-operatives, dairies and dairy farming associations could be traced quite reliably to the 1880s and 1890s, the journal's authors were especially keen on demonstrating their respect for a centuries-old tradition. Thus, they pointed out that dairying was an activity as old as humankind and that it had an especially long tradition in the Alpine foothills region of Bavaria. 'We see how much and how little the way of working has changed since olden times' (Bilder aus der ..., 1929).

In the course of preparations for the anniversary issue, many members of the dairy farming association proved to be conscientious amateur historians, collecting images, publications, and quotations which served to portray an age-old economic activity that could be linked to a young industry which had spread throughout Europe. While authors writing in the early nineteenth century perceived commercial dairying as the manifestation of limited agricultural options, all subsequent writers were aware of the fact that milk production could serve the interests of at least all farmers taking into account recent findings of science and technology and to position themselves in response to new market realities. Thus, dairy farming had acquired an entirely new goal: whereas traditional dairy farming had served agriculture, the modern milk industry directed its position in the service of the nation's economy.

However, on growing markets transactions became impersonal and pursuing new opportunities forced dairy farming associations to identify their products from that of other competitors. For this purpose, history became the vehicle to localisation of authenticity and quality. Apparently, in no other period were the roots of European dairying sought with such zeal than in the decades between 1870 and 1930 (Martiny, 1895). And it is this link between local tradition and quality of food that was adopted for the marketing of most modern dairy products. Even today, labels for *high quality* products refer to the history of a local dairy tradition. *Emmentaler, Greyerzer, Gouda, Edamer, Parmigiano* or, more generally, Swiss, Dutch, French or Italian cheese. Signifying typical products by certifying the geographical origin is a common instrument to guarantee high prices and fair competition on national and international markets. A lot of brand marks use figures of landscape, animals and plants as memory anchors that illustrate desired traditions. As anthropologist Cristina Grasseni put it in her study on dairy and breeding practices in an Italian valley of the Alpine mountains, the multiplicity of tastes 'is equated with a guarantee of multiplicity of *saperi* (knowledge, wisdom) as a reservoir of diversity of identities and traditions. Mountain valleys are seen as repositories of antique recipes and tastes, and eco-tourism feeds on local foods as a resource for local development' (Grasseni, 2001, chap. 7: 1).

II. Nationalisation of landscapes – internationalisation of knowledge

Traditions that claim to be old are more often invented, as Eric Hobsbawm and Terence Ranger pointed out in their *Inventing Traditions* (1983). To establish continuity with a suitable past naturalises change; it makes proposed novelties appear to be the logical outgrowth of past achievements. In this sense, the discourses very often deal with the future, rather than with the past by itself. The future appears as the inevitable fulfilment of a historically legitimated destiny. Thus, invented traditions occur more frequently at times of rapid social transformation, when *old* traditions are disappearing. As Swiss historian of economics Hansjörg Siegenthaler (1992) has emphasised, such learning processes accelerate in times of crisis. The disposition to learn increases in a context of uncertainty, because crisis and uncertainty devaluate general rules, which have been self-evident up to now. Quite similar to the observations of *Inventing Traditions*, Siegenthaler points out that references to history become important in the communication and production of the new. Familiar interpretations will be abandoned, if people loose their trust in the future. One talks about things of the everyday life, but now with a new meaning. Although it is not possible to benefit from history, historical narratives themselves become a major topic in communication.

In fact, the notion of milk as an ancient food, and the constant referring to traditional practices of dairying since the end of the 19th century, can be precisely read in this sense. All forms and institutions of milk production as well as the kind of milk consumption and dairy products we are used to, are quite recent in origin. They were constructed and formally institutionalised throughout the industrialised world with great rapidity since the 1870s. Farm economies and dairy manufacturing methods became standardised and – due to the internationalisation of scientific and technological determined farming practices – butter and cheese products were no longer of local origin. Even before 1914, British cheddar cheese was produced in Argentina, Canada and Denmark, and sold in the cheese stronghold Switzerland. Similarly, the technique of producing *Emmentaler* (Swiss cheese) was employed in the Ukraine as well as the USA (50 Jahre Schweizerische Milchwirtschaft, 1937: 168).

However, it is important to recognise that the process of agricultural change lasted much longer. The history of the modern dairy industry started at least at the end of the 18th century and it ran parallel with the dissolution of regional structures in favour of nationally defined territories of agriculture. Discourses on *old* and *new* referred to this process, in which free trade, delocalisation and dissemination of knowledge initiated a complex process of globalisation of agricultural products. Against this background, they can be read as a communication strategy which attempt to fill the gap between constant change in an industrialising world and the assumed character of unchanging and invariant peasants traditions.

In this chapter, I will examine the seamless transitions from region to nation, from rural to national economy, from local to broad markets and from landscape to technology by studying the shift from *old* to *new* forms of milk production. In general, these transformations are well known, however we do not know in detail how they passed of.[1] How did cultural landscapes and contextualised dairy practices become part of national economic areas? How did processes of nationalisation frame the modern milk business? Which effects had the integration into broad markets on local farm practices? How did farmers learn to produce for a national/international, in any case widely competitive economy instead of working under the conditions of narrow markets? What was the impact of anonymous producer/consumer relations on the perception of food and food patterns?

Beside those specific questions regarding the history of the modern milk business, one could raise more general questions: What benefit can be derived from studying dairy history in combination with the processes of nationalisation/internationalisation? What are the benefits of such an attempt to the history of technological change? Of course, there exists a broad range of scholarship on nationhood, nation building and national identity.[2] Since Benedict Anderson's path-breaking analysis of the way in which modern nations have concealed their historicity behind a carefully crafted facade of naturalness and eternity, it is beyond debate that nations are modern constructions and

[1] On the history of milk and dairy, see DuPuis, 2002; Lysaght, 1994; Ottenjann and Ziessow, 1996; Vatin, 1990; Teuteberg, 1986.

[2] Classical texts are Anderson, 1991; Hobsbawm, 1992; Gellner, 1991.

must be explored as such. In opposite to this trend, European histories of technology mostly have been written as a collection of national histories or, at best, as a comparison of different nations. However, if the overall aim is to explore processes of linking and de-linking of regional/national infrastructures, then comparing technological systems and practices in different countries can only be the first step to reveal patterns that remain hidden when countries are examined in isolation.

One of the rare investigations which reflects national economies and technological development as expression of the concept of *imagined communities*, is Gabrielle Hecht´s Radiance of France (1998). By national identity, which shaped the technological development, Hecht means 'the ways in which people imagine the distinctiveness of their country and define uniquely national ways of doing things' (p. 10). She asks, for example, how technologists define their niches in national policymaking and how they enact policy choices in technical practices and artefacts, how local communities situate themselves within a nation. Hecht, too, observes that debates on those topics typically refer back to the past. As she summarises, this entails engaging in various political, cultural, and technological acts, many of which derive legitimacy by invoking the relationship between one country and the rest of Europe.

In the case of milk production and dairy technology, the problem of nationalisation and internationalisation must be explored historically from two directions. Firstly, the development of an internationally standardised milk production system must be approached from a territorial and environmental perspective. Agriculture as a site-bounded branch of production has always been influenced by the perception of land resources as well as the redefinition of regional and national borders. Thus, it was not only the world wide economic and political restructuring of markets that has accelerated the modernisation processes in the European agriculture and food systems. As I will argue, one of the most significant indicators of change that brought the modern milk production on its way, was the innovation of dairy as an area-independent branch of agricultural production. Whereas ancestors judged dairy farming as a strategy of land use, narratives of an industrialised dairy farming system usually neglect landscape and soil but start with the cow as productive force. Only because market values were placed on authenticity and locality, new ways of envisioning the landscape were also required.

Secondly, the development of the modern milk production system must be explained from a trans-European perspective of exchange in knowledge, technology and visions of perfect farming.[3] Economically or politically motivated barriers to cross-border trade of commodities never touch industrial visions. The cross-border exchange of technological innovations, scientific knowledge and marketing strategies is, in fact, most often supported by firms and states while national markets in terms of labour and products are protected, and brands signify national provenance. National tariff systems protect at best the outcomes of production. Production methods and knowledge usually flourish the borders and launch different countries business.

Thus, it is only the combination of both approaches that makes clear in what way the symbolic intensification of region and nation interacted with the technological improve-

---

[3] M. DuPuis (2002) tells the story of the US dairy industry from the perspective of a *Perfect Food Story*. In her path-breaking work she explains how milk was constructed as 'nature's perfect food' and how this ideal shaped patterns of consumption as well as the implementation of a new production system.

ment and standardisation of dairy practices. In order to make this evident, the intersection of national/international food policies and technological change must be taken seriously by reconstructing the parallel processes of nation building and the definition of food markets and food production systems.

III. Dairy zones – Animal husbandry as a strategy of land use

For centuries, milk production was interconnected to the regional conditions of living. It was part of local farming systems, depending on the perception of the landscape and its potentialities. Different ways of envisioning land feasible for the purposes of agriculture existed in direct relation to the soil condition. For pre-industrial farmers, it was nature that determined to a great extent how land could be used. Some land was never meant to be farmed, others prohibited grain production. To balance the equilibrium between agriculture for grain production and animal husbandry was one of the fundamental rules of all pre-modern farming systems.[4] Unlike today, there existed no market for animal feed. The available fodder was instead dependent on the way the land was being used, i.e. the relationship among all types of land management such as arable farming, permanent meadows, forests, pastures, orchards, vineyards, etc. Today, subsidies and market access have resolved this issue. In earlier times, farmers needed to adjust more to the variety of land areas which had developed through the symbiosis of rock formations, elevation, climate and vegetation.

Against this background, it becomes comprehensible that all over Europe the saying *Dairy farming is land use* (Schneider, 1916: 7) was just as self-evident as the idea of a *dairy zone*.[5] With this phrase, people reflected the long-standing rule that animal production for human food has been linked to soil. In order for a region to be able to produce extraordinary dairy products, it was first necessary to find a solution to the omnipresent problem of finding adequate fodder. An old farmers' rule holds that 'cows give milk through their mouths' (Rüger, 1851: 7). All mountainous areas were characterised as potential *dairy zones* due to their steep cliffs, rocky terrain and difficult climatic conditions. In addition, coastal wetlands, river meadows, highland and lowland moors, which could only be made arable through large-area cultivation work, seemed like obvious locations for cattle husbandry and dairy farming. Besides the coastal countries of Holland (the Belgian nickname for the Dutch is *kaaskoppen,* or 'cheese heads'...), Denmark, Ireland, Sweden etc., it was particularly the northern Alpine countries, at first Switzerland, which had a pronounced and often highly developed dairy farming system. In France, besides the Alps, especially the Massif Central, Les Vosges, Les Pyrénées and Corsica have been the *old* cheese regions.[6]

---

[4] This problem is explored in detail by Riemann, 1953.

[5] I took the phrase from S. McMurry (1995: 12–15) who describes how dairying families in Oneida County, NY, thought this way when they started an intensive milk production in the middle of the 19th century. Even today, Italian farmers in a small Alpine village think in terms of dairy zones (Grasseni, 2001).

[6] To the industrialisation of milk in France, see Vatin, 1990. The complexity of rural economies of pre-industrial cheese making is also stressed in Whittaker and Goody, 2001.

Though people drew agricultural boundaries in different ways, dairy zones were conceptually well bounded in the sense that it was believed to possess definite natural limits.[7] This can be read also from data on different types of cheese production. Although, for instance, there existed a large choice of cheeses in France, most of them have been very small, because the herds were mostly restricted to one or two animals and usually employed for draught, which meant low milk production. The same goes for the widespread making of sheep and goat cheese. In opposite to these, only cheese of the Alps was larger and generally stored for a longer time to mature (Pinard, 1995).

In the more arable-farming-oriented regions, the milk of cows, goats and sheep were at best a by-product of animal husbandry, available in pitifully insufficient quantity and only on a seasonal basis, and therefore used only on a subsistence level.[8] According to limited soil fertility, crop farming and cattle-raising functioned in no way independently from one another, and no balance existed between the two production systems. The historian of agriculture Wilhelm Abel was one of the first who mentioned that for a long time animal husbandry had been displaced in support of grain growing. In the late Middle Ages, the meat consumption averaged at about 100 kg per capita and per year (Abel, 1978: 124; Henning, 1994: 316). In a permanently growing population during centuries, it declined to 16 kg until 1800 (Achilles, 1993: 69).[9] This phenomenon, named by historians of agriculture the *graining* of the European agriculture, illustrates that livestock farming in comparison with grain production by no means had the same relevance.

The French historian Fernand Braudel confirmed these observations by emphasising invisible boundaries of European fat geographies. Butter had been consumed only in Northern Europe – with the exception of North African zones up to the Egyptian Alexandria. The rest of Europe preferred pig lard, bacon or olive oil as a fat source. Due to the amount of milk amounts available, those regions with a highly developed fatty cheese culture reduced the production of butter.[10] Other milk products like yoghurt – nowadays appreciated throughout Europe – were restricted on single zones, at first Turkey and some parts of the Orient.

Beside this 'localness' of dairy production, a general rule was that most dairy products were not consumed in fresh fluid form. In regard to regional studies on dairy food habits throughout Europe, the anthropologist Patricia Lysaght concluded that in some countries 'fresh milk was considered a luxury food, while in others it was thought to be unwholesome and unhealthy. In fact, it seems that the drinking of fresh milk was the least important aspect of milk utilization overall' (Lysaght, 1994: introduction). Because of lactation cycles and feed restrictions, yearly-round fresh milk drinking even for cow-holders was

---

[7] Up to now, agricultural history has not yet mentioned this perception of dairy zones. The only exception is McMurry (1995). For the history of dairy as part of agrarian subsistence economy in various European countries, see Lysaght, 1994; Pirtle, 1973. For Germany, see Ottenjann and Ziessow, 1996; Teuteberg, 1986. On the history of dairy in the Netherlands, see De Vries, 1974: 156–162, 166–167; Bakker, 1992.

[8] As such it has been an important part of farmer's women work. What is remarkable about the 19th century, however, is the uniformity of change that occurred in different countries. Worldwide, men assumed responsibility for this area of work. Cf. McMurry, 1992; Shortall, 2000.

[9] Only after the end of the 18th century the situation changed completely. Until 1913 the consumption of meat per head increased again to 51kg (ibid., 257).

[10] For the situation in pre-industrial France, see Pinard, 1995.

simply not possible. Farmers in grain growing regions did not have extensive winter storage systems for fodder. To keep the life stock through the winter was an unsolved problem, and reports on the weakened state of animals by spring widely spread through the literature of agricultural reformers since the end of the 18th century (Schwerz, 1816: 123).

Only in the light of particular conditions, there existed decisive exceptions. Already Adam Smith had mentioned that under two spatial conditions farmers would start to produce milk for the market: Firstly, if they face problems with grain production, and secondly, if the farms are located nearby big cities (Smith, 1978). Johann Heinrich von Thünen in his *Isolated State Model* from 1829, too, described the need for fresh milk as a somehow natural need of the town.[11] To him, dairy products were luxury goods that only find demand from wealthy town citizens. Two decades later, in 1841, another author described the surrounding of a town as its 'natural area of nutrition', where one could find gardening and 'milk meadows' (Kohl, 1841: 48).

All authors remarked that trade between city and country influenced the profitable efficiency of animal husbandry and reduced the cost of transporting milk and dairy products to urban markets. From a farm perspective, increasing milk production meant shifts from arable farming to permanent meadows and pastures, an investment that would be profitable only if enough people were willing to pay the price. In fact, this perception of land use coincided with reports from several urban markets.[12] Older descriptions about the town of Hamburg show that inhabitants kept cows as well as other productive livestock. Places where people rounded up cows in order to milk them were named like *cow bridge*, *milking hill*, *milk street* or *cows mill* (Voigt, 1903). Since the beginning of the 17th century several estates in Schleswig-Holstein sent fresh milk and butter by boat to Hamburg (Hanssen, 1880: 417). It was the so-called *Koppelwirtschaft*, the coupling of field and field rest on fenced paddocks with the exclusive use of the fallow land for milk production that advised owner's of estates to do so. They engaged people from the Netherlands to build up dairies in order to produce butter. For that reason, early dairies in the northern German states were named *Holländereien*.

Sometimes, even both factors – closeness of an urban market and poorly yielding land – merged, as the story of the village Neuholland, close to Berlin (about 48 km), illustrates.[13] Until the 17th century, the damp, swampy and sedgy land at the course of the river Havel was not reclaimed. Then, local authorities motivated Dutch people to settle down, reclaim and drain the land, protect it against the water, cultivate meadows and start dairy production in order to accommodate the demand of the Berlin gentry for excellent butter. In 1659, the first Dutch family from Brabant settled down; within a generation further colonists followed and for a long time an important dairy centre of Berlin could exist.

To sum up, an extensive production of dairy goods, in the sense of butter and cheese making, has a long commercial tradition in Europe, but cows, goats and sheep were employed for this purpose only under specific regional conditions. Making milk, butter, and cheese depended on spatial perceptions of agriculture. Agricultural boundaries were

[11] The Thünen model is discussed in Cronon, 1992.

[12] The pre-industrial dairy near cities is adressed by Dieterichs, 1856; see also Fenton, 1995. The milk trade in Munich is examined by Spiekermann, 1994.

[13] The story of this village is told in Peters et al., 1989.

drawn in different ways, animal husbandry as a precondition of milk production was perceived as having clear-cut natural limits. Yet in 1853, one agrarian writer noted that most plants and animals have their natural habitat, outside of which they do not exist.[14] However, this view of the landscape was neither static nor fateful. In fact, dairy farming was a situated knowledge and activity. Under the conditions of near-by markets, farmers produced and stabilised *dairy zones* in a way quite similar to the way they developed skills, techniques and products. They produced land as adaptation of crop cultivation methods to the come across landscape. In this sense, milk production was both, a question of *natural* landscapes and one of *imagined* landscapes.

IV. Decline of the 'old' dairy tradition

Although the European economic development between the 16th and 18th century was based on territorial countries, they had only slight influence on the perception of regional landscapes, especially *dairy zones*. Whether in Switzerland, the Netherlands or Sweden, all European *dairy zones* were to a lesser extent defined by the political system then by problems of land management. If this changed in the 19th century, then because of the transformation of a rural perception of the landscape to one of a nationwide agricultural territory with open markets. Across Europe and without any regard for soil conditions or vegetation zones, agricultural regions, which had been used for grain production, were being converted to intensive dairy farming and a new system of milk processing and trade. The dynamics that caused the rise of an industrialised and highly engineered milk production were based upon two crucial preconditions: separating arable farming and animal husbandry and thinking in terms of different branches of production.

The first challenges to the old agricultural systems occurred in what Paul Bairoch called the *organic phase* [15] of agricultural modernisation, beginning in the late eighteenth century and lasting until the 1870s. In this period the traditional relationship between land, fodder, cattle, and dairy production had been torn apart. In the discourses that marked agricultural reforms, natural spaces were re-evaluated, but not with the specific purpose of improving animal production or dairy farming. Around 1760, when proponents of the new Physiocratic School of economic thought began to argue for reforms,[16] the idea was to accomplish a fundamental land reform, with a view to using the natural nitrogen cycle more effectively. To reach this goal, more dung had to be produced. Consequently, all of the Physiocrats' proposals and measures for agricultural reform were based on four innovations: 1) abolishing the old style of meadow farming in all arable farming

[14] In regions with deciduous forest, he wrote, the farmer would be well advised to breed goats, rather than cattle or sheep, in oak and beech tree forests he should prefer pig breeding. Dry and spacious areas should be allocated to sheep, fertile gardens and orchards to bees, lush meadows to cows (Stamm, 1853: 372).

[15] Following this model the first 'organic' phase started in the late 18th century and ended during the major depression that hit European agriculture from 1875 to 1890. The second phase is described as the mechanical phase, lasting until the Great Depression. The third phase is classified as being in the period of the welfare state and thus the post-WWII period. (Bairoch, 1976; Pfister, 1995: 176).

[16] For this whole section, see Pfister, 1995: 175–202; Teuteberg, 1986: 163–184.

regions, i.e. no longer allowing land to lie fallow and parcelling out communal pastureland; 2) planting fodder plants, such as clover and sainfoin; 3) summer stall feeding, to make it easier to collect cattle dung and liquid manure. The added quantities of cattle dung could be used to 4) insure that fields and meadows were fertilised more intensively.

The implementation of these ideas was very controversial and took at least several decades. Nevertheless, since the 1840s, livestock husbandry, long regarded by many central European farmers as useless and burdensome, suddenly seemed profitable. Developments in international grain markets created the harsh framework for this change of heart. Europe's last major famine occurred in 1846/47; in the years that followed, the liberalisation of the grain trade and the effects on distribution of an increasingly wider and more closely linked network of railroad lines offset an under-supply of grain, and not just at the local level. National and international markets established themselves for the long haul, and were soon followed by price wars. Agriculture based traditionally on grain production became unprofitable and needed to be replaced with other sources of income.[17]

Suddenly many farmers in the grain-growing regions hoped to increase their incomes through what they called *artificial meadows* or *artificial feed*, and expanded cattle husbandry.[18] Formerly neglected fallow land attracted increasing interest; [19] schools for land improvement were founded; and the profession of the so-called *Wiesenbaumeister* ('master of greenland') was launched.[20] Regions, where, despite all the natural advantages of abundant vegetation, farmers had stubbornly adhered to the practice of planting grain, also reacted. One example was the Allgäu region in the area at the foot of the Alps, where, around the turn of the 18th and 19th centuries, 'farmers, having received foreign ideas, thought of the natural possibilities of (...) making the 'yellow' and 'blue' Allgäu (flax growing) into the blossoming 'green' Allgäu' (Jahn, 1955: 20). This reorientation was so radical that today, in the Bavarian part of the Allgäu, the farm landscape is said to have been completely *'greened'*, and local products are advertised with the image of Germany's traditional dairy art.

When the feeding situation improved, the next step, logically speaking, was to bring dairy farming into the former grain production areas. Although the primary intention was not to produce a dairy cow, for the first time the notion of a milk-producing cow as an isolated natural entity became a real possibility. The separation of land and animal

---

[17] For the agrarian crisis in 19th century see Bairoch, 1976; Teuteberg, 1986; Achilles, 1993; Pierenkemper, 1989.

[18] Data on the size and composition of cattle herds demonstrated the new priorities. In the canton of Berne, around 1760, the grain-growing areas still had a large number of draft animals, particularly oxen, whereas in the mountainous regions the cows were in the majority, commensurate with their significance for cheese making. Transitional areas, also called field grass areas, were also already heavily oriented toward dairy farming. After 1790 (until 1911) the cow population began to outgrow that of horses, oxen, and sheep in all parts of the country to an incredible, yet varying extent (Pfister, 1995: 189).

[19] In Prussia in 1840, about 20 percent of agricultural area counted as fallow land, in 1867 it was only 10 percent, by 1913 the percentage of fallow land had decreased to 2,7 percent (Teuteberg, 1986: 166).

[20] See on this greenland movement Häfener, 1847. The term *greenland* in German means all agricultural area used for the production of fodder.

permitted the construction of a self-evident form of animal production. A system of animal breeding, feeding, and husbandry was emancipated from its cultural roots and imagined as a separate branch of production.

In all those places, where one used to say that 'cattle guarantee true pure yield only in those open and fertile pasture areas not capable of arable farming, but not in general in usual arable farms where cattle are and must be held out of sheer necessity',[21] now farmers were beginning to grow interest in cattle and its products as a goal of production.

V. Urbanisation as a component of agricultural change

This change of heart among farmers, which did not take place in an uninterrupted fashion, was accelerated by different influences. First, I have to mention those developments that have been the same, more or less, for all farmers. Thus, it seems necessary to describe some elements of the *new* dairying structure, which developed all over Europe between 1870 and 1930. In my description I will mainly concentrate on the German market.

All across Europe, the growth of cities exposed the limits of the traditional system of food supply. Due to a steady rise of population growth the processes of urbanisation and industrialisation by itself provoked a dramatic increase of milk production. In Germany between 1873 and 1907 the number of cows increased from scarcely 9 million to more than 11 million while the population increased from 41 to 62.6 million people. However, the constant growth of milk production had not been enough to satisfy the growing demand on the milk markets. Compared to head per population, in the same time the number of cows decreased from 21.8 cows per 100 inhabitants in 1873, to 17.5 in 1907 (Handwörterbuch der Staatswissenschaften, II, 1910: 701). Although farmers successfully enhanced milk output per cow, the amount of milk produced in Germany was insufficient to serve the needs of all branches of milk production. Especially the production of butter suffered from the lack of milk. In 1896, for the first time, the amount of imported butter exceeded that of exported butter. In the next three decades the German Reich became one of Europe's biggest butter importers. In 1931, 21 percent of all butter and 15 percent of all cheese consumed in Germany was imported.[22] In 1912 the surplus of imports was 196.2 million RM, or 6.5 percent of the German production. Only because in 1913 a custom of 20 RM per 100kg of imported butter had to be paid, the import of cream increased instead of butter (Hittcher, 1913: 12).

Whereas relatively durable foodstuffs such as grain were quite adaptable to the new sales requirements under the condition of urbanised food systems, the market for perishable fresh products reacted through price increases. Especially milk became a rather lucrative product. Converted into the *Reichsmark*, between 1855 and 1865 milk cost between 9 and 10 pennies per litre when purchased from a farmer in the city of Berlin. By 1875 the price had gone up to some 14 to 15 pennies, and between 1880 and 1890 it held firm at around 18 to 20 pennies, of which the producer pocketed 12 to 15

[21] *Theoretisch-praktisches Handbuch der größeren Viehzucht* (1810), quoted in Abel, 1986: 66.

[22] 70 percent of this imported cheese came from the Netherlands. F. Mendelson, 'Bedeutung, Lage und Aussichten der deutschen Milchwirtschaft', in Sering, 1932: 504.

pennies (Martiny, 1891: 10). Until World War I these high prices did change scarcely, whereas substantial range of prices up to 5 pennies per litre between big and smaller cities were calculated (Arnold, 1911: 599).

Yet, the constantly high prices do not lead back to radical changes in consumption patterns. Again, I have to mention that the increasing urban demand at first was an effect of the general population growth, and not a result of new milk diets. Due to the high prices fresh milk, butter and cheese remained, what economics call, inferior goods. With the exception of families who had to nourish babies, and despite of a steady rise of the dairy industry, fresh milk only slowly became a nation-wide beverage. Until World War I Germany faced highly regional distinctions in the consumption of milk and dairy products. While in regions with a long tradition in animal husbandry milk hold a regular part in nutrition (big cities like Hamburg or Munich with their old milk trade traditions also showed quite high consumption levels), other regions like Saxony, Thuringia or Silesia attached no importance to milk production. Even on the verge of the war, statisticians and economists wondered about the wide differences in milk consumption. In 1913, an account of the milk consumption per capita in several German cities showed a high consumption level in Munich (0.44 litre), Stuttgart (0.50 litre), Flensburg (0.50 litre), Ulm (0.50 litre) and Heidelberg (0.58 litre), and a low level in Eastern German cities like Hindenburg (0.09 litre), Gera (0.11 litre), Cottbus (0.14 litre) or Görlitz (0.15 litre) (Westphal, 1931: 120–121).

Those impressing differences make evident that diet in general does not change as fast as retail systems do. In fact, one has to take different dispositions to learn into account, respectively those determined by social and economic factors. For instance, if there existed enough economic pressure, then nutrition habits changed very radically. This is the lesson one can learn from the case of farm families that took part in the milk boom. Parallel to the rise of the modern dairy industry, several doctors entertained suspicion that farm families did not give enough milk to their children because they sold every drop of it. Paediatricians like Ignaz Kaup claimed that in rural regions milk consumption declined dramatically.[23] The fast development of the cities and the single respect for money should have caused malnutrition on the countryside, he argued, and for him the children were the mourners. His arguments were widely debated (Berg, 1912: 131–134). The same was reported about countries like Denmark and Ireland. Despite its importance for butter production, milk and milk products were little eaten by farm families and labourers, it was noticed.[24]

These facts imply some of the fundamental problems of the modernised milk market. In the pre-industrial era, traffic systems prohibited the transportation of dairy products apart from butter and cheese. Except for the dairy farmers in and around big cities, hardly any milk was sold in liquid form. According to a limited amount of land, most farmers had to make a decision, regarding the composition of the farm animals and the cattle herd. Should animals be fed through the winter or purchased in the spring? Were more cows for milking or calves desired? In regions in which mainly livestock farming (milk cows, bulls or draft oxen) was conducted, milk was needed almost completely to

[23] I. Kaup, *Ernährung und Lebenskraft der ländlichen Bevölkerung. Tatsachen und Vorschläge* (Heft 6 der Schriften der Zentralstelle für Volkswohlfahrt), Berlin, 1910.
[24] For Ireland see Crawford, 1995: 226.

suckle young cattle, with little left over for butter and cheese. If the focus was on butter and particularly marketable cheese, then, conversely, farmers decided against extensive cattle rising. But even if farms specialised on dairying, in most cases they did not have enough milk to produce fresh milk products (cream), butter and fat cheese at the same time. As relative small units they needed any kind of co-operation to bring together enough milk. That was the reason why even in *dairy zones* once more butter and another time more cheese was produced. Often, even those specialised regions faced the problem of not having enough butter to sell it on the local markets. Since the 16th century the Swiss authorities repeatedly banned butter exports in order to guarantee the domestic supply of butter.[25]

Yet under industrialised conditions the urban market required all known milk products at the same time and in good quality. Fresh milk as a regular commodity and as raw material for dairy products, no longer processed on the farm, animated farmers to orient their production much more to the demand efforts then to landscapes and soil. This remove from the soil enabled new areas of specialisation and the diversification of products. However, this did not necessarily happen for the benefit of the farmers, as will be shown below. Moreover, a reorientation in production targets was evoked by industry's and residential areas' hunger for land. Agriculture more and more was pushed from central locations to the fringes, and trade in foods which were transported across large distances became much more important. To examine the consequences of these developments, one must study certain regions. For the period between 1870 and 1910, general trend statements present the enlargement of agricultural and forest areas in the Ruhr Valley. The picture looks very different if one explores the development on a smaller level. The southern part of the Ruhr Valley, the first area that was grasped by industrialisation (e.g. around the city of *Essen*, originally a zone of distinct agriculture and a lot of forests), lost its fields and forests nearly completely (Reif, 1990: 337–345).

VI. Emergence and Competition on a National Level

Farmers, based in the immediate vicinity of large consumer centres, were understandably the first to react to the observation that the price of fresh milk was rising faster than that of other agricultural products. In the Ruhr Valley many sold their farmland as construction sites, often involuntarily, and only practised pure dairy farming on their remaining property. In extreme cases, these farms had no independent agriculture anymore but instead purchased nearly all the necessary fodder on the fodder markets, which were growing fast due to imports. Without their own fodder base, they could only purchase highly pregnant or freshly lactating cows, milked them for a lactation period and then resold them for slaughter (Reif and Pomp, 1996).

Such highly specialised *milking-only* farms were not the only consequence of the increasing urban demand for milk, though. The growing urban milk markets produced specialisation on very different levels: product diversification went parallel with a diversification of dairy farming practices, a new complexity in milk retail and – most important – the formation of new structures of milk processing. Especially the professionalisation

[25] For conflicts about the supply of butter in Switzerland, see Anderegg, 1898, II: 534–535.

and institutionalisation of dairies as intermediate between milk producers and consumers became a highly contested terrain in urban areas. With the spread of railways and transport improvements even the more remotely located farmyards were able to share in the milk boom, and their milk was distinctly cheaper than that of the milking-only farms.[26] Moreover, new system levels became necessary because of inherent technical problems that resulted from the *nature* of a highly sensitive, perishable good.

It is self-evident that milk transportation and selling in the cities no longer could be done by the farmers themselves. Most notably, in urbanised regions the milk distribution systems underwent a radical change. In lieu of milk production within the cities and direct selling of milk by farmers and milk dealers, the delivery of milk, mostly by train, increased.[27] 'Close connections between producer and consumer have been disrupted. Milk like all other commodities became a marketable good', was an often voiced observation (Arnold, 1911: 588). In all big cities of Germany, the author amplified, one could consider similar developments. Within only two decades, between 1890 and 1910, a dramatic increase of milk was delivered by train. In Dresden, for example, in 1886 6.9 million litres of milk were transported by rail, in 1895 already 20.1 million litres arrived by train. Until 1910 the amount of milk transported by train once again was doubled to 40.6 million (Arnold, 1911: 588). Approximately the same amount was brought into the city by horse-drawn carriage; however, this amount could only be estimated (Martiny, 1895: 13).

The milk transport system by railway expanded like a spider web in all directions, as a consideration of the Bavarian milk trade pointed out. Researchers had divided the receipt of *Railmilk* (as it was named by contemporaries) into the magnitude of cities on the one hand and zones of distance on the other. Thus, they found that as bigger the city, as further the distance to the farmers, who produced the milk. For big cities like Munich and Nuremberg one had calculated that in the year 1904 3.2 percent of the sold milk had been brought into the towns from a distance wider than 100 km., and 26.3 percent were delivered from a distance between 50–99 km (Arnold, 1911: 589; see also Spiekermann, 1993).

The flourishing milk business in the neighbourhood of cities appealed farmers to produce as much milk as possible and to sell a great deal of it in liquid form. As a consequence, the question of further processing, especially of unsold milk, evolved into a constant conflict between farmers and milk dealers. While some farmers understandably wished to sell the whole amount of daily produced milk to the trade, dealers preferred to buy milk on sale or return. In the beginning of the milk boom, fixed-term contracts with a term of one year were common. Under these contracts, milk dealers took the annual yield and processed the surplus of unsold milk for own account. In doing so, however, they needed a shop with cellar and facilities (Scholz, 1949: 12).

Other farmers only unwillingly sold milk. For them residues of the butter production remained an important part of farm surplus to feed different farm animals. Especially pigs were fed with the residues of butter or cheese production. In general, farmers did not like long-term contracts because of seasonal fluctuations in milk production. Thus, long-term contracts were difficult to maintain, wherefore farmers were interested to

---

[26] See Reif and Pomp, 1996: 96.

[27] Until 1900, the number of milking-only farms within the city of Berlin increased, too. In 1864, Berlin counted 5 cow holders, in 1893, their number was increased to 397 (Martiny, 1985: 14).

Figure 11.1 ***Railmilk*-flow to Berlin, 1903 and 1927**

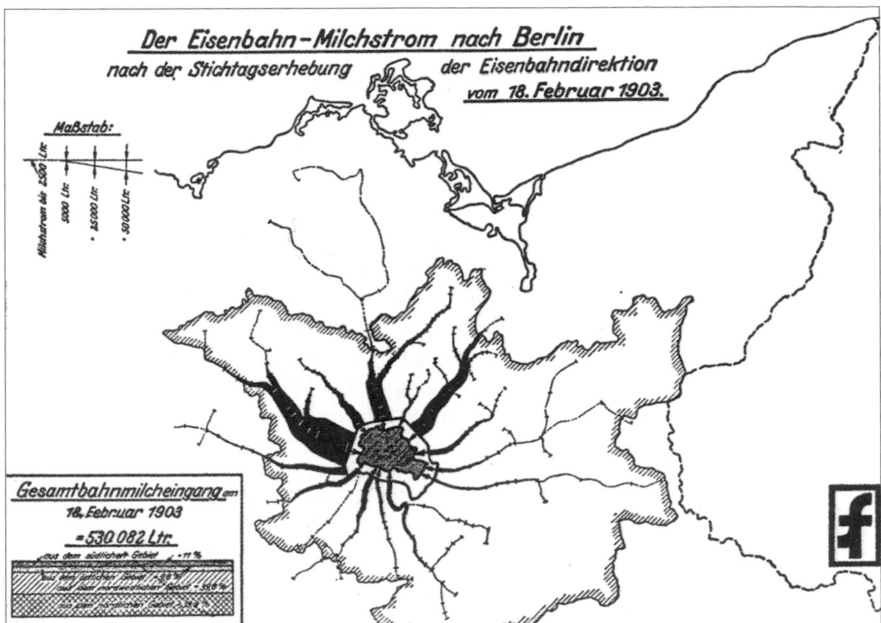

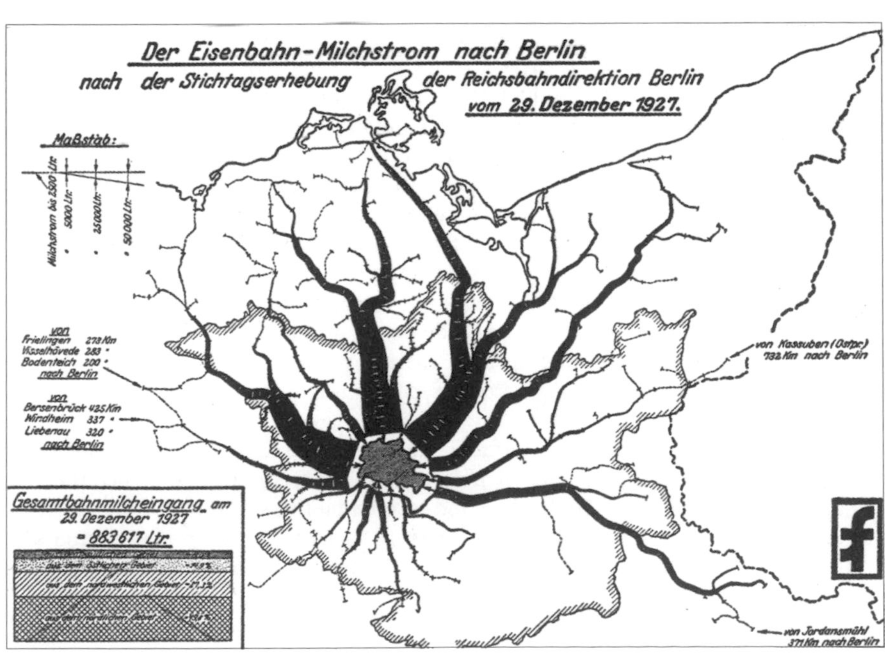

Source: Brandt, 1928: fig. 14.

minimise the dependency to urban dealers as much as possible. Depending on the location and the amount of milk produced, it finally could be much more attractive for a village or community to organise the milk transport by themselves.

Against this background it was only a question of time until the first farmers' co-operatives developed a system of milk processing and trade. Directed by a group of farmers in 1883, the first German farmers' dairy co-operative was set up on the countryside in Schleswig Holstein, an idea that soon spread all over the country. Operating as collecting point for the milk of all participants, the surplus of unsold milk was processed or returned to the farms. Because of inherently technical questions farmers preferred those dairy co-operatives. Cooling units, storage rooms, cream separators and other equipment to guarantee hygienic manufacture could be financed collectively. Yet, the most important reason was that for the first time sufficient quantities of milk enabled work on a profitable level.

This model of dairy co-operatives became very successful. At the end of the year 1894 all over the German Reich 1,145 associations of farmer's co-operatives were registered (Beschreibung der Milchwirthschaftlichen, 1895: 7). Dairy experts judged these co-operatives as a farmer's training for economic thinking and book keeping. 'In former times, proceeds of a sale came in only in small proportions. With the participation in a co-operative the situation changed at once. Even without commercial knowledge the member of a co-operative receives an overview about his profit because of the monthly settlement. On the occasion of milk payment farmers draw a comparison and stimulate one another to increase the profits of the milk business' (Berg, 1912: 52).

However, farmer's co-operatives together with private dairy enterprises and single milk traders forced a scaling up and steady specialisation. At the turn of the century, the business was split up into individual markets. Main parts of the German dairy industry had the milk supply in liquid form on their agenda. To a lesser extent farmer co-operatives with no access to liquid markets or as a result of surplus processing produced cream, butter or another durable dairy product. But, milk for the manufacturing market had not commanded comparable prices with liquid milk outlets. Until World War I, when German milk policies started to regulate the milk market, there was no one to alter this situation.

Already in 1904 agronomists warned against this imbalance of milk production: 'There exists an oversupply of liquid milk and an undersupply of livestock production. I argue that farmers loose substantial amounts of money, only because they have been misled to specialize in 'milking-only' farms, even far away from cities' (Ring, 1904–1905, II: 16). The decrease of domestic butter prices since the introduction of the cream separator around 1885 was alarming, too.[28] Europe-wide butter production increased dramatically, because cream separators speeded up the curdling process. By then, cream had been separated from the first milk through prolonged standing in open containers only by the aid of gravity.

It was a doom loop. Because manufacturing of durable dairy products was not that profitable, a run on the liquid milk market was the consequence. Because the liquid milk supply was higher than the demand and large quantities of milk were sent back to the producers, farmers started to invest in milk processing. Although farmers' co-operatives produced scarcely more then cream, butter, sour milk and non-fatty-cheeses,

---

[28] In the year 1877 the German engineer Wilhelm Lefeldt announced the invention of a milk centrifuge (Martiny, 1904: 176).

they nevertheless competed with creameries, cheese or condensed milk factories and dairies in possession of urban dealers, philanthropic and consumer societies and finally with imports from foreign countries (Scholz, 1949: 13).

Needless to say that urban dealers faced the problem of the milk surplus as well. For them, too, it was advisable to know how to make butter, soured milk, cured cheese etc. The skilful preservation of unsold liquid milk was indispensable. Even butter and cheese could not be stored in a cellar for more then a few days or at least weeks. Thus, urban milk dealers, started to invest in the processing of milk, too. With the diffusion of the cream separator the bulk handling of milk came in use so that even urban dealers were no longer interested to sell on commission and return the unsold milk to the farmers. What happened can at best be shown by example. In Berlin, Carl Bolle, who had started his horse drawn carrier service in 1882, was one of the first, who built up a large dairy in the urban area of Berlin.[29]

Firstly, he was very innovative in questions of milk distribution. In 1890 already 116 milk coaches left the dairy's courtyard every morning. At more than 3,000 stops throughout Berlin the milk was sold. In 1891 Carl Bolle already had more than 700 employees, nine years later they were about 2,000.

Figure 11.2 Development of the Carl Bolle Dairy, in numbers of horse-drawn cars and sales volume in million of litres

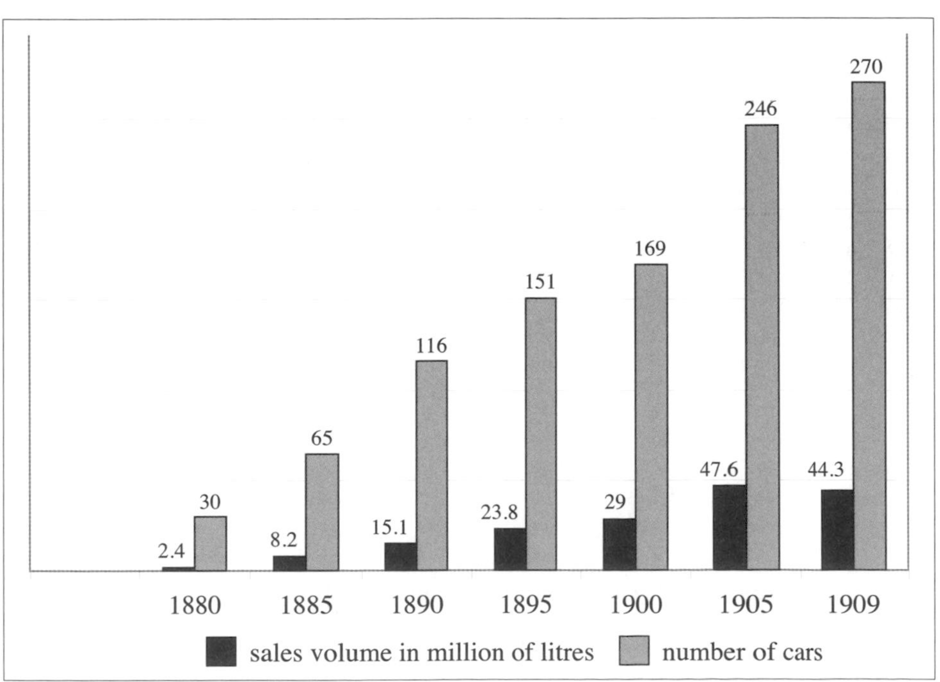

Source: Arnold, 1911: 612.

[29] To the history of Bolle see Arnold, 1911: 612–615.

Secondly, he became very innovative in milk processing and dairy technology. Bolle did what every firm undertook to assure its position. He constantly enlarged his facilities, endeavoured to persuade the consumers about the quality of his products, and undertook first steps to integrate supply- and direct distribution functions in order to advance his competitive position. Bolle bought his milk predominantly from farms all over the Mark Brandenburg. Already in 1887 he possessed a plant with stables and different technical and commercial equipment, altogether an area of 11,000 square meters. To a large extent the milk arrived on the railway to Berlin. In the Berlin factory the milk cans were seized quantitatively and milk filtered qualitatively by sample analysis in the laboratory. Since 1900 Bolle started to pasteurise, followed by cooling to 2–4 degrees Celsius. The unsold milk was centrifuged to cream, which was sold directly or used for the butter production. Residues in the centrifuges were sold as skimmed milk and/or processed to curled cheese, and also buttermilk, lactose, lactic acid and salts were produced and sold to pharmaceutical, paper and textile industries. The range of products was enlarged constantly, later kefir, yoghurt, and even *milk champagne*, a foaming mix beverage, completed the program.

Such kind of business activities destabilised the Berlin milk market: Bolle's factory system made him independent from the farmer as well as from the consumer. The smaller milk dealers, who worked by lease contracts with farmers nearby Berlin, could prevent a Bolle monopoly only with the instrument of dumping prices. However, this happened to the disadvantage of the farmers to whom the loss was passed on. The decline of producer prices, as a consequence, forced farmers and farm co-operatives to stop selling milk to Berlin wholesalers, and process it with own facilities. In 1900, for the first time the term *milk war* appeared in the newspapers, when farmers' associations announced their own sales and distribution organisation for the Berlin market. Bolle and other wholesalers reacted with the cancellation of lease contracts and the foundation of a new and even larger dairy. These economies of scale, however, forced the small milk dealers to associate too in order to organise milk from remote areas.

I stop telling the story here. In the end, all milk branches were antagonised. Big wholesaler, small milk traders, single farmers and farm co-operatives and – not to forget – more and more consumer organisations[30] battled against each other. Consequently, the Berlin trading area for milk became larger and larger, and dairies sprang up like mushrooms.[31] The city of Berlin, for years, built the arena of a *milk war*, a battle which likewise took place all over the country. Between 1900 and 1914 Hamburg, Hannover, Dresden, Düsseldorf and Munich reported about similar battles (Scholz, 1949: 14).

[30] They grew up for mainly two reasons: Because of the high prices for liquid milk the milk retailers started to sell different types of milk, the whole milk (natural fat content) or skimmed milk which quite often was adulterated with water. A second reason to organise a kind of consumer control was the question of hygienic production and processing methods. In the milk wars at the turn of the century hygiene became a most favored argument to attack a competitor.

[31] Thus, it became common to publish the conditions of milk delivery so that customers could get an idea about the feeding and care of cows on the farm, the milk collection system, transportation and processing of milk (Zitzen, 1915: 293).

VII. National Promotion of the Dairy Industry

According to the historical narrative of the modern dairy industry, the rise of the modern dairy industry in Europe was the work of a few forward-looking men who started the promotion of dairy in the 1870s.[32] Among them, two Germans became quite well-known, especially Benno Martiny, an agronomist who had been the secretary-general of an agricultural society between 1862–1874 in Eastern Prussia, and later became a professor of dairy sciences in Berlin. In 1871, he published *Die Milch, ihr Wesen und ihre Verwertung*, judged by contemporaries as the opus magnum of the dairy industry.[33] Together with Wilhelm Fleischmann, another dairy expert,[34] he was said to have founded the institutional and scientific basis of the modern dairy industry in Germany.

Within the community in which these two men, among others, acted as *Founding Fathers* of the new milk business, dairymen were described as a new generation of farmers and scientists (Beschreibung der Milchwirthschaftlichen ..., 1895: 5). One talked about a movement in progress, a movement that tried to achieve the common purpose of scientific and technological advancement in agriculture. The self-proclaimed dairy experts described themselves as savers of agriculture in a new world of globalised economics. The past, to which they referred, imposed fixed and inflexible practices, while the modern scientific and technological system proposed progress. By developing the paradigm of *old* dairy traditions they ought to bridge the gap between an out-dated agriculture, inflexible, inefficient and not market-orientated and a modernised agricultural industry.

However, the 1870s were not the first decade when proponents of progress brought the dairy business on the political agenda. Since the days of the economic school of the Physiocrats in the late eighteenth century, countries like Switzerland, France or even the German States debated the question of agricultural modernisation and dealt with the milk question on a regional level (Im Hof, 1983). Also, market orientation and scientific and technological improvements of the dairy were by no means new. Already in 1868 the Prussian statistician Viehbahn could point out that about half of the German milk production was traded in form of butter and cheese, because 'the selling of milk and the manufacture of butter and cheese increases in bigger estates' (quoted in Spiekermann, 1994: 91).

This was in part due to some technical innovations that circulated already at that time. For example, the Swedish owner of a large estate, Gustav Swartz, became well known with his 1863 invented method to separate cream and milk with icy water, which allowed to produce a sweet butter in less than 36 hours.[35] In the 1860s the equipment in cheese factories changed insofar as it became standard practice to use walled-off cheese

---

[32] A critical discussion of this tendency can be found in Spiekermann, 1994.

[33] One year later, in 1872, he edited the first number of the German dairy newspaper *Milch-Zeitung*. He also was responsible for the first dairy exhibitions, organised 1874 in Danzig and 1875 in Königsberg. Last but not least, in 1874 he was one of the founders of the German dairy union *Milchwirtschaftlicher Verein* in Bremen.

[34] Wilhelm Fleischmann had studied agricultural chemistry at the laboratorium of Justus von Liebig. Since 1863, he had worked at several agricultural experimental stations. In 1868, he became the director of the milk control in the town of Lindau. After he was engaged in building up a dairy in 1872, he decided to stay in the dairy business and become a dairy teacher. Later on, the university of Göttingen awarded him a professorship of dairy sciences (Schuler, 1942).

[35] A detailed description can be found in Altrock, 1936: 617–619.

cauldrons with outer covers, an iron grate, ventilation, and a direct link to the chimney instead of open fires.[36] Also condensed milk became an immediate success in urban areas where fresh milk was difficult to distribute and store. In 1856, the American inventor Gail Borden was awarded a patent for condensed milk. However, it was not until 1861 that financing was secured and the first plant was operational, soon followed by an introduction of condensed milk on the European market.

Yet, technology in the sense of mechanical facilities has not been the only reason for debates surrounding modernisation of dairy farming. For political reasons, the extensive agricultural reforms of the late 18th and early 19th century, motivated by the desire to make farm production more efficient, took very different directions, in particular in the old dairy zones. The interplay between politics and economics becomes particularly clear with a closer look at the case of Denmark.

At the beginning of the 19th century, Denmark was helplessly caught in the conflict between Napoleon and the rest of Europe (Pedersen, 1973; Frandsen, 1994). The loss of Norway in 1814 meant that the former dual monarchy, which geographically had stretched from the North Cape to the Elbe, was reduced to Denmark itself and the German duchies. Until the Danish-Prussian war in 1864 almost a third of the nation's greenland area was German. Holstein and Lauenburg belonged to the German Confederation, while Schleswig was nationally divided. Shortly after the loss of the German speaking population and area in the war with Prussia, the Danish parliament passed several regulations to promote a large-scale shift from the cultivation of plants to livestock farming. 'What has been lost outside, must gain influence inside', was the slogan (Altrock, 1936: 633).

This meant, that the redefined nation Denmark switched to a more intensive type of agriculture based on importing grains and growing fodder crops and feeding both to livestock for the production of bacon, butter, cheese, eggs, and meat.[37] Moreover, the idea of co-operative enterprises at a local level was invented in Denmark and spread from here via Holstein to Germany. Danish farmers were not only joining together in dairy co-operations but also in companies and associations dealing in fertilisers, foodstuffs, the marketing of agricultural products and the promotion of agricultural technology – from plant and animal breeding up to the control of milk-yields. In contrast to German farmers, it had become routine for Danish farmers very early on to compare feed costs with the prices they got for milk and butter.

---

[36] In the 1880s a type of cart-borne fire was developed in which, unlike open fires and closed fires, the heat could be discharged using a portable cart, called the fire cart; when not in use, this cart could be moved away from the cauldron and to a water heater. Attempts had been made since the 1860s to use steam heating methods but this innovation found little support because the steam often found its way into the milk, and customers claimed to be able to detect an aftertaste in the cheese. The first working steam cheese factories were not built until after the turn of the century. In 1913 there were five cheese factories with open fires and hanging cauldrons in the canton of Thurgau, 132 cheese factories with cart-borne fires, and 40 steam-operated cheese factories (Gutzwiller, 1923: 98–100).

[37] The results of this intensification quickly made themselves apparent in the trade field. Butter exports amounted in 1865/66 to about 43.000 casks or 4.85 million kg; by 1873/74 the figure had risen to about 103.000 casks or 11.5 million kg. Around 1900 exports of butter had reached about 61 million kg, over 90 percent of which went to England that was buying most of the Danish agricultural exports (Pedersen, 1973: 5).

Finally, agricultural modernisation in general had motivated reformers to get more information about cattle breeding and dairy practices in the *old* dairy zones. Long before the 1870s, there existed the idea to make the cow a 'milk machine' (Orland, 2003). Since the beginning of the century and intensified in the 1870s, a lively travel activity took place. This was directed to the North (Denmark, Netherlands, South of Sweden and Northern Germany (e.g., Steinmüller, 1802/1804; Schatzmann, 1870; Das Molkereiwesen, 1882; Schwarz, 1908), and to the Alpine dairy zones in Switzerland or the Allgäu (von Oppenau, 1886; Lindner, 1955). Last but not least, dairymen and cheese makers from the *old* dairy zones were most welcome experts who brought the knowledge of cheese making throughout Europe. Around 1820 Swiss dairymen founded the first cheese factories in the German Allgäu region (Aufsberg, 1913: 13–15). And even in Russia, it was a Swiss who mediated the knowledge of making fat cheeses. When a sovereign named Meschtschersko built up a dairy on his estate in 1815, he employed a Swiss dairyman (Milch-Industrie, 1885: 1). As famous as the dairymen have been the cows. In the 17th and 18th centuries, cows from the North Sea coast and from Switzerland were already well known because of their milk.

What was at stake in the 1870/80s, in fact, was the construction of a public sphere, which already existed in principle and which now demanded recognition in the political sphere on a nationwide, better to say, European level. In the 1870s a movement was intensified that started in different regions at different times, and it was merely a matter of time until nearly every region in Europe faced agricultural societies that were engaged in promoting dairy farming.[38] In respect to the proliferation and professionalisation of the dairy industry, numerous professional dairy associations of all kind were founded. More important, a new type of professional institution grew up that, state-financed, performed operational, scientific and technological tasks, which served particular interests within the dairy industry.

First of all, it was the issue of milk hygiene that intensified national interest during the 1870s. For many decades, the prohibition of milk adulteration and the general set-up of milk safety and quality regulations were placed at the centre of urban reform discussions (Du Puis, 2002). A countless number of medical reformers, public health experts and urban policy makers joined the milk reform debate. They were not only engaged in the infant mortality problem or suspected milk to be a cause of spreading epidemics; more general, the milk question turned out to be one crucial factor in building up the food provision system of urbanising societies. In fact, the question of milk reform was part of a broader food safety and city sanitation movement, which was itself part of political movements addressing the living conditions and inequalities of the economic system.

The ability of sellers to survive on a market that – as a result of the proliferation of different types of milk trade – often varied considerably from one town to the next, was depending on their power of persuasion to offer a healthy food. However, this was not an easy task. It needed decades, until the different countries passed nationwide milk regulations. For instance, the German milk legislation was not passed before 1930. Up

[38] Even a country like Great Britain, which was known for its dependency on food imports (e.g. butter and cheese), dairy farming attracted more attention since the end of the 19th century. And of course, Great Britain too had its *Founding Fathers* (Enock, 1943).

to this date, there existed no nationwide binding scientific definitions for 'good' or 'pure' or 'clean' milk, nor did any standardised method of setting such norms exist. Thus, chemists, engineers, agriculturists and a lot of entrepreneurial visionaries were just as able to set up rules for the hygienic treatment of milk as were doctors or bacteriologists. Especially the new breed of milk experts, whose presence was enough to trigger structural change in agriculture with dairy farms as a hitherto unknown link between farmers and consumers, felt obliged to create a set of rules for milk hygiene.

The cumulative growth of the sector together with different health questions indicated a growing attention being paid to the field by different national agencies. In order to act as a neutral instance of marketing co-ordination, policy makers, however, needed methods of valuation and control instances for supervision. For this reason, more and more organisations were founded that were financed by the public sector, neutral to business, and generally populated with academically trained scientists. Up to the Weimar Republic, all over the country numerous technical dairy schools were established. They were guided by nine Federal dairy teaching, research and experimental facilities or scientific institutions for milk issues, all of which dated back to the 1880s.[39] From their point of view, however, the solution to the safety question and the institutionalisation of a functioning dairy industry required technical skills of a high order, efficient equipment and scientifically informed competence on all levels of producing, processing and selling milk. For them, the rise of an academic milk science as a force for the benefit of the consumer was the answer to the overall *milk question*.

VIII. Integration into international markets: the Swiss case

Before I continue to discuss the national utilisation of scientific benefits, a closer look on the old dairy systems and their integration into the new economics of dairying seems indicated. In earlier times the *old* dairy zones had been the centre of butter and cheese trade, after 1870 those regions suddenly stood at the periphery or were at best part of a diversified trans-national milk business. The increase of national dairy industries was hostile to regional peculiarities and local identities; all producers became part of a dense network of dairies. One may thus ask how the emergence of the milk industry did integrate the knowledge and experiences of the *old* dairy zones into the policies of trade promotion.

To answer this question, one should look at the Swiss case. While the European milk markets slowly began to change, the Swiss animal husbandry and dairy system underwent some important changes.[40] Within a few decades and long before the 1870s the Swiss mountain dairy farming lost its territorial roots, which was – as said above – a result of the fact that farmers in the grain-growing regions hoped to increase their incomes through the expansion of cattle husbandry. One effect of this tendency was a change of the old

---

[39] At the beginning, national institutes for research in dairying only had food chemical laboratories where the chemical composition and the nutritive value of milk and milk products were analyzed. Starting in the mid-1880s they were joined by comprehensively equipped bacteriological laboratories (Riedel, 1936: 217–222).

[40] For more details, see Orland, 2004.

division of labour between valley and mountain farming. Until the 1820s milk and cheese were primarily produced on the mountain meadows in the summer. But then, the first cheese factories were established in the valleys, not by farmers but by dairymen with business acumen, and cheese traders.[41] The latter were primarily concerned with shortening their transportation routes when they attempted to convince valley farmers of the virtues of dairy production. In addition to cheese factories, they created warehouses where they could store unripe cheese purchased from the Alpine meadows, and allow it to age. Gradually non-farmers began to acquire expertise in cheese making, and to refine it as they pleased. Land that was once used to grow crops was now used for grazing.[42]

These cheese factories of the early 19th century presented farmers with a new type of competition. Soil quality, fodder, care of animals and Alpine meadow farming were no longer relevant. Cheese traders naturally were not interested in those issues. Their sole concern was cheese as a raw material, which – as people were slowly beginning to realise – could easily be produced in large quantities and sufficient quality in low-lying areas, too. However, a sensible method of organising the procurement of the quantities of milk needed to produce fatty cheese was needed, since no single valley farmer had herds the size of those brought together in the Alpine meadows. Joining forces to form co-operatives seemed to be the answer, and not just to this problem. United, co-operative farmers could resist the prices forced upon them by cheese traders and – if they could reach an agreement – run cheese-making operations themselves.

Despite initial objections and resistance among farmers and cheese traders, the co-operative model spread quickly from its origins in western Switzerland. The constant improvements in the transportation infrastructure from the 1840s onwards and the reduction (and, in some domestic transportation routes, total abolishment) of customs duties served to convince the last doubters that cheese could also be manufactured in the valleys. In 1847 an agricultural census in the canton of Berne counted 217 valley cheese factories; a decade later this number had already swollen to 355.[43] Only mountain cheese production failed to benefit from the general upswing; its share of Alpine cheese production steadily declined.[44]

Beside an expanding emigration of Swiss knowledge in cheese production, general developments within Swiss agriculture until the beginning of World War I differed not very much from those in other countries. Urbanisation and industrialisation increased milk prices since the 1860s, changing price levels and the thrust of development advanced

[41] Many of the cheese dealerships that had begun to develop in Switzerland since the 1760s had been established by former dairymen or cow herders. In the 1810–1820 decade many of them emigrated to the Bavarian Allgäu region and there tried to manufacture the same quality of heavy and fatty cheese from their home region. The first valley cheese factories were established there in 1815, on farmsteads in Hofwyl and Kiese. Following their initial success, such valley cheese factories spread throughout the rest of Allgäu and Swabia. Another cheese trader from Allgäu, Karl Hirnbein (1807–1871), imported another type of soft cheese from Belgium: Limburg cheese (Flad, 1989: 26–32).

[42] On changes in land use within the Canton Fribourg, see Walter, 1980.

[43] This period of growth slowly came to an end in the 1870s. Exact figures are in Pfister, 1995: 198.

[44] It is estimated that in 1870 in all of Switzerland there were about as many valley cheese factories (2,600) as Alpine dairy farms (2,800) (Anderegg, 1894: 87).

the milk production according to quantity. Further on, the foundation of the Anglo-Swiss Condensed Milk Co. in Cham (1866) and Henri Nestlé's infant food factory in Vevey, founded in 1867, opened up broader markets, because condensed milk and dry milk products could be easier transported over long distances (Pfiffner, 1993).

Figure 11.3 Development of producer prices for fresh milk in Switzerland, 1800–1914

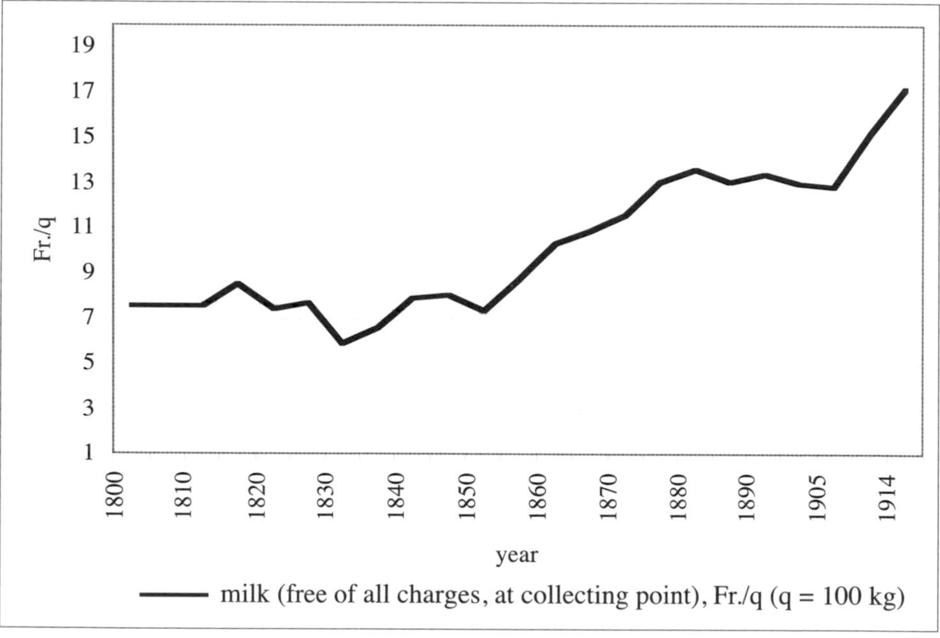

Source: Ritzmann-Blickenstorfer, 1996: 482–483.

Like other countries, Switzerland very soon got a dense network of fluid milk production enterprises. The new preferences were followed by bulk-buying contracts, and the collecting and selling of milk to diversifying factories and private consumers was done by milk delivery co-operatives. In 1920, 3,519 local co-operatives were counted, thereof 2,134 had been founded before 1900 (before 1850: 345, 1850–1879: 870, 1880–1899: 919) (Stocker, 1982: 14). Switzerland, too, experienced the conflicts of interests between the different branches of the dairy industry and regions. The only difference to other countries has been that cheese factories and cheese traders (who traditionally were engaged in cheese export) played a more important role as in other areas. Until the beginning of Word War II local milk prices resulted from the profit cheese traders realised on the international markets (50 Jahre Schweizerische Milchwirtschaft, 1937: 152). While in Denmark, for instance, the butter market dictated the conditions of price policy, business regulations and the control of the liquid milk trade, in Switzerland the cheese market played this role.

Figure 11.4 Development of the milk production in Switzerland, 1837–1930

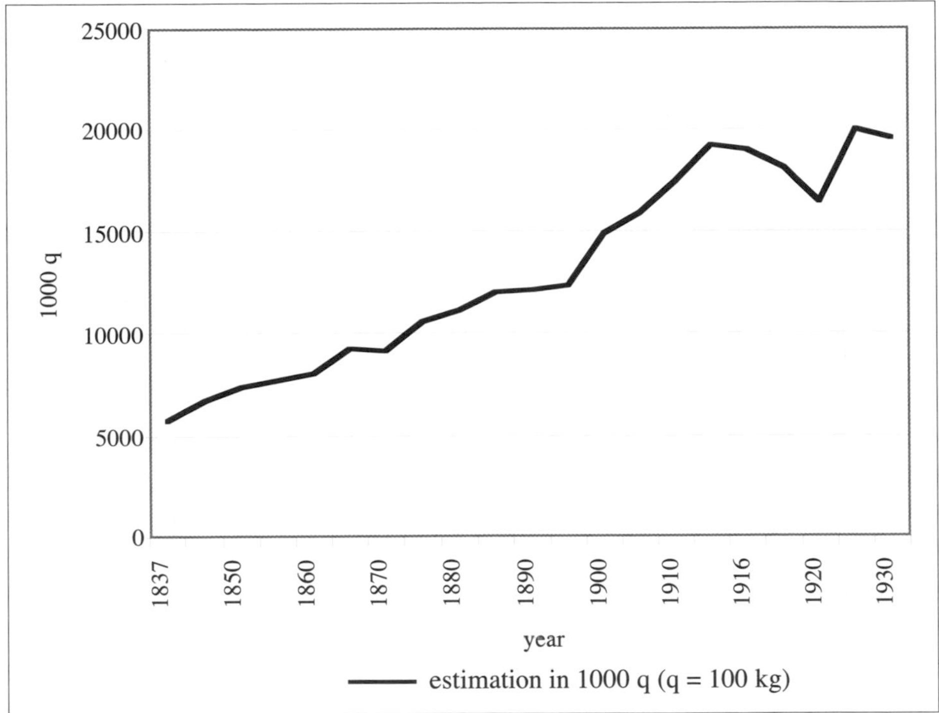

Source: Ritzmann-Blickenstorfer, 1996: 552.

Since the 1870s (with the only exception of World War I, which entailed a drastic decline in milk production), the Swiss struggled with constant overproduction and sales problems. A first and successful reaction to stagnating markets and emerging competition in the 1890s has been the formation of strategic alliances. At the end of the 19th century, all over the country regional dairy organisations were founded with the aim to reduce competition and achieve synergy in nationwide and trans-national positioning. Enhancement of innovations, concentration of resources and the acquisition of State subsidies (at first for dairy cattle breeding) became major tasks of dairy co-operatives (Senti, 1982: 18). In 1907, the formation of the *Zentralverband Schweizerischer Milchproduzenten* (ZVSM) completed this politics of co-ordination. However, all strategic alliances could not insulate from fluctuations in demand, political influences on business cycles or the increasing international competition. In a long-term perspective, only arrangements between the farmer's unions and the government could prevent, that since 1920 the milk prices for producers collapsed. A long story of pegging and agricultural subsidy began (Stocker, 1982: 41).

IX. Designing national authenticity

How did nationality help to solve the economic problems of everyday life? What pushed the farmers/milk traders to give up their local traditions and develop hope that they would benefit of the integration into national federations? I will argue in the following that the production of national milk symbols, which played a central role in the transition of the *old* Swiss dairying practices into an industrialised milk business under the conditions of an emerging international trade, was a kind of crisis management. First it helped to integrate the different local identities within the country. Secondly, the prediction of a typical Swiss product was aimed to improve the competitive position of Swiss products, 'because we have to confess, that since the 1870s it [i.e. the Swiss dairy industry] in its quality evolution did not withstand the competition of foreign countries'.[45]

'Milk is the fundamental material of our *nation, to build and support its ability to work'*. When Rudolf Schatzmann, a Swiss country pastor, formulated this sentence in 1872, he was active on behalf of Alpine dairy farming for more than a decade (Schatzmann, 1872, quoted in Kollreuther, 2001: 22).[46] As a member of a mostly bourgeois, urban reform movement which felt compelled to take matters of the mountain dairy farming into their own hands, he was used to ride a wave of national sentiment. The Swiss nation should awake about the fact that Alpine dairying farming was passing away. On 25 January 1863, some 30 men from all parts of the country had convened to form the Swiss Alpine Farming Association (Schild, 1865: 3–17). Their job was to unite activities to improve the Alpine landscape and agriculture. In the founding document, 'activity in the pure Alpine air and a diet of cheese, butter, milk and meat' was praised as a source of health, physical strength, and the spirit of freedom of the Swiss people. The shepherd, who embodied these Swiss virtues, could not be allowed to fall into oblivion, nor could his regions be allowed to suffer under the indifference of modern man. It was said that there were 'immense treasures to be had' in the mountains (Schild, 1865: 3–17).

Thus, Schatzmann and others acted into two directions: On the one hand, the aim was to catch up to the economic boom in the valley by improving Alpine farming. People like Rudolf Schatzmann,[47] thought that science and technology were to provide the tools to that end. They used lectures, essays, courses, exhibitions and inspections of cheese factories to tirelessly preach the gospel of the latest achievements in the science and technology of dairy farming to the mountain inhabitants. Thinking that in order to save Alpine farming, dairy farming needed to be modernised, his greatest triumph was the opening of the first Swiss milk testing station in Thun on 1 September 1872, later joined by another station in Lausanne.[48]

On the other hand, promoters of the Alpine dairy farming system used their certain activities to persuade the urban population about the healthy and tasty products of the mountain meadows. In his *Alpwirthschaftliche Volksschriften* of 1873, Schatzmann praised

[45] As one dairy expert put it in retrospect: Widmer, 1937, 23.

[46] Since 1859 Schatzmann edited the *Alpwirthschaftliche Monatsblätter*, published annually altogether six times until 1866 by publisher J.J. Christen in Aarau.

[47] For his personal history, see Wahlen, 1979.

[48] These were modelled on Alpine testing stations already in place in several locations in the German and Austrian Alps which conducted botanical, soil and fodder production experiments ('Die Alpenversuchsstationen' 1867: 129–133, 137–40).

'the aromatic Alpine herbs and the fresh, healthy Alpine air' as the important conditions 'for healthy, strong and beautiful cattle and tasty butter and cheese' (Schatzmann, 1873: 3). Schatzmann, like his companions, made no secret of the fact that every innovation in the valley made the soil there more attractive, thus exposing ever more painfully the disadvantages of higher elevations. Nevertheless, he never got tired to remind his fellow citizens that the Alpine nature was the trump of the Swiss national economy.

At the same time, when people like Schatzmann kept working hard on saving the mountain dairy farming, another part of the Swiss agriculture tried to define the national authenticity of the Swiss cow. Until the beginning of 19th century, in Switzerland like in all other countries, local cattle lacked distinguishing any kind of specific characteristics. Cows in dairy zones as well as in other regions were a motley lot; they varied in size, shape, and colour. They were the result of unintended cross breeding within herds, or because they roamed, and bred freely, they reflected geographical characteristics. In any case, they were not uniform. At best, agricultural writers differentiated between so-called lowland and highland cattle.[49] In a country like Switzerland all breeds counted as highland animals, regardless of other traits.

When European farming regions began realising the economic potential of dairy farming, agricultural reformers in Switzerland tried to promote dairy cattle production in different Swiss cantons. For centuries Swiss livestock farming concentrated on breeding bulls. Cows had been part of the reproduction cycle; their milk was needed almost exclusively to suckle young cattle, with little left over for butter and cheese. The 19th-century changes in international cattle trade and the new recognition of the Swiss cow, forced agricultural farmers to think in terms of specialised breeds and more over to concentrate on the breeding of milk cows.

At first, agricultural reformers started to explore the standards of animal husbandry in Switzerland. Since the very beginning of 19th century, Physiocratic societies had founded commissions in order to evaluate local cattle. Such commissions sent out authorised agents to visit farmers, and distributed questionnaires to local authorities (Duerst, 1923: 21–23; for the canton of Glarus, see Hösli, 1948: 43). Commissions were specifically interested in ascertaining what role the cattle trade played in a community, inquiring, whether communities held enough cattle to turn a profit, which stocks were doing well or poorly, or whether farmers had enough fodder for the winter. One result of these activities was the prohibition of cattle imports. In other words, the social construction of a breed of cattle was initiated by the demarcation of geographical boundaries.

A second influential activity should become the organisation of cattle exhibitions in order to initiate competition within local farmers. One of the first was organised by the small council of Bern in 1806/1807. Prizes were given for the best stock bull and the best cow (Duerst, 1923: 22). In the next decades, exhibitions developed their momentum and farmers from different villages, communities, and regions competed with one another. The rivalry here was different from that which characterised the old cattle markets. The aim was not to achieve a good price then and there. Awards were not short-dated bills but rather prospects for future profits. Above this, cattle exhibitions were educational measures that intensified local and regional competition. Before long, other cantons followed suit. In 1811 the city fathers of Luzern passed regulations governing the

[49] On the next paragrah, find more details in Orland, 2003.

scheduling of cattle exhibitions at regular intervals. In 1818, for the first time the local government of the canton Luzern awarded prizes for sires, and in 1837 the law on cattle exhibitions (*Gesetz über die Schau von Zuchtvieh*) was passed (Duerst, 1923: 22). When the first cattle exhibition in the canton Appenzell was planned in 1846, the organisers hoped to profit from the long-standing experience of other cantons (Duerst, 1923: 37).[50]

Until the middle of the 19th century, all these measures to improve dairy cattle production were not dominated by nationalistic ideas (Weishaupt, 1991). Although the nationalistic idea of a farmers and herdsmen state was already developed among the literacy elite at the end of the 18th century (those who by means of socio-economic positions have been no farmers),[51] cattle exhibitions at first functioned as local or regional events. Only by and by, the idea of a *national cow* came into business, and in part this had been an immediate consequence of the processes of classification. As said before, one of the first activities to initiate the social construction of a breed of cattle was the demarcation of geographical boundaries. In the long run, cattle exhibitions became the common place where people negotiated the terms to distinct the animals in question from other breeds.

However, there remained an important question to be resolved: Which prominent features could one identify, and how should they be conserved? What was the *true* type cow of a specific region or breed, or a dairy animal at all? To answer these questions one needed methods of evaluation; it was the herdbook that should become the most prominent instrument that directed the work of breeders. With the expansive use of herdbooks, however, the cultural and intellectual climate changed. While several descriptive terms distinguishing animals were in common use, e.g. scrub or mongrel, cross-bred or pure-bred, the system of herdbooks formulated scales of points that designed to aid in acquiring the skills needed to select cows by bodily conformity. The first Swiss brown cow herdbook of 1879 (*Verzeichnis edler Thiere der Braunviehrasse*) included not less then 23 bodily characteristics and bodily positions that had to be evaluated (Weishaupt, 1998: 38). Every animal had to be measured and weighed.[52] The results were coded in registers; each item got its own score.

Within the herdbook, the classification of dairy cattle was no longer a question of local traditions. As well as methods of measurement and rating required a process of standardisation, national standards of appearance, height, weight, and performance became more and more important. Although only a small fraction of the national herd consisted of animals that were recognised as belonging to definite breeds, by the way of measuring in the 1880s the quasi-official Swiss cow was defined: The *Schweizer Braunviehrasse* (Swiss Brown Race). All other breeds, such as the Appenzeller, Haslitaler, Prättigauer, and many more, continued to exist on farms, in villages, and in the minds of people.[53] But the breed that was sent to exhibitions, existed on paper, and was thought to be the origin of the first Swiss herdbook of 1879, was the Swiss Brown Race.

---

[50] In Germany the first cattle exhibitions with nation-wide relevance took place 1863 in Hamburg, 1868 in Mannheim, 1874 in Bremen.

[51] Before the 18th century the highlanders of Switzerland did not form a distinct group (Guggenbühl, 1998).

[52] Several instruments have been developed for the process of measurement (Pressler, 1886).

[53] Likewise in other regions: when the German Society for Agriculture *Deutsche Landwirtschaftsgesellschaft* organised its first cattle exhibition in 1887, one of the organisers worried about the

If European nations were invented by the construction and acceptance of one language, one past, one *ethnic* group, then one should add to this list *one landscape*, *one botanic* and *one animal* too. During the 1880s herdbooks sprang up all over Europe (Hansen, 1921: 144–402). They all performed the same task, to wipe out difference and create difference at the same time. It is important to recognise that all of today's famous dairy cattle breeds were defined and designed in the second half of the 19th century: Shorthorn, Ayrshire and Jersey for Great Britain, the Swiss Brown Cow for Switzerland, the Holstein Frisian breed for Netherlands and Nothern Germany, the Danish Red for Denmark (Politiek and Bakker, 1982: 89–126). Nationalism claimed for cultural homogeneity. Thus, cattle-breeders were among the first who developed genealogical myths and discourses about heredity. They divided and counted out, they looked for natural entities and placed them instead of subtle differences in merging and unselected, in any case difficult to identify groups of animals. Thus, it was just a small step to arrange plants and animals according to the nation they belonged to. Swiss breeders were among the first to ascertain external traits which marked the already famous dairy cow and which could be transformed into representations of the Swiss nation.

Finally, there existed a third context that helped to integrate the work of agricultural reformers, cattle breeders and dairy farmers into a wider domain of national policy. Likewise, national authorities and bureaucracies mediated the invention of other national traditions, the representations of the good cow, and the healthy milk. Since the 1880s, national exhibitions took place at regular intervals. Originally planned as exhibitions of trade, under the patronage of the federal government they soon became events of national self-portrayal: Switzerland *en miniature* (Hettling, 1998). 'Get to know yourselves!' was the overall slogan of 1883 (Gugerli and Speich, 1999). According to the concept of the organisers every exhibitor was invited to present its own economic potential by comparing it with those of the competitors in business.

In Zurich in 1883 the first exhibition took place, in which the nation exhibited itself. Geneva in 1896 and Bern in 1914 went on with this tradition, which was peaked out in the fourth exhibition in Zürich in 1939. In this year the double-faced representation of the Swiss dairy industry was abundantly clear. On the one hand, the exhibition served the purpose to demonstrate the economic potential of the Swiss dairy industry (Zollikofer, 1939–1940: 562–571).[54]

For this purpose, a complete cheese factory had been built up, representing the state-of-the-art in dairy science and technology. A sparkling milk fountain symbolised the daily processing of milk on its way from the cow via collecting points to the consumer and the different processing operations. Another section designed in 'a cheerful green' was dedicated to demonstrate the healthiness of milk and cheese, 'this ribbon of life' (Zollikofer, 1939/1940: 565), which was said to be the most important food of the Swiss nation.

On the other hand, the organisers exhibited the *old* traditions of Swiss dairy farming, showing pictures of the mountain cheese manufacturing, of the herding up of cattle to

diversity of the German races. Only Bavaria counted 28 cattle stocks at this time. During the next 30 years they were reduced to twelve, in the year 1903 the society found ten, in the year 1925 eight and in 1948 today's four *Fleckvieh*, *Frankenvieh*, *Braunvieh* and *Pinzgauer*. (Diener, 1980: 79).

[54] Zollikofer, 1939/1940: 562–571.

the Alpine meadow elevations and the return of the animals to the valley in autumn. A set of pails, bowls, cloths, wooden stirring instruments, storage and transporting vessels and butter kegs used in the alpine huts in summer time represented the archaic working conditions in the mountains. Finally, some figures were already completely detached from the dairy context so that they could function as artful symbols and most prominent features of the nation.[55] Already in 1883, a larger-than-life sculpture of a herdsmen boy decorated the art section of the exhibition (Gugerli and Speich, 1999). In 1939, a wooden sculpture of a farmers family stood in front of the buildings harbouring the facilities of dairy industry.

Altogether, the exhibitions operated with the contrast of old and new, of out-of-use technology and state-of-the-art working conditions in modern dairy industry. Thus, with the experience of international competition, a strong relationship between national self-images, state of mind, landscape aesthetics and food habits was constructed. Milk, cows, dairymen and cheese became formative rhetorical elements of the physical environment, which formed the myths of the nation. In this manner, agriculture did not only bring information from the periphery to the centre of the modern national economy, it also participated in constructing and stabilizing the state's defining power of national products.

X. World trade with dairy products

Eric Hobsbawm (1992) reminded us of the fact that the paradox of the nation as an economic unit in liberal capitalism is that it has no location. The national economy is no site-bounded economy; the basic modules are not places or regions but enterprises. However, enterprises normally do not follow national interests. For them, the general rule is to make as much profit as possible. Therefore, markets are not limited and defined by national borders. Depending on the prevailing terms of the market firms will neglect national borders and operate worldwide. However, with industrialisation and economic growth the newly created firms competed with each other on the market of durable dairy products to an increasing degree. The extraordinary transport problems could be minimised with pasteurisation technologies; new products like condensed milk or milk powder could keep stable for a long time. With the exception of the fresh milk markets, the dairy industry like others could go for a business on large scale.

Several displacements on the butter and cheese market exemplify the consequences of the creation of larger markets. The best-known case is that of the competition between Dutch and Danish dairy trade on the British market. Parallel to the processes of industrialisation, England encouraged imports of butter and cheese, because local dairies were left heavily dependent on the sale of fluid milk (Taylor, 1987). Yet, in 1860 the Netherlands had exported 16,868 tons of butter to Great Britain, while from Denmark only 589 tons were sold on the British market. In 1890 these figures had been reversed completely. Now, the Danish exported 42,000 tons to Great Britain, the Dutch only about 8,000 (Bakker, 1992: 108). Dutch butter exporters, however, were lucky to find a compensation in Germany, so that for decades a peaceful coexistence could exist between the two competitors (Smidt, 1996).

[55] About the common use of dairy symbols in Switzerland see Marchal and Mattioli, 1992.

Switzerland too faced more and more trouble with the cheese export. Up to the 1880s, fatty cheeses like the Emmentaler, Gruyère or Appenzeller had been the export hits of the Swiss nation. But then, in 1885, cheese exporters for the first time faced a strong trade crisis. At this time about 13 to 15 percent of the world cheese trade consisted in cheese coming from Switzerland (Kollreuther, 2001: 37). For the next decade, the amount of exported cheeses still reached a high level, but it was no longer possible to dictate the market conditions. And although Swiss fat cheese asserted its position as one of the most popular cheeses marketed in the world, since World War I the total exports from Switzerland steadily decreased. In 1913 about 36 million kg fat cheese left the country. After a complete collapse within and after World War I, and a partial recovering since 1922, the cheese market for decades did not reach its pre-war level (Altrock, 1936: 83).

Figure 11.5 Development of export/import of cheese in Switzerland. Export 1851–1939, Import 1851–1913

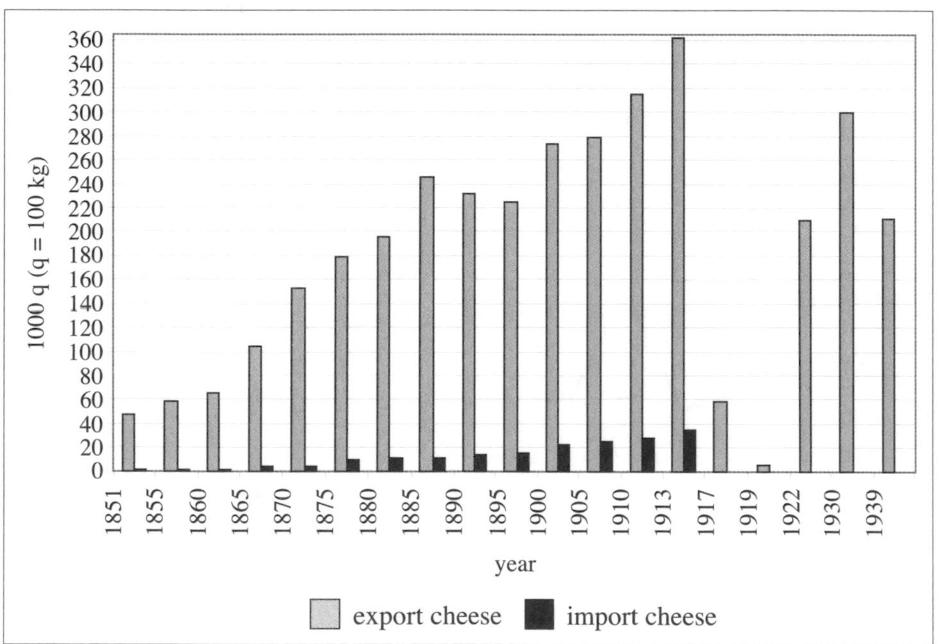

Source: Ritzmann-Blickenstorfer, 1996: 662, 666, 682.

Contemporaries very soon had caught up the culprit. 'Foreign imitation products' were called to be responsible for the dramatic decline of the Swiss cheese trade (50 Jahre Schweizerische Milchwirtschaft, 1973: 165). Farmers were said to be the victims of international manipulations. A wave of patriotism flow through the country and demanded for the complete relief of foreign commerce. Further on, dairy organisations claimed that the domestic demand should be developed to the maximum. Dairy products advertisement was intensified, all over the country people were reminded that consumption of dairy products would improve the nation's health (Kollreuther, 2001).

What was neglected, however, was the fact that Swiss people had been the ones who brought the knowledge of Swiss cheese making into other countries. For example, at the turn of the century the United States had become a strong business rival in Swiss cheese making. In Ohio as well as in Wisconsin, Swiss emigrants had settled down in the 1860s and 1870s. The knowledge of making Swiss cheese was their capital to survive and find a new subsistence. However, these two regions furnished the principal output of Swiss cheese for the country, and – on a low level – exported *Swiss* cheese to Europe. Especially throughout the war and post-wartime (1916–1922) they could increase their exports to Europe (Pirtle, 1973: 114).

The same is to say for another Swiss speciality, the cattle trade. The first so-called Brown Swiss cow set foot on American soil in 1869. Next shipments followed in the 1880s. Although altogether only several hundreds of animals were directly imported from Switzerland, American breeders as fairly clever businessmen started new breeds and – as they claimed for – improved the type of the Brown Swiss cow. As early as 1880 the first Brown Swiss breeders association was founded, and because Brown Swiss were noted for their longevity and high yields, they became one of the most popular dairy cows in United States (Pirtle, 1973: 42).

Figure 11.6 Export rates Swiss cattle, 1851–1913

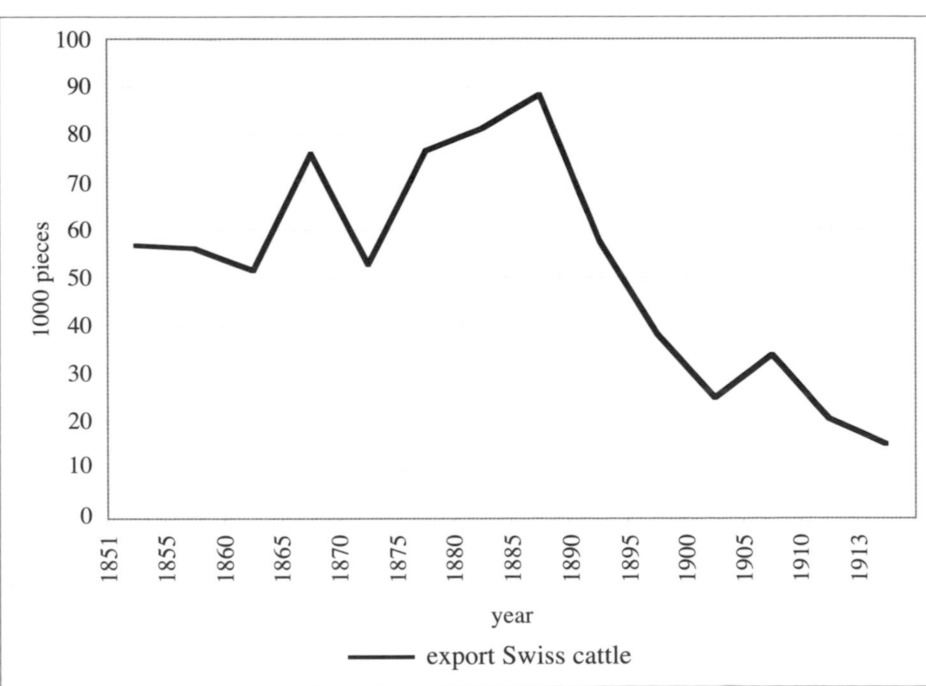

Source: Ritzmann-Blickenstorfer, 1996: 666.

General export rates of Swiss cattle confirm these observations. While the milk boom of the 1870s for a few years increased export rates, such developments soon were shaped by the construction of Swiss Brown breeds in different countries. Although the breeding business was among the first Swiss agricultural branches that received a substantial amount of State subsidies, no continuous boom could be generated, and the export rates declined.[56]

Last but not least, one has to mention that already at the turn of the century the European market was the destination for milk products from all over the world. Depending on the differences of the dairy branches, products came from the United States, Canada, New Zealand, Uruguay, Argentina, Russia and Australia (Altrock, 1936; Pirtle, 1973; Lewthwaite, 1980). Together with the already mentioned European leaders in dairy business (Denmark, the Netherlands, Switzerland, Ireland, Germany, Sweden, Finland, Norway and France)[57] about 25 countries counted as the biggest world traders. Consequently, these national dairy industries acted as a globalised community.

As said above, since the 1870s–1880s professional associations sprang up like mushrooms all over Europe. This trend found its expression on a global level as well. A scaling up of organisations took part that already in 1903 resulted in the foundation of the International Dairy Federation (IDF) based in Brussels. Every two years (with the exception of the wartime), this institution organised world congresses of the dairy industry till this day.[58]

Founded on the occasion of the first world congress of the dairy industry, the world organisation was aimed to overcome tensions between the national industries. Thus, it should inspire national governments to pass legal regulations that facilitate the international trade in dairy products. Its specific responsibilities included administering the federal government direct subsidy to farmers and providing leadership and undertaking research on behalf of the national milk supply organisations. Further on, the international organisation should work together with the private sector to import and export milk products, and it had the task to promote research in the sciences and technologies of dairy on a general level. Thus, the IDF served as an advisory body to governments and as a centre of knowledge and expertise for the dairy industry.

XI. Standardisation of national trademarks

With the diversification and globalisation of the dairy market, people involved could no longer be contented with a local view. While during the process of building up a new division of labour, the ideology of the nation played a crucial role for the integration of local and regional traditions into a wider context of national economies, after World War I it increasingly lost its relevance. This does not mean, however, that economic policy measures of the national governments declined as well. In opposite to this, governmental

[56] In 1879, for the first time the Swiss federal state granted financial aid to cattle husbandry and breeding. Since this date, government support constantly increased till this day (Ritzmann-Blickenstorfer, 1996: 564).

[57] France is a somewhat special case. Although about 400 different cheeses were registered around 1930, they practically played only a minor role on the international markets. Between 1909 and 1930 France always imported twice as much cheese as it exported (Altrock, 1936: 37).

[58] See the homepage: http: //www.fil-idf.org/.

actions started by now. Though national economies had been undermined by the world economy, they still existed and were very active in regulating national markets.

So far, most federal governments had been very discrete in regulating the milk markets. Of course, public authorities were well aware of the *milk question* as it was discussed since the 1880s. Several countries passed food legislation in order to fight the problem of food adulteration, as in Germany, for instance, in the year 1879. Special milk regulations existed too. However, most of them acted only on a regional level. In 1876, the German government founded the Imperial Public Health Department that over years was engaged and tried to find solutions for the complex struggles on the milk markets, with only slight success. Already in October 1882, a commission hold a meeting in order to pass 'technical materials for the preparation of an imperial order, regarding the control of milk by policy'.[59] However, this, like other regulations, has never been passed. The paper of the Public Health Department only served its purpose as template that was used by several towns and communities to tighten local regulations (Zitzen, 1915: 294).

Up to World War I, it was always the same argument that hindered federal public authorities to regulate the milk market on a national level. The urban milk markets were said to be too non-transparent and too complex to be regulated on a top-level. In administrative respect this meant that public authorities relied on the old regulations of the medical and veterinary police that worked more or less since centuries. Although police officers since the 1880s received assistance by chemical examination, this kind of food control, however, was limited.

World War I brought a radical change. Great parts of the European agriculture that meanwhile earned a good deal of its income from animal husbandry, imported much food from foreign countries (Germany, e.g., about 30 percent) (Spiekermann, 2001: 7–9; for Denmark, Pedersen, 1973: 7). This was far too much in the perspective of war economies. The immediate lack of feeding stuff resulted in harum-scarum slaughter of cattle, favoured also by boom prices for fat stock of all kind. Until 1916 the German Reich had enough meat, but to a lesser extent butter, cheese and liquid milk. A centralised governmental control of food production was inescapable. Since 20 July 1916, in Germany milk production and processing was under strict control by the federal government.

After the end of the war, diverse war regulations of the market remained in existence. Revolutionary drifts, the rise of social democracy, and the relevance of communal administration started a debate about the question whether milk production as a fundamental need of life should stay under centralised control (Böhret, 1966). Many towns went even further and built up dairies, creameries and milk selling points under the auspices of local authorities (Brandt, 1928). However, in a long-term perspective all these activities did not survive, mostly because the dairy organisations set out to overcome the antagonisms of the different branches. Quasi by self-control the dairy industry tried to prevent the state intervention. This, however, went along with a new defined role of the state.

From the mid-1920s on, public authorities experienced a decisive role as a kind of master controller of regulations set up by the dairy associations in collaboration with scientific institutes. Until 1914, there existed no social or legal conventions in a way that all commercial partners in the milk business felt constraint to team up. Every part

[59] 'Technische Materialien zum Entwurfe einer Kaiserlichen Verordnung, betreffend die polizeiliche Kontrolle der Milch' (Reinsch, 1903: 8–9).

of the business worked for its own benefit. Individual opportunism had produced crass asymmetries in the information about commodities, and milk wars have been the result of this keen competition.

After the war experience of rigid control systems, all participants of the business, from the farmer and his co-operatives, the butter- and cheese factories, wholesalers and small milk traders, consumer organisations and philanthropic societies, placed the responsibility for a nation-wide quality control of dairy products with the government. For all of them, it seemed obvious that the dairy industry needed to provide confidence with the aid of trademarks or guaranties. The past had shown that it was practically impossible to check the quality of milk and dairy products by organoleptic tests. It was quite easy to sell the consumer a pup. But it was also clear that one needed a neutral instance in order to gain the confidence of the consumer. Thus, it became the task of the government to design selection procedures and control labels. In his role as a neutral or even objective controller the state grew up to the big national advertiser for a healthy good. Through government intervention, the deficiencies within the dairy industry, including a substantial variation in the quality of produce and marketing should be eliminated.

The 1920s became the period of quality control, standardisation processes and the administration of product characteristics in the history of the modern dairy industry (Altrock, 1936: 473–540). Although quality control was by no means new, it now changed its character. Before World War I, only export countries like Denmark or the Netherlands had used quality control procedures and trademarks. The first trademark for butter was invented in Denmark 1901, the Netherlands followed with a trademark for cheese in 1904 (Dold, 1937: 65–66). Testing results, produced by deputies of the government, butter wholesalers and dairy experts, in Denmark had been the basis for negotiation of prices.

Exhibitions, that played an important role as the locations of testing and judging, were not new, too. In Germany, dairy exhibitions took place since the 1870s. The German dairy association had organised its first exhibition in Danzig in 1874, afterwards nearly every year an exhibition was organised at another place (Molkereiwesen in Dänemark, 1876: 81). But while these events were industrial fairs and only secondly events of quality inspection, the new type of exhibition of the 1920s was organised for just one purpose, the state control of dairy products. For this reason, most of the exhibitions of the 1920s and '30s took place near by or directly in dairy science institutes that operated as a kind of governmental executive.[60]

The first and overall goal was the evaluation and unification of existing higher, medium or cheaper grades. Several descriptive terms distinguishing butter or cheese were in common use, e.g. for cheese during the years 31 grades. 1932 these grades had been rearranged and reduced to only 14 grades, divided in six main groups (Claus, 1938: 325). A renewed system of qualification formulated scales of points that designed to aid in acquiring the skills needed to select products by conformity. This was seen as indispensable because 'the character of a nationwide testing requires that the samples send in from several regions despite differences in the distances need to receive the same testing conditions' (Claus, 1938: 326).

[60] Since 1927 the testing of fresh milk products were completely managed by dairy science institutes. The main reason was the debate about pasteurisation technologies, which were said to destroy the natural quality of the milk. The operating expense was immense. In 1935, for example, 1300 samples of liquid milk were tested (Claus, 1938: 328).

Yet it was not only the constant and invariable quality that should be achieved by unification. It is the 'advertising effect of the huge German State exhibition and the voluminous industrial fair of the German dairy industry which was sent out far beyond the German border' (Claus, 1938: 330). In fact, from its beginning in 1925–1926, the quality control actions were combined with both the developments of trademarks and brand advertising. The Schleswig-Holstein ministry of agriculture had been the first. Under steady influence of their Danish neighbours, in 1925 the ministry published a butter trademark in combination with a new ministerial butter testing system. Soon other provinces of Germany followed suit. Pommern, Oldenburg, Mecklenburg in 1926, Eastern Prussia, Saxonia, Hannover, Hessen in 1927, interestingly all provinces that were among the first to pass a butter trademark had not been traditional butter regions.

In opposite to this, Bavaria was very averse to use a butter brand. The ministry of agriculture declared in 1927 that brands are an improper method to improve the product quality (Altrock, 1936: 331). In Switzerland too the introduction of the butter brand *Floralp* in the year 1928 did not met common consent. Several farmers associations refused the use of the brand as a needless method of advertisement (50 Jahre Schweizerische Milchwirtschaft, 1937: 353). In fact, it looks like as if all regions with a young dairy industry hoped that brands would help to industrialise an *authentic* experience of skilful dairy practices. Such regions, too, had no problems with the scaling-up of branding. For them, the production of national brands for milk, butter and cheese, assumed a definite form that minimised existing problems with differing qualities. Those regions, however, which judged their products as *really* high quality products according to the vision of longstanding dairy traditions, refused the nationalisation of trademarks. This might explain why only 6 of 16 federal dairy control stations accepted the national trademark of the *Deutsche Markenbutter*, when it was registered in 1931, the year of the Federal German Milk Law.

XII. Conclusion

The imagination to construct 'new' communities can be read as a result of the emergence of new technologies allowing the transcendence of space. In other words, space, landscape, agricultural territories etc. became uniform and abstract categories, widely unstressed with local specificity, especially those of the 'natural' environment. Pre-industrial societies, too, produced images and ideas about the landscape. As has been shown in this paper, the idea of *dairy zones* functioned in a way quite similar to the way farmers developed skills, techniques and products. Land use was, in fact, the adaptation of crop cultivation methods to the come across landscape. Milk production was both, a question of *natural* landscapes and one of *imagined* landscapes. It was based on decisions about land management, forms of collaboration and the division of labour on a regional basis. In this sense, what W.J.T Mitchell taught us about the characteristics of landscapes had the same status for pre-industrial agricultural societies, too. 'Landscape is a natural scene mediated by culture. It is both a represented and presented space, both a signifier and a signified, both a frame and what a frame contains, both a real place and its simulacrum, both a package and the commodity inside the package' (Mitchell, 1994: 5).

However, there existed some decisive distinctions towards later perceptions. Authors describing the conditions of dairy farming in the early nineteenth century perceived commercial dairying as the manifestation of limited agricultural options. For them, dairy farming had become a local tradition merely because it was the most profitable form of agricultural land use in regions with specific natural conditions. Only under the condition of near-by markets, farmers neglected conditions predetermined by nature. Thus, pre-industrial views of the landscape were not static but flexible. In fact, dairy farming was a situated knowledge and activity.

Even though the material basis of regional agriculture was eroded by modern markets, dynamics in technological change cannot be explained as exogenous factors that initiated change. In opposite to most historians of agriculture, I argue that the modernisation of dairy technology and the emergence of the dairy industry since the 1870s can be read as a result of the re-evaluation of landscapes, animals and plants as it took place since the end of the 18th century. Spatial perceptions of landscape and agriculture disappeared, not only as a result of urbanisation processes that broadened markets and forced people to find technological solutions for transportation problems. Far beyond this, processes of agricultural modernisation, initiated since the end of the 18th century with very different goals, caused a radical re-evalutaion of agricultural territories and local rural perceptions. Across Europe and without any regard for soil conditions or vegetation zones, agricultural regions that had been used for grain production were being converted to intensive animal husbandry and dairy farming. Long before cream separators, milking machines and pasteurisation technologies engineered dairying, the separating of arable farming and animal husbandry and the thinking in terms of different branches of production became influential. The centuries old relationship between land, fodder, cattle, and dairy production was dissolved and new forms of division of labour could be established. Against this background, the institutionalisation of a new system of milk processing and trade was almost inescapable.

Fundamental land reforms, including new perceptions of landscape and animals, usually do not count as technological developments. However, rural experiences and productions methods most often are linked. From this perspective, landscape is both, cause and result of agricultural development, while technology is much more a mediator between cultural values, social groups, and institutions on the one hand, and the natural environment on the other. Rather than a simple productive force, technology itself can be the outcome of environmental conditions. It is not always technology that changes the environmental attitudes and practices of land use. More often the attitudes and practices of a society change technology.

To neglect the spatial dimensions in agricultural production methods and technology draws the curtain over some phenomena that are from some interest for today's debates on agriculture and technology. If today a still increasing number of people in an almost completely urbanised Europe reject the modern technological and corporatist model of agri-food production they mostly refer back to a broader territorial vision of rural development. As David Goodman and Michael Redclift point out in their chapter in this volume, the *Quality Turn* broadly focuses on the local. 'That is, a transition from the 'industrial world', with its heavily standardised quality conventions and logic of mass commodity production, to the 'domestic' world, where quality conventions embedded in face-to-face interactions, trust, tradition and place support more differentiated, localised and 'ecological' products and forms of economic organisation'.

This revival of the local appears in another light, if one takes into consideration that the local and the region did never completely disappear from the scene. As has been shown in this paper, for different reasons its boundaries have been actively reconstructed by symbolic means. In particular, the issue of a proper naming and of publicity had become crucial for the economic survival of dairy production under the conditions of growing markets and international competition. The assumption of the priority of originality, authenticity and continuity of local traditions became part of producing dairy products as standardised commodities. Even national policies encouraged farmers and manufacturer to publicise 'local products' in the name of the nation. Thus, the transition from tradition to modernity was not a change produced by economics and technology. It was and is, still today, an inherent part of the material practices. In other words: as local rural experiences dissolved, tradition was invented.

Bibliography

50 Jahre Schweizerische Milchwirtschaft, 1887–1937. Festschrift unter gefälliger Mitwirkung einer Anzahl von Fachleuten (1937) Schaffhausen.

Abel, W. (1978) *Geschichte der deutschen Landwirtschaft vom frühen Mittelalter bis zum 19. Jahrhundert*, 3rd ed., Stuttgart.

Abel, W. (1986) *Massenarmut und Hungerkrisen im vorindustriellen Deutschland*, 3rd ed., Göttingen.

Achilles, W. (1993) *Deutsche Agrargeschichte im Zeitalter der Reformen und der Industrialisierung*, Stuttgart.

'Die Alpenversuchsstationen im landwirthschaftlichen Bezirke Westallgäu' (1867) *Landwirtschaftliche Blätter für Schwaben und Neuburg*, VI, pp. 129–133 and 137–140.

Altermatt, U., Bosshart-Pfluger, C. and Tanner, A. (eds) (1998) *Die Konstruktion einer Nation. Nation und Nationalisierung in der Schweiz, 18.–20. Jahrhundert,* Zurich.

Altrock, W. [et al.] (1936) *Milchwirtschaftliche Betriebslehre, part II*, Handbuch der Milchwirtschaft, III, Vienna.

Anderegg, F. (1894) *Allgemeine Geschichte der Milchwirtschaft*, Zürich.

Anderegg, F. (1898) *Illustriertes Lehrbuch für die gesamte schweizerische Alpwirtschaft*, Berne.

Anderson, B. (1991) *Imagined communities: Reflections on the origin and spread of nationalism*, London.

Arnold, Ph. (1911) 'Zur Frage der Milchversorgung der Städte', *Sonderabdruck aus den Jahrbüchern für Nationalökonomie und Statistik*, Jena, pp. 585–642.

Aufsberg, Th. (1913) *Bausteine zur Geschichte der Milchwirtschaft im Allgäu*, Weiler.

Bairoch, P. (1976) 'Die Landwirtschaft und die Industrielle Revolution 1700–1914', in C. Cipolla and K. Borchardt (eds), *Europäische Wirtschaftsgeschichte*, vol. 3, Stuttgart–New York, pp. 297–332.

Bakker, M.S.C. (1992) 'Boter', in H.W. Lintsen [et al.] (eds), *Geschiedenis van de techniek in Nederland. De wording van een moderne samenleving 1800–1890,* vol. 1: *Techniek en modernisering landbouw en voeding*, Zutphen, pp. 103–134.

Berg, G. (1912) *Die Milchversorgung der Stadt Karlsruhe unter besonderer Berücksichtigung der Produktions- und Preisverhältnisse*, Munich–Leipzig.

Beschreibung der Milchwirthschaftlichen Verhältnisse im Deutschen Reiche (1895) edited by the Deutschen Milchwirthschaftlichen Verein, Berlin.

'Bilder aus der milchwirtschaftlichen Vergangenheit, vom Geheimen Regierungsrat Dr. Theodor Henkelsen' (1929) *Süddeutsche Molkerei-Zeitung*, 4 June 1929, pp. 5–19.

Böhret, C. (1966) *Aktionen gegen die "kalte Sozialisierung" 1926–1930*, Berlin.

Brandt, K. (1928) *Der heutige Stand der Berliner Milchversorgung*, Berlin.

Claus, W. (1938) 'Entwicklung der Reichsprüfung für Milch und Milcherzeugnisse in Deutschland', in *Der Reichsminister für Ernährung und Landwirtschaft, Wissenschaftliche Berichte des XI. Milchrtschaftlichen Weltkongresses, 22. bis 28. August 1937 Berlin*, vol. II Berichte der Sektion II, Hildesheim.

Crawford, M.E. (1995) 'Food retailing, nutrition and health in Ireland, 1839–1989: One hundred and fifty years of eating', in A. den Hartog (ed.), *Food, Technology, Science and Marketing*, East-Linton, pp. 221–238.

Cronon, W. (1992) *Nature´s metropolis. Chicago and the Great West*, New York–London.

Diener, H. (1980) 'Förderung der deutschen Haustierzucht und der tierischen Produktion im 19. und 20. Jahrhundert durch staatliche Maßnahmen', *Bayerisches landwirtschaftliches Jahrbuch*, p. 78–120.

Dieterichs, I.F.C. (1856) *Ueber Milch- und Kuhwirthschaft im nördlichen Deutschland in Nähe grosser Städte*, Berlin.

Dold, D. (1937) *Untersuchungen über Grundlage, Aufbau und Entwicklung der deutschen Milchwirtschaft mit Vergleichsangaben aus der Weltmilchwirtschaft*, Munich.

Duerst, U.J. (1923) *Kulturhistorische Studien zur Schweizerischen Rindviehzucht*, Bern-Bümplitz.

DuPuis, M. (2002) *Nature's Perfect Food. How Milk became America's Drink*, New York–London.

Enock, A.G. (1943) *This Milk Business. A Study from 1895 to 1943*, London.

Flad, M. (1989) *Milch, Butter und Käse. Ein Beitrag zur Geschichte der Milchwirtschaft in Württemberg*, Stuttgart.

Frandsen, S. (1994) *Dänemark – der kleine Nachbar im Norden: Aspekte der deutsch-dänischen Beziehungen im 19. und 20. Jahrhundert*, Darmstadt.

Gutzwiller, K. (1923) *Die Milchverarbeitung in der Schweiz und der Handel mit Milcherzeugnissen*. Schaffhausen.

Fenton, A. (1995) 'Milk and milk products in Scotland: The role of the Milk Marketing Boards', in A. den Hartog (ed.), *Food, technology, science and marketing*, East-Linton, pp. 89–102.

Gellner E. (1991) *Nationalismus und Moderne*, Berlin.

Grasseni, C. (2001) *Developing vision, developing skill: Locality and identity in rural Northern Italy,* Ph.D. thesis, Manchester.

Gugerli, D. and Speich, D. (1999) 'Der Hirtenknabe, der General und die Karte. Nationale Repräsentationsräume in der Schweiz des 19. Jahrhunderts', *WerkstattGeschichte,* 23, pp. 53–73.

Guggenbühl, C. (1998) 'Biedermänner und Musterbürger im "Mutterland der Weltfreyheit". Konzepte der Nation in der helvetischen Republik', in U. Altermatt, C. Bosshart-Pfluger and A. Tanner (eds), *Die Konstruktion einer Nation. Nation und Nationalisierung in der Schweiz, 18.–20. Jahrhundert,* Zurich, pp. 33–47.

Häfener, F. (1847) *Der Wiesenbau in seinen ganzem Umfange nebst Anleitung zum Nivellieren, zur Erbauung von Schleussen, Wehren, Brücken etc.*, Reutlingen.

Handwörterbuch der Staatswissenschaften (1910) Jena.

Hansen, J. (1921) *Lehrbuch der Rinderzucht. Des Rindes Körperbau, Schläge, Züchtung, Fütterung und Nutzung,* Berlin.

Hanssen, G. (1880) *Agrarhistorische Abhandlungen*, Leipzig.

Hecht, G. (1998) *The radiance of France, nuclear power and national identity after World War II*, Cambridge MA.

Henning, F.-W. (1994) *Deutsche Agrargeschichte des Mittelalters – 9. bis 15. Jahrhundert*, Stuttgart.

Hettling, M. (1998) 'Die Schweiz als Erlebnis', in U. Altermatt, C. Bosshart-Pfluger and A. Tanner (eds), *Die Konstruktion einer Nation. Nation und Nationalisierung in der Schweiz, 18.–20. Jahrhundert,* Zurich, pp. 19–32.

Hittcher, Dr. (1913) 'Neuere Erfahrungen und Fortschritte in der Milchwirtschaft Deutschlands', *Internationale Agrartechnische Rundschau*, p. 4.

Hobsbawm, E. (1992) *Nationen und Nationalismus. Mythos und Realität seit 1780*, Frankfurt am Main.

Hobsbawm, E. and Ranger, T. (eds) (1983) *The invention of tradition*, Cambridge–New York.

Hösli, J. (1948) *Glarner Land- und Alpwirtschaft in Vergangenheit und Gegenwart*, Glarus.

Im Hof, U. (1983) *Die Gesellschaft im Wandel. Mitglieder und Gäste der Helvetischen Gesellschaft*, Frauenfeld.

Jahn, W. (1955) 'Die allgemeinen physischen Faktoren der Landwirtschaft', in K. Lindner (ed.), *Geschichte der Allgäuer Milchwirtschaft. 100 Jahre Allgäuer Milch im Dienste der Ernährung*, Kempten-Allgäu, pp. 15–38.

Kaup, I. (1910) *Ernährung und Lebenskraft der ländlichen Bevölkerung. Tatsachen und Vorschläge*, Berlin. (Schriften der Zentralstelle für Volkswohlfahrt; 6).

Kleinpeter, F.X. (1879) *Allgemeine Betrachtungen über das Molkereiwesen in Deutschland im Vergleiche zu anderen Ländern*, Bremen.

Kohl, J.G. (1841) *Der Verkehr und die Ansiedlungen der Menschen in ihrer Abhängigkeit von der Erdoberfläche*, Dresden–Leipzig.

Kollreuther, I. (2001) *Milchgeschichten. Bedeutungen der Milch in der Schweiz zwischen 1870 und 1930* (Lic. Phil.), Basel.

Lewthwaite, G.R. (1980) 'New Zealand milk on the map', *Annals of the Association of American Geographers*, 70, pp. 475–491.

Lindner, K. (ed.) (1955), *Geschichte der Allgäuer Milchwirtschaft. 100 Jahre Allgäuer Milch im Dienste der Ernährung*, Kempten–Allgäu.

Lysaght, P. (ed.) (1994) *Milk and milk products from Medieval to Modern Times*, Edinburgh.

McMurry, S. (1992) 'Women's work in agriculture: Divergent trends in England and America, 1800–1930', *Comparative Studies in Society and History*, 34, 2, pp. 248–270.

McMurry, S. (1995) *Transforming rural life. Dairying families and agricultural change, 1820–1885*, Baltimore.

Marchal, G.P. and Mattioli, A. (eds) (1992) *Erfundene Schweiz. Konstruktionen nationaler Identität*, Zürich.

Martiny, B. (1891) *Die Versorgung Berlins mit Vorzugs-Milch. An Hand der Geschichte erzählt*, Bremen.

Martiny, B. (1895) *Kirne und Girbe: Ein Beitrag zur Kulturgeschichte, besonders zur Geschichte der Milchwirthschaft*, Berlin.

Martiny, B. (1904) *Vor hundert Jahren. Darstellung der Milchwirtschaft Gross-Britanniens um das Jahr 1800. Ein Vorbild für die gegenwärtige Entwicklung der deutschen Milchwirtschaft*, Leipzig.

Mendelson, F. (1932) 'Bedeutung, Lage und Aussichten der deutschen Milchwirtschaft', in M. Sering (ed.), *Die deutsche Landwirtschaft unter volks- und weltwirtschaftlichen Gesichtspunkten*, Leipzig, pp. 500–524. (Berichte über Landwirtschaft, N.F., Sonderheft Nr. 50).

Milch-Industrie (1885) *(Organe für das Molkereiwesen in Beziehung auf Technik, Wissenschaft und Handel)*, Berne.

Mitchell, W.J.T. (1994) 'Imperial landscape', in W.J.T. Mitchell (ed.), *Landscape and power*, Chicago, pp. 5–34.

Das Molkereiwesen in Dänemark, Schweden und Schleswig-Holstein (1876) Reiseberichte von B. Vissering, G. Küster und H. v.d. Hessen, nebst Berichten über die Molkerei-Ausstellung zu Frankfurt a.M., Celle.

Das Molkereiwesen und die landwirthschaftliche Thierzucht in Dänemark, Schweden und Norddeutschland. Bericht über eine im Jahre 1881 ausgeführte Studienreise (1882); erstattet an das kgl. Staatsministerium des Innern, Abtheilung für Landwirthschaft, Gewerbe und Handel vom Molkereiconsulenten und Wanderlehrer für landw. Thierzucht, Prof. Feser in München, Munich.

Niemann, A. (1823) *Die holsteinische Milchwirthschaft*, 2nd ed., Altona.

Orland, B. (2003a) 'Turbo-cows: Producing a competitive animal in 19th and Early 20th century Switzerland', in S. Schrepfer and Ph. Scranton (eds), *Industrializing organisms: Introducing evolutionary history*, New York–London, pp. 167–198.

Orland, B. (2003b) 'Cow's milk and human disease. Bovine tuberculosis and the difficulties involved in combating animal diseases', in *Food and History*, 2003, 1, pp. 179–202.

Orland, B. (2004) 'Alpine milk. Diary as a strategy of land use in premodern times', *Environment and History*, pp. 327–364.

Ottenjann, H. and Ziessow, K.-H. (1996) (eds), *Die Milch* (Arbeit und Leben auf dem Land, 4), Cloppenburg.

Pedersen, E.H. (1973) *The Danish agricultural industry 1910–39*, Copenhagen.

Peters, J., Harnisch, H. and Enders, L. (1989) *Märkische Bauerntagebücher des 18. und 19.*

Jahrhunderts, Selbstzeugnisse von Milchbauern aus Neuholland, Weimar.

Pfiffner, A. (1993) *Henri Nestlé (1814–1890). Vom Frankfurter Apothekergehilfen zum Schweizer Pionierunternehmer*, Zürich.

Pfister, C. (1995) *Im Strom der Modernisierung. Bevölkerung, Wirtschaft und Umwelt im Kanton Bern 1700–1914*, Berne–Stuttgart–Vienna.

Pierenkemper, T. (ed.) (1989) *Landwirtschaft und industrielle Entwicklung. Zur ökonomischen Bedeutung von Bauernbefreiung, Agrarreform und Agrarrevolution*, Stuttgart.

Pinard, J. (1995) 'The development of cheese consumption in France in the past 150 years', in A. den Hartog (ed.), *Food, technology, science and marketing*, East Linton, pp. 117–126.

Pirtle, T.R. (1973) *History of the dairy industry* (reprint of 1926 edition), Wilmington.

Politiek, R.D. and Bakker, J.J. (eds) (1982) *Livestock production in Europe. Perspectives and prospects*, New York.

Pressler (1886, orig. 1854) *Neue Viehmesskunst*, 3rd. ed., Leipzig.

Reif, H. (1990) *Konkurrenz um den Boden – Die Landwirtschaft zwischen Verdrängung und Anpassung*, in Köllmann [et al.] (eds), *Das Ruhrgebiet im Industriezeitalter. Geschichte und Entwicklung*, Düsseldorf, I, pp. 337–345.

Reif, H. and Pomp, R. (1996) 'Milchproduktion und Milchvermarktung im Ruhrgebiet 1870–1930', *Jahrbuch für Wirtschaftsgeschichte*, 1, pp. 77–108.

Reinsch, A. (1903) *Die gesetzliche Regelung des Milchverkehrs in Deutschland, insbesondere in den größeren deutschen Städten*, Hamburg.

Riedel, W. (1936) 'Milchwirtschaftliche Unterrichts-, Versuchs- und Forschungsanstalten', in W. Winkler (ed.), *Handbuch der Milchwirtschaft*, vol. 3, *Milchwirtschaftliche Betriebslehre, Zweiter Teil: Organisationen der Milchwirtschaft*, Vienna, pp. 217–222.

Riemann, F.-K. (1953) *Ackerbau und Viehhaltung im vorindustriellen Deutschland*, Kitzingen-Main.

Ring, E. (1904–1905) 'Die Versorgung der Grossstädte mit Milch und der Kampf um den Milchpreis. Vortrag gehalten in der Ökonomischen Gesellschaft im Königreiche Sachsen, Dresden am 11. November 1904', *Mitteilungen der Ökonomischen Gesellschaft im Königreiche Sachsen*, Dresden, pp. 1–34.

Ritzmann-Blickenstorfer H. (ed.) (1996) *Historische Statistik der Schweiz*, Zürich.

Rüger, D. (1851) *Die neue chemisch-praktische Milch-, Butter- und Viehwirtschaft*, I, Löbau.

Schatzmann, R. (1870) *Die Weide- und Milchwirthschaften von Schweden, Dänemark, Holstein und Holland. Ein Reisebericht*, Aarau.

Schatzmann, R. (1872) Die Milchfrage vor der gemeinnützigen Gesellschaft des Kantons Bern, Aarau, quoted in I. Kollreuther (2001) *Milchgeschichten. Bedeutungen der Milch in der Schweiz zwischen 1870 und 1930* (Lic. Phil.), Basel).

Schild, J. (1865) 'Bericht über die Aufgabe des schweizerischen alpwirthschaftlichen Vereins und dessen bisherige Arbeiten', in R. Schatzmann (ed.), *Schweizerische Alpenwirthschaft*, No. 6, Aarau, pp. 3–17.

Schneider, I. (1916) *Die schweizerische Milchwirtschaft mit besonderer Berücksichtigung der Emmentaler-Käserei*, Zürich–Leipzig.

Scholz, B. (1949) *Mein Leben für den Milchhandel*, Hamburg.

Schuler, A. (1942) *Wilhelm Fleischmann. Der Begründer der Milchwirtschaftswissenschaft. Seine Lebenserinnerungen und sein Lebenswerk*, Hildesheim.

Schwarz, R. [et al.] (1908) *Bericht über die Studienreise nach Norddeutschland und Dänemark. Im Auftrage der Deutschen Sektion des Landeskulturrates für die Markgrafschaft Mähren*, Brünn.

Schweizerische Landesausstellung 1939 (1939), Zurich.

Schwerz, J.N. (1816) *Beobachtungen über den Ackerbau der Pfälzer*, Berlin.

Senti, A. (1982) 'Die Milchwirtschaft seit dem 18. Jahrhundert', in *75 Jahre. Die schweizerische Milchwirtschaft zu Beginn der achtziger Jahre*, Berne, pp. 14–42.

Sering, M. (ed.) (1932) *Die deutsche Landwirtschaft unter volks- und weltwirtschaftlichen Gesichtspunkten*, Leipzig. (Berichte über Landwirtschaft, N.F., Sonderheft Nr. 50).

Shortall, S. (2000) 'In and out of the milking parlour: A cross-national comparison of gender, the dairy industry and the state', *Women's Studies International Forum*, 23, pp. 247–257.

Siegenthaler, H. (1992) 'Hirtenfolklore in der Industriegesellschaft. Nationale Identität als Gegenstand von Mentalitäts- und Sozialgeschichte', in G. Marchal and A. Mattioli (eds), *Erfundene Schweiz. Konstruktionen nationaler Identität*, Zürich, pp. 23–36.

Smidt, H.A.R. (1996) 'Dutch and Danish agricultural exports during the First World War', *Scandinavian Economic History Review*, 44, 2, pp. 140–160.

Smith, A. (1978) *Der Wohlstand der Nationen. Eine Untersuchung seiner Natur und seiner Ursachen* (nach der fünften Auflage von 1789 übersetzt und mit einer umfassenden Würdigung des Gesamtwerkes herausgegeben von Horst Claus Recktenwald), Munich.

Spiekermann, U. (1993) 'Milchkleinhandel im Wandel. Eine Fallstudie zu München 1840–1913', *Scripta Mercaturae. Zeitschrift für Wirtschafts- und Sozialgeschichte*, 27, 1/2, pp. 91–145.

Spiekermann, U. (1994) 'The retail milk trade in transition: a case-study of Munich, 1840–1913', in P. Lysaght, *Milk and milk products from Medieval to Modern Times*, Edinburgh, pp. 71–93.

Spiekermann, U. (2001) '"Fleisch giebt Fleisch!". Zur Geschichte der Tiermehlverfütterung in Deutschland vor dem Zweiten Weltkrieg', in *Zeitschrift für Ernährungsökologie,* 2, 2001, pp. 7–9.

Stamm, D. (1853) *Die Landwirthschafts-Kunst in allen Theilen des Feldbaues und der Viehzucht. Nach den bewährten Lehren der Wissenschaft, der Erfahrung und den neuen Entdeckungen in der Natur, gründlich, faßlich und ermuthigend erläutert*, Prague.

Steinmüller, J. (1802/1804) *Beschreibung der schweizerischen Alpen- und Landwirthschaft*, 2 vols, Winterthur.

Stocker, Th. (1982) *Geschichte des Zentralverbandes schweizerischer Milchproduzenten 1907–1982*, Burgdorf.

Taylor, D. (1987) 'Growth and structural change in the English dairy industry (1860–1930)', *Agricultural History Review,* 35, 1, pp. 47–64.

Teuteberg, H.-J. (1986) 'Anfänge des modernen Milchzeitalters in Deutschland', in H.-J. Teuteberg and G. Wiegelmann (eds), *Unsere tägliche Kost* (Studien zur Geschichte des Alltags, no. 6), 2nd. ed., Münster, pp. 163–184.

Trendtel, F. [et al.] (1931) *Die Milchversorgung der Städte und grösseren Konsumorte* (Handbuch der Milchwirtschaft, ed. by Willibald Winkler, vol. 2, part 1), Vienna.

Vatin, F. (1990) *L'Industrie du Lait. Essai d'histoire économique*, Paris.

Voigt,F. (1903) *Geschichtliches über die Versorgung Hamburgs mit Milch*, Hamburg.

von Oppenau, Fr. (1886) *Die Hebung der kleinbäuerlichen Milchwirthschaft in Elsass-Lothringen. Im Auftrage des landwirthschaftlichen Bezirksvereins Unter-Elsass*, Strassburg.

Vries, J. de (1974) *The Dutch rural economy in the Golden Age, 1500–1700*, New Haven–London.

Wahlen, S. (1979) *Rudolf Schatzmann 1822–1886. Ein Bahnbrecher der schweizerischen Land-, Alp- und Milchwirtschaft und ihres Bildungswesens*, Münsingen.

Walter, F. (1980) 'Cadastre et histoire rurale. Contribution à l'étude de la petite propriété familiale en Pays de Fribourg (milieu du XIXe siècle). Questions et méthode', *Schweizerische Zeitschrift für Geschichte*, 30, 1, pp. 29–58.

Weishaupt, M. (1991) *Bauern, Hirten und „frume edle puren". Bauern und Bauernstaatsideologie in der spätmittelalterlichen Eidgenossenschaft und der nationalen Geschichtsschreibung der Schweiz*, Zürich.

Weishaupt, M. (1998) '"Viehveredelung" und "Rassenzucht". Die Anfänge der appenzellischen Viehschauen im 19. Jahrhundert', in M. Fuchs (ed.), *Appenzeller Viehschauen*, St. Gallen, pp. 11–48.

Westphal, W. (1931) 'Statistik des Milchverbrauches', in W. Winkler (ed.), *Handbuch der Milchwirtschaft, Bd. 2, Erster Teil: Die Milchversorgung der Städte und grösseren Konsumorte*, Wien, pp.112–145.

Whittaker, D. and Goody, J. (2001) 'Rural manufacturing in the Rouergue from Antiquity to the Present: The examples of pottery and cheese', *Comparative Studies in Society and History,* 43, pp. 225–245.

Widmer, A. (1937) 'Aus der Vorgeschichte', *50 Jahre Schweizerische Milchwirtschaft*, Schaffhausen.

Zitzen, E.G. (1915) 'Die Milchversorgung der Städte', *Milchwirtschaftliches Zentralblatt*, p. 44.

Zollikofer, E. (1939/1940) 'Die Schweizerische Milchwirtschaft', *Schweizerische Landesausstellung, 1939, Die Schweiz im Spiegel der Landesausstellung*, Zürich, pp. 562–571.

'Zum Geleit' (1929) *Süddeutsche Molkerei-Zeitung*, Kempten-Allgäu, 4 June 1929, p. 1.

12 Fast food and slow food. The fastening food chain and recurrent countertrends in Europe and the Netherlands (1890–1990)

Anneke H. van Otterloo, University of Amsterdam

I. Introduction

At the start of 2003 problems affecting the McDonald's *fast food* chain were hot news. For the first time in its existence the hamburger giant struggled with profit losses at the stock exchange, in which it had participated since 37 years. The company had to close down 517 loss-making outlets all over the West. These outlets should be added to the number of 200 outlets already closed by the end of 2002. The chain of fast food restaurants increased only in China, where it remained very profitable. The expectation for the future in the West was not favourable at all, and the closing of many other establishments seemed to be imminent.[1] This was fully contrary to the profitable expectations of market researchers who analysed the willingness of seven European markets (Britain, France, Germany, Italy, Spain, the Netherlands, and Belgium) for four American fast food chains (McDonald's, Burger King, Kentucky Fried Chicken, and Pizza Hut) in the 1990's (*Fast Food in Europe,* 1990). The deterioration of the McDonald's market position followed some years of severe critical notes on the methods of production and serving, and on the quality of the products. This may or may not be a coincidence (Schlosser, 2001; Kincheloe, 2002).

The Golden Arches seem to have reached a point of diminishing returns. This decline occurs after a period of fifty years of unprecedented and uninterrupted growth in the USA and the Western hemisphere (Ritzer, 2002; Schlosser, 2001; Alfino, 1998; Love, 1986; Kroc and Anderson, 1977; Boas and Chain, 1976; Baltesen, 1996). Successes continue in the East. For Western countries, though, it may be asked if a sudden change of consumers preferences is at stake (Watson, 1997).[2] If food choices were made the way some McDonald's critics would like them, the appetite for *Fast Food* even might turn into its opposite: a taste for *slow food*. Indeed, the Slow Food movement expresses culinary ideas and practices, opposing the handling and eating of hamburgers in McDonald's restaurants and other *junk food* (Miele and Murdoch, 2002). Slow Food advocates Taste, and tries to disseminate knowledge and preferences for 'traditionally' and regionally produced foods, while mass-produced fast food like Big Macs and French fries are standardised McDonald's items, having nothing to do with local or even national, French or Italian, culinary traditions.

[1] This was reported by the Dutch papers *NRC/Handelsblad*, 24 January 2003, and *Volkskrant*, 1 February 2003.

[2] The question of changing consumers' preferences is also tackled by Richardson and Agular, 'Consumer change in fast food preference', http: //www.ifama.org/conferences/2004Conference/Papers/Richardson1004.pdf.

The aim of this chapter is, first, to explore the culinary fields of Fast Food and Slow Food at a national, a European and an international level, and next, to put these opposite trends into a long-term historical perspective.[3] This is done by linking both trends to the 20th-century development of the food chain and its resulting assortment in modern-industrial (Western-European) societies, and in particular the Netherlands. In their recent study on Slow food \ Fast Food, Mara Miele and Jonathan Murdoch have conceived of these phenomena as two contrastive contemporary food cultures, condensed within 'culinary networks', having a recursive relationship with each other (Miele and Murdoch, 2003: 26). *Fast Food* stands for the American hamburger culture, indicating 'food as fuel', while *Slow Food* means 'cultural heritage', 'emblematic of local, regional and national ways of life'. They locate both cultures in the 'context of new food consumption trends' in post-industrial societies. David Goodman and Michael Redclift, this volume, also deal with changes in the modern-industrial food system at an international (European) level, and ask whether these changes must be seen as re-articulation or resistance. They situate the trends I want to discuss into the broad scale and scope of the modernisation and (future) development of the (Western) European agricultural system as a whole. This means cheap mass food production and, along the thinking of Ulrich Beck, the resulting risks for health and environment (Beck, 1992). The authors see the forms of resistance, of which Slow Food might be but one example, as *Alternative Agro Food Networks* (AAFN's), which have taken various shapes since the 1980s. AAFN's have risen in the wake of environmental and health food movements in Western Europe, criticising intensive agricultural production and its risks. Guthman (2003), finally, contrasts Fast Food with Organic Food, and puts the dichotomies between fast and slow, reflexive and compulsive, or even good and bad eaters, under critical scrutiny.

From the point of view of the history and sociology of technology, both food trends may be considered to be part of the development towards a large-scale modern-industrial food chain in the 20th century (Schot, 2000, 2001; Van Otterloo, 2000a). The food chain or food system appears to expand from a small-scale regional and national to a large-scale European and global level. Meal patterns and culinary cultures are shaped by consumers at all levels, choosing from the assortment of food products, supplied by the food chain. Consumers do so, according to the rules of their eating habits, comprising criteria of preference, time, money and identity shaped by class, gender, ethnicity, status, nation, religion or ideology and perhaps knowledge (Mennell et al., 1992). As the industrialisation of food during the twentieth century implied ever more technological processes and processing, I may state that the resulting choices and meal-patterns originate in mutual interaction between technology and society over (a long-term period of) time. In this chapter the following interrelated questions are dealt with: What do the culinary trends of Fast Food and Slow Food represent? How did the modern-industrial food chain develop and expand in the 20th century with both Fast Food and Slow Food as its corollaries? If Fast Food had its precursors, what then were Slow Food's predecessors?

[3] See for the use of Pierre Bourdieu's (1984) term 'cultural field' to the culinary sphere, Ferguson, 1998.

II. Fast food and slow food: the field

The contemporary labels *Fast Food* and *Slow Food* have various connotations. Until further notice we conceive these twin concepts as loose indicators of two opposite food trends, or culinary cultures: one dominant or main stream, connected to large-scale networks; the other, its counterpart, alternative, marginal and connected to small-scale networks. The opposites have various connotations to be discussed below. In a moral sense they symbolically even stand for *bad food* and *good food*, at least in the eyes of different groups of actors, in their role as producers and consumers. The actors and actor groups take positions at the various locations in the agro-food system or, the *food chain* (Tansey and Worsley, 1995; Van Otterloo, 2000a: 238–240). Both trends deserve more explanation. Although Miele and Murdoch (2003) have done pioneering work on the clarification of these 'binary' contrasts, to my opinion their analysis does not go far enough and has to be extended in time and scope. The authors situate both opposite consumption trends in 'post-industrial' abundant western societies, and describe them 'as cultures, condensed within particular culinary networks' (pp. 26–27). They limit themselves to the cases of McDonald's chains and the Slow Food movement. In my view, Fast Food and Slow Food may represent much broader phenomena than the two cases they deal with. The trends discussed are less clear in meaning and contrast, less limited to the consumption context of eating out and, finally, less time-bound and restricted to the period since the 1980s.[4] In this section I will illustrate these mutually related points.

To get a fuller impression of the opposite phenomena discussed, we need to know the meaning of other opposites, resonating in the daily-life use of the terms Fast Food and Slow Food. Except for 'mainstream' food versus 'alternative' food, other meanings of related opposite concepts play a part. Such are globalisation versus localisation or regionalisation, the USA versus Europe, large scale industrial production versus small scale production, resulting into large quantities of cheap, standardised and tasteless mass products of poor quality, versus products of high quality, tradition and variety, and finally sustainable versus wasting agricultural methods. The list of opposites is not exhausting. What stands out in these connotations, is the proneness of food to associate itself with value-bound social criteria of identity and social status. Consumers cohere and distinguish themselves with and from each other by food preferences and taste, experiencing their own group as 'higher' classified and the others as 'lower' (Bourdieu, 1984). Group status and feelings of group superiority and inferiority thus are linked to many of the opposites just mentioned. Producers and consumers, having different positions in the food chain, are involved in a power struggle of supply and demand, but are also competing among themselves for the leading positions in the market and in the status hierarchy of food consumption. When studying Fast Food versus Slow Food as opposites, it is important to keep the different connotations and the social forms they may take in mind. This social character of food production and consumption complicates the description of the field, for which reason it is not easy to make clear-cut distinctions between trend and countertrend. Julie Guthman, in dealing with the rise and long-term development of organic food versus main stream fast food in California, points out that

---

[4] Julie Guthman (2003) describes the history of organic food in California and puts the 'binary opposites' Fast Food / Organic Food into perspective.

the *yuppy* social context of this culinary trend has been very important in its diffusion (Guthman, 2003: 51–53). In the same vein, this is probably true for the various critical noises, taking on McDonald's food I discussed above. Cultural and economic capital are unevenly distributed among classes and selectivity in eating habits has often an elite character. What is more, the same consumer may perhaps switch from a McDonald's party with the children to a high status vegetarian dinner at home with friends.

The trends indicated as Fast Food and Slow Food need in my opinion not be limited to the sphere of eating out, as Miele and Murdoch maintain (2003: 7). Both types of food may also be eaten in the household, although the McDonald's chain and perhaps also the Slow Food movement began as outside locations of preparation and consumption of certain types of food. McDonald's and similar hamburger restaurants and their products have got their counterpart at home. Fast Food covers also the various types of industrially produced *convenience foods*, distributed by supermarkets and other outlets to eat at home (or outside, elsewhere than in hamburger restaurants), with as little effort as possible (Nestle, 2002: 19–20). Convenient foods comprise snacks of all varieties, readymade dishes from canteens, take away meals from Chinese-Indonesian or Italian restaurants, and complete frozen or fresh and cooled meals from the supermarket (Thoms, 2003; Oddy, 2003; Gabaccia, 1998: 149–175; Knop and Schnitz, 1983). Convenience food is highly industrially processed fast food, which means to consumers that they may prepare, if necessary at all, and eat it, out or at home, *quick* and *easy*. Cookery time spent in the kitchen, nowadays has been much reduced or even has become absent. More specifically relating to snacks and snack food it also means inexpensive, *cheap* food, eaten in *informal* ways and places, as streets, but also at home (Albert de la Bruhèze and Van Otterloo, 2003: 318).

Since the 1990s Fast food, embodied in the American hamburger, even stands as a model for a general social process, the so-called *McDonaldization*. George Ritzer has indicated four dimensions characterising the organisation and products of McDonald's restaurants all over the world: efficiency, predictability, calculability, and control (Ritzer, 1993: 9–12). This rational organisation method has contributed to the, in his eyes deplorable, success of the fast food chain. McDonaldization, he maintains, increasingly pervades (post)modern industrial society to its farthest corners. The debate resulted in an international flood of books, which identifies ever more incarnations of McDonaldization in contemporary society (Ariès, 1997; Ritzer, 1998, 1999, 2000; Dagevos, 2002: 48–50). This discussion in social science and history went far beyond the field of food, considering McDonaldization as an icon of the American way of life and its excessive (ir)rationalisation. Ulrich Beck's earlier characterisation of modern-industrial society as a *risk society* is more often used as a theory analysing problems related to the food chain.[5] This concept has evoked at least as much as scholarly discussion as Ritzer's did (Adam et al., 2000). The critical debates and studies about 2000 may also be inspired by Beck's 'reflexive modernization'. As far as in the literature mentioned, it comes near to the critical noises in the 1990s, focused on aspects of large-scale modern-industrial food production and its excrescence. The problems that came to light caused a series of food scares. Cases in point are BSE (mad cow disease) and other animal diseases, toxic accidents in production lines, genetic modification of plants and animals, and the absence (or presence) of regulating European food laws.

[5] Beck (1986) and Goodman and Redclift (this volume), for instance, use his vocabulary.

Some critical noises are clustered under the banner of Slow Food, a movement established in Bra, Italy in 1986, by 'a group of food writers and chefs' (Miele and Murdoch, 2002: 250). This group has expanded to a network at a world-wide scale ever since, and presents itself on the Internet from all the four corners of the world, advocating *taste*. Such critical networks may represent various *new social food movements* or, as they more specifically are labelled AAFN's, *alternative agro-food networks* (Goodman and Redclift, this volume). Social movements rally around a commonly defined problem, a wrong thing in society, and try to change this abuse from an ideological point of view of how the situation should be. New food movements, e.g. AAFN's, are striving at obtaining various types of *food quality*. A famous contemporary example of producers' actions against the main stream food trend of fast food and industrial mass production is that of José Bové and François Dufour. These French *farmers* (sheep-breeders, possessing however a lot of cultural capital) reached world paper headlines in 1999 by demolishing an establishment of McDonald's in construction in Milau. They did this as a protest against American trade-barriers against Roquefort cheese which, in their turn, were measures against the European refusal to the import of hormone-beef and they were joined by urban consumers (Bové and Dufour, 2000: 11–14; Miele and Murdoch, 2003: 25). Another example consists of a recent Austrian network, labelled *ARGE Fast Food Slow Food*. Its information on the Internet says that this is a co-operation between various expert institutions, doing governmentally commissioned research on food and landscape in Austria. The Dutch *Innovatienetwerk Groene Ruimte en Agrocluster* ('Innovation-network Green Space and Agro-cluster') is characterised by a similar mixed composition of experts in various positions, even belonging to the authorities of the national government.[6]

When considering the literature on Fast food and Slow Food, it is striking that most authors restrict their reports to the final two decades of the 20th century. Julie Guthman (2003) is an exception to this. She points out that 'organic food' and its culture in California dates back to at least the 1970s, and some elements of the culinary Bay Area climate even go back to the '20s and '30s. 'Salad mix' was an item within this climate that got new culinary meaning with new generations (pp. 46–47). Although the American food history, especially on California, differs from that in Europe, her approach is markedly near, though in an entirely dissimilar way, to what I suggested earlier. Both trends, Fast Food and Slow Food, might turn out to be two sides of the same coin, and no sudden shift in preferences is the matter. The rise of these trends goes together and forms part and parcel of the development of the modern-industrial food chain already from the 1890s. The case of the Netherlands is described below as an example and for a provisional lack of other cases. It might be illustrative for varieties of (similar or different) developments in European countries and in Europe as a whole.

[6] See for *ARGE Fast Food Slow Food*, http://www.klf.at/html/prj/klf2_03.html, and for *Innovatienetwerk Groene Ruimte en Agrocluster*, http://www.agro.nl/innovatienetwerk/.

III. Fast Food and the development of the food chain (1890–1990): the case of the Netherlands in a European context

The recurring rise of the multiple trends of Fast Food and its ditto countertrend of Slow Food and its predecessors both are related to the growth of the food chain and the resulting assortment of food products in the 20th century. The concept of the food system or food chain, comprising the phases from the field to the plate (production to consumption), is widely used in contemporary literature (Tansey and Worsley, 1995). To my knowledge, the actual (technological) history of the food chain as a whole, in its social and cultural context, has been described for the Netherlands only.[7] Specific studies for parts of the chain or for certain products have taken place, at the European and at the global level (e.g., Mintz, 1985). In the Dutch study mentioned, the food chain was specified as 'a large system of mutually interconnected phases, links and locations', where *phases* stand for *production, distribution, preparation, consumption and processing of waste* of food. *Links* refer to the locations, where *transformation* of (raw) materials into food products takes place and other adding of value. The locations are also the places where individuals, social groups and organisations are active in handling food. It appeared, that unstable mutual power relationships existed over time between the actors and actor groups, located at the several phases and links in the chain. That is, the actors and actor groups involved maintained mutually variable relations of power and dependency with each other. The food chain, and its *assortment* from which consumers may choose is to be considered as a system of phases, links and locations, organised by people, through which the stream of foods circulates (Figure 12.1).

This scheme of the food chain is an abstract model, intended to make sure that no relevant aspect of the food supply and demand is left out. It is not a picture of a real contemporary or historical situation. Studying the history of the chain as a whole demands the attention simultaneously to as much aspects as possible, at several levels of society over a certain period. The assumption is that modes of (primary and secondary) production, the resulting products, and modes of consumption, are closely bound up with other phases in the food chain and its development. All of the phases must be noticed to get a full description and explanation of what happened in the food history of a certain region, country or continent. The different links (locations), where people are (or were) active in transforming food in society during a certain period, have been the concrete places for research in the case of the Netherlands.

It is clear that this method requires divergent types of literature and (source) materials. In this way the principle of the mutual shaping of technology and society in the field of food was given shape. The study of the Dutch food chain is obviously too limited to derive precise statements on the development of the food chain in Europe as a whole, or on the history of European-American co-evolutions of food production and consumption, let alone the circulation of foods (artefacts), food knowledge and technology, and the

[7] Van Otterloo, 2000a: 238–240; Schot, 2000; 2001. This research was done in the framework of the interdisciplinary project, dealing with the history of technology and its social context in the twentieth century in the Netherlands and described in seven volumes.

Figure 12.1 The food chain

Phases	Links: Locations
1 Primary production, *arable farming and stockbreeding, horticulture, fishing*	1 Farm, *greenhouse, trawler*
2 Secondary production, *food processing*	2 Factory, company, *foods and ingredients*
3 Distribution, *transportation, wholesale trade and retail trade, storage*	3 Market place, *store, means of transport, entrepôt*
4 Preparation, *buying, cooking, serving*	4 Kitchen, *catering, private households, institutions*
5 Consumption, *eating in different social settings*	5 Table, *food and meal patterns, eating out (street, restaurants) and at home*
6 Waste disposal	6 Garbage can, rubbish

circulation of people involved.[8] Finally, the growth of the food chain and the circulation of foods at a global level have also been studied only fragmentary until now. Therefore I limit myself to the Dutch case, when relating the trends of Fast Food and Slow Food to the 20th century history of the food chain. Some main lines in the development may also hold for other European nation-states, although differences in degree, time and scope are to be expected. From the Dutch study it appeared, that the food chain and its assortment went through some radical changes between 1890 and 1990, the time span studied. These changes in the making up of the food chain are summarised as processes of *lengthening, differentiation and condensing*, which happened in three periods, 1890–1920, 1920–1960, and 1960–1990, to be described very briefly below (Van Otterloo, 2000b: 249–309). These periods, I suppose, may also be indicative for some recurrent stages in the rise of Fast Food and Slow Food as trends and countertrends. The contemporary Slow Food movement took only one shape of contemporary countertrends. Slow Food stands here also for the different countertrends to the increasingly industrialised fast food assortment, produced by the *lengthened and differentiated* chain in the past.

[8] 'The circulation of foods (artefacts), food knowledge and technology and people' was an indicator for the research done in the project *Tensions of Europe* (ToE) and the Barcelona Conference on Agriculture and Food to which this chapter originally contributed.

III.1. The food chain in the Netherlands

The decades between 1890 and 1920 were characterised by a powerful industrialisation of food and the growth of many new links/locations in the phase of primary and secondary production (Van Otterloo, 2000c: 263–269). American imports of grain, lard and canned meat already had revolutionised European agriculture and food markets. Innovations in food preservation and shipping technology, followed by an ongoing series of technological and organisational changes in the world food supply since 1900, caused an increasing entanglement between competitive American and European agricultural and food realms. This entanglement between food companies and markets was going to increase during the twentieth century. In the 1890s the Netherlands had recovered from a great agricultural crisis. The production and delivery of raw materials to the industrial making of bread and cake, meat, dairy products, milk, vegetables and fruits expanded enormously. New food factories started to process these materials into new products, while striving to improve production processes and products through new food knowledge and technology on nutritional value and hygiene. An example of the then *quality turn* is the slow and intermittent process of learning how to reckon with the bacteria in dairy production, which finally became institutionalised as a general daily rule. The food shortages at the end of the First World War urged to measures from the national and local governments, that caused a circulation among consumers of new factory foods, such as canned products, milk powder, sugar, bread, margarine and sausage. These products may be seen as the first industrially produced types of Fast Food, bought ready to eat without further preparation or effort.

In the next period, between 1920 and 1960, first economic upsurge and crisis, next World War II and the reconstruction afterwards, put a stamp on the development of the food chain (Van Otterloo and Sluijter, 2000: 280–295). Industrialisation and urbanisation created larger food markets, as citizens no longer could grow their own ingredients. Food factories as Verkade (biscuits and chocolate) and Van Nelle (coffee, tea and tobacco) used ever more brand names for their products to win and keep the confidence of the consumers. The geographical distance between production and consumption was already growing, and food advertisements began to play an important role to overcome it. At the phases and links/locations of preparation and consumption, new ideas and activities were in the make. Ideals on living a modern (if possible, *electrical*) and comfortable life were omnipresent in society.

The food chain, until then, mainly consisted of three groups of actors: producers, distributors, and consumers – with a limited role of the government to sanction and enact rules. In this period the chain was enriched with new actor groups and organisations, striving at advancing and modernising eating and living. Educational organisations and advice institutions in the field of food, cookery and household activities are examples of it. They jointly began to shape a *new intermediate field* between production and consumption, purposely or unintentionally, focused at attuning both ends of the chain (Van Otterloo and Sluijter, 2000: 293–295). For the lengthening and differentiating processes in the chain had caused an increasing distance between producers and consumers. The growth and expansion of the intermediate field meant at the same time the increase of the process of *condensing,* the third general regularity in the evolution of the chain, seen from hindsight. The condensing of the chain is a corollary of lengthening and differentiation, opening possibilities for attuning and co-ordination. During this period optimistic ideals to improve the *quality* of food, cooking, housing and living were shared by industrial food companies, local and national government authorities, new educational institutions and groups of scientists. Cookery schoolteachers and other women's organisations, for instance, were active in diffusing 'modern, rational' knowledge on food quality, choice and preparation. The American dream of the good life shimmered already in the period between the wars; it could be realised only, however, by the well to do. The cookery ladies taught women how to cook with aid of electrical equipment and with less water to keep the vitamins. The necessity of such vitamins for food quality and health was stated by scientists who based their statements on empirical research in new laboratories. Standards of food quality, though, were themes of heavy and serious discussion among scholars and other parties in and outside the food chain, such as industrial companies and government authorities. New food educational and food research institutions were established, partly financed by the local and national government authorities.

The occurrence of the economic crisis and the World War II caused a decline in the optimistic ideals and in the standard of living. A new governmental organisation was set up to distribute food and to advise people how to optimally utilise their food portions. After the war efforts were made for reconstruction, partially with help of the Marshall Plan, but otherwise life and eating very much continued as it was before 1940. America was popular as a cultural model since the inter-war period, but this did not result in the borrowing of special dishes. It was a time of the chewing gum, not yet for hamburgers and other mass-produced fast foods.

In the Netherlands small snacks were produced by a variety of small entrepreneurs (Albert de la Bruhèze and Van Otterloo, 2003). Efficient industrial organisation and marketing departments, however, captured their places in the factory and business links/locations in the food chain. Other new links/locations started to develop in the phase of distribution: the first self-service shop in the Netherlands opened its doors in 1948, and soon others followed. In the 1950s, though, many housewives continued to preserve their meat, vegetables and fruit at home, in stead of buying them readymade, canned or frozen. The period between 1920 and 1960 is most importantly characterised by the rise of a varied *intermediate field*, bridging the growing geographical and experiential distance between production and consumption, and attuning both ends of the chain. An important point in this attuning process was the development of a set of ideas on *quality standards* of good food, which in the vocabulary of Goodman and Redclift might be seen as a set of 'quality turns' (Goodman and Redclift, this volume).

Many new changes in the food chain and its assortment took place between 1960 and 1990, most aptly characterised by a strong expansion. The era of mass food consumption began for the Dutch in the 1960s, accompanied by variety and convenience. A similar era dominated the American food supply and demand already decades earlier. New technologies, applied to mass food production, opened possibilities to a real food revolution. Canned food was completed by frozen and dried (dehydrated) food, soon from a compound character, that is composed of particles of different (raw or processed) materials. Dry soup, instant pudding and coffee were at the start of this development, while complete frozen meals were at the temporary end. The Unilever concern and other food companies started a programme of diversification and product development. The programme demanded special departments, to accommodate its employees of many different food disciplines, educated at new food technology departments at universities. Addition of various types of preservatives, flavourings, aromatic substances, colours, stabilisers, thickeners and anti-clotters to foods, became a normal practice in production lines. Bio-industry (calves, chickens, pigs) expanded enormously and artificial meat, named TVP (Texturized Vegetable Protein), was composed at the same time. Some consumers began to complain about the excess of improper substances, additives and contaminants in food. The expansion of the self-service shops at the distribution links, followed by supermarkets, heralded a new era. A complete new way of purchasing and storing emerged in combination with the family-car, the fridge at home and industrially frozen and cooled food products (Van Otterloo, 1995; Van Otterloo and Sluijter, 2000: 281–293, 295–310).

When considering the history of the Dutch food chain and its assortment in the 20th century, I may conclude that the development has resulted in a turn of supply (and demand) of an increasing quantity of highly processed, industrialised food. From hindsight the evolution of an industrialised assortment may look self-evident, but that is only appearance. This industrial type of food generally does not need much preparation time before consumption, at home or outside, and therefore we may look upon it as a kind of *Fast Food*. McDonald's hamburgers relate to this general category of Fast Food as one variety only. The trend to the production and consumption of Fast Food in this general sense appears to have been part and parcel of the development of the food chain as a whole. As mentioned earlier three general changes turned out to be the most important: *lengthening, differentiation* and *condensing*, explained below and put in European context. These general changes might be the outcome of the food chain's evolution in other European societies, for America and for the global food chain/food system as well, though details and pace in these other time/space contexts may be completely different. Yet, a similar final outcome of chain changes, chain trends and assortments may be plausible. The processes that turned out to be regularities in the development of the Dutch food chain, might with varieties in time and pace correspond with food chain changes in other European nation-states, and perhaps even with the evolution of the food system at a European and global level as a whole.

III.2. The food chain and fast food trends in the European context

In the course of the food chain development, similar differences are to be found in the *lengthening* of the chain, that is the growth of the number of links and locations, and the increasing of the geographical distance between producers and consumers. The process of lengthening implies an international expansion and entanglement of national chains at a European and a global level. The diffusion of McDonald's outlets through Europe and the world, for instance, is an aspect of this lengthening of the chain at a global level.[9] Focusing on consumption these outlets belong to Phase 5 and Location 5, though the other phases and locations are important as well (see Figure 12.1). The geography (and culture) of Europe has divided food production and consumption into distinct North-Western and South-Eastern parts and patterns (Pujol, this volume). The food chains, the assortments, the meal patterns and the trends of Fast Food and Slow Food may have developed differently in pace, scope and content in the various North-Western and South-Eastern European countries. It might, therefore, be not by accident, that the Slow Food movement originated in Italy with similar protests following in France. Both countries have an eminent culinary history, and they stick to their traditional meal patterns and taste. At the same time globalisation and the growth of a *world cuisine* may have an integrating

[9] See Fast Food in Europe (1990).

influence in Europe (Kalb et al., 2000; Featherstone, 1995; Goody, 1982). The McDonald's global fast food chain may have such integrating consequences. The international interweaving of food companies (Phase 1, Location 1) is another case in point. An outcome of a long-term lengthening of the European and American food chains, to wit the mix of leading food companies, is shown in Table 12.1 (Leonhäuser, 2002: 20). The lengthening of the chains at the consumption side may also be seen as an aspect of globalisation of food and the growth of a world cuisine.

Table 12.1 The ten leading world-wide food companies in 1997

		Food sale (1997 in Millions US$)	Food sale (in % of the total)
1. Nestlé	Switzerland	45,380	95
2. Philip Morris Inc.	USA	31,890	44
3. Unilever PLC/NV	NK / UK	24,170	50
4. Con Agra Inc.	USA	24,000	100
5. Cargill Inc.	USA	21,000	38
6. PepsiCo Inc.	USA	20,910	100
7. Coca-Cola Co.	USA	18,860	100
8. Diageo Guiness + Grand Metropolitan	UK	18,770	93
9. Mars Inc.	USA	13,500	100
10. Danone	France	13,970	94

Source: Leonhäuser, 2002: 20.

A second very general change in the case of the Dutch food chain in the twentieth century is an increasing *differentiation*, especially accelerating in the second half of this era from the 1950s and 1960s. This means, that the chain's links and locations, where the food-activities take place, have become more complex in themselves. Production, the assortment and consumption as a whole develop into much more types and varieties than ever before. Differentiation in consumption is mutually developing in a similar direction. Differentiation also means the increasing of the number and varieties of products into a huge pile of foods. From the 1960s onwards the share of Fast Food of various types – to eat out and at home – in the chain's food assortment has continuously increased. Raw materials (that is to say, without any processing) to prepare foods and meals were no more available by the 1990s. An example of a similarity in food chain *differentiation*, is the production and consumption of meat in the different parts of Europe, to be seen in Table 12.2 (Leonhäuser, 2002: 27).

In this table, the differentiation between European countries in the production and consumption phases of the food chain appears in the figures. In Greece and Spain the meat consumption today is higher than in the other European countries mentioned, except for Luxembourg. This situation is a recent one, which dates back only from about two decades. It has at least to do with economic affluence. Western-European countries before had a higher meat-consumption pattern than the Mediterranean countries for reasons of earlier economic growth. Meat production means a differentiation in the phase of

Table 12.2 Average availability of meat by type (g/person/day) in 1998

	Greece	Ireland	Luxembourg	Norway	Spain	United Kingdom
Pork	15	9.8	32	20	24	12
Beef & veal	52	25	43	26	31	19
Poultry	37	31	30	12	58	34
Other meat & meat products	46	72	81	71	65	73
Total	150	138	186	129	178	138

Source: Leonhäuser, 2002: 27.

agricultural production between agriculture and cattle breeding (Figure 12.1: Phase 1, Location 1). A Dutch example of the influence of the rise of income on consumption (and thus production) of meat and dairy products over time, is to be found in Figure 12.2, showing expenditures over a long-term period. The historic turning point in Dutch consumer expenditures around 1960 is illustrated.

Figure 12.2 Private consumption in the Netherlands, 1925–1985, millions of guilders

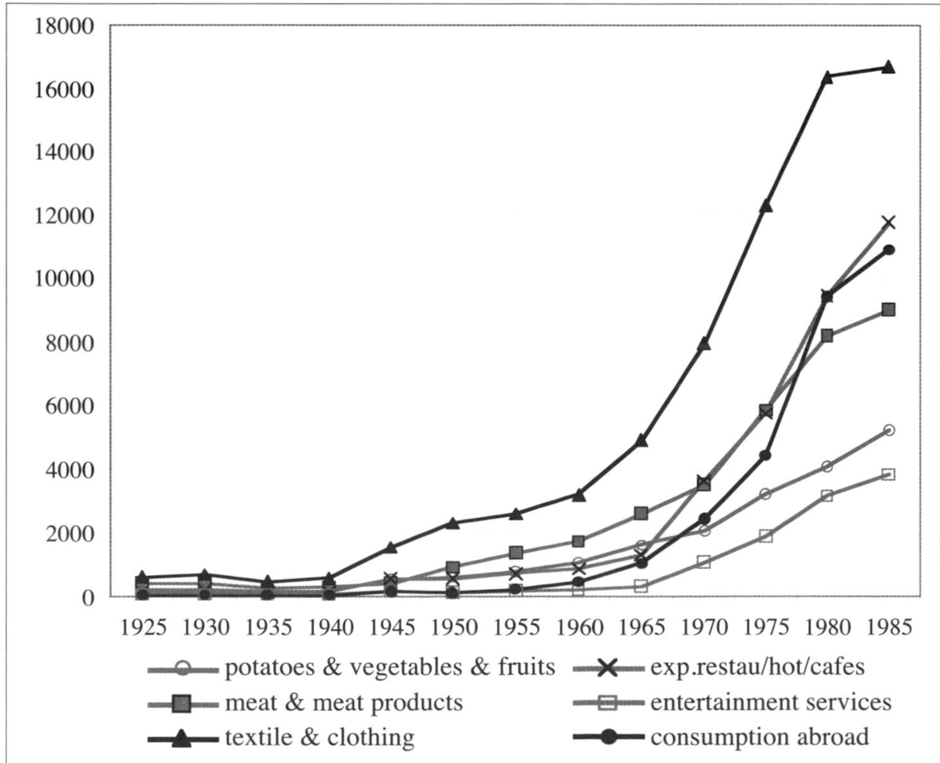

Source: Van Driel and Boon, 1991, table 5.

Meat (chicken) and readymade dairy products, (new and exotic) vegetables and fruits, (alcoholic and soft) drinks, luxury foods and many processed, dried and frozen products were diversifying the meal at home (see the two relevant symbols in the graphic explanation chart at the bottom of Figure 12.2). The food chain increasingly produced a vast field of industrially processed foods, to be designated as convenient foods, including fast food, ready from the shelve to the table without further preparation than heating, later by microwave. This meant a differentiation process in the links and locations of the food chain.

A final example of the process of lengthening and differentiation in the food chain, among others leading to Fast Food production, is the evolution of the cultivation and use of the potato. The production and processing of *potatoes* and its many by-products and derivatives, demand own technology and organisation. Main products and by-products have exchanged places over time, so potatoes now mainly are produced and consumed in the form of fast food (crisps and French fries, mainly). Potatoes in the 18th and 19th centuries became an essential survival food and component of the European meal, consumed at home. The shape this tuber has taken in the European market for consumers, however, has radically changed in the last three decades of the 20th century. The contemporary provision and consumption of potato products that are ready to eat outside or at home, have taken the place of the former use of fresh and relatively unprocessed potato, to be peeled and cooked or mashed at home. The tubers increasingly are supplied and consumed as types of convenience foods (Camaggio and Lagioia, 2002: 61–67). The separate and at the time interconnected development of the frozen French fries chain has been an important part of this process. The appearance of French fries and potato chips has several consequences for producers (more potatoes needed) and consumers (more fats eaten), and possibly also for the European meal pattern, which might take a more uniform fast food shape. At the same time it means that lengthening and differentiation of the food chain, in the case of the potato, goes along with an evolution into a fast food assortment.

A third important long-term general change in the evolution of the Dutch food chain in the twentieth century may be indicated as *condensing,* that is the condition of mutual tuning of phases and links, enforced by the common situation in which actors and actor groups in the contemporary Dutch food chain find themselves since the 1980s. The growth of an *intermediate field* between production and consumption since the 1920s has played an important role in the process of condensing. Where actors and actor-groups in the past strove for their own goals at dispersed locations in the production, distribution, preparation and consumption of food, they now have learned to attune their behaviours to each other. Their efforts resulted over time in the development of a technologically and organisationally complex food chain, controlled by none of the individual actors alone. Unbound and regionally dispersed producers, distributors and consumers about 1900, found themselves into an intertwined network about 2000. This new situation posed restrictions to the behaviour of these actor groups in the food chain, because they had now become interdependent to each other. The various interconnected links and locations in the chain, and its actors, mutually had to reckon with each other. What is more, they had to reckon with the parties and organisations in the newly developed intermediate field, at least consisting of groups of scientists, the media, the local and national governments and more actor groups that got involved in matters of food over time. Comparable developments might have taken place in the European chains from elsewhere. Part of the condensing process might consist of the recurrent countering of the processes of rationalisation, that is to say the lengthening and differentiation of the food chain over the 20th century and the ' fastening' of its assortment. The three general developments just described, *lengthening, differentiation and condensing*, that have been characteristic for the development of the Dutch food chain and probably for other European and global food chains as well, did not take place without tensions and conflicts. Recurrent protests by farmers and critical consumers are to be described further.

IV. Slow Food and its predecessors, 1890–1990

In the above sections it appeared that Fast Food and its vast variety of manifestations over time and place, have been the (un)intended corollary of the development of the food chain in the 20th century. Competing economic and political power forces compelled to rationalisation of the chain, which contributed to its lengthening, differentiation and diffusion from the local and regional to the national, European and global levels in a relatively short period. Now I turn to Slow Food and its equally vast variety of manifestations over time and place. In its restricted meaning *Slow Food* is the movement, originally founded in 1986 in Bra, Italy, to counter the feared flood of McDonald's and similar manifestations of Fast Food in that country. Ever since that point in time, this movement expanded over the globe, operating via the Internet (Goodman and Redclift, this volume; Miele and Murdoch, 2002; Murdoch and Miele, 2004). Slow Food organises meals in dinner clubs and tasters' meetings of *traditional* cheeses or wines. The Slow Food's network may be considered to belong to the set of recent initiatives that opposes the Fast Food assortment in its many varieties, to eat out and at home. These groups and initiatives propose alternatives to the main stream food chain's way of producing and handling highly processed and standardised types of food. The characterisation of the increasing process of McDonaldization in modern industrial and consumer societies by George Ritzer, powerfully go for the development and mode of operation of the contemporary food chain itself. The four dimensions he discerns in this process (efficiency, predictability, calculability and control) might be considered as aims of the chain's producers, distributors and consumers, as far as these actors and actor groups consciously strove for them (Ritzer, 1993: 9–12). The elements of McDonaldization are part of the rationalisation process of the chain and, to be clear, they imply its growing lengthening, differentiation and complexity. The standardisation of the food assortment, produced in the chain, belongs to this process also. My hypothesis is that the rise of the modern, industrial food chain between 1890 and 1990 has been accompanied by recurrent manifestations of discontent and protest of a similar, yet also different character as the contemporary Slow Food movement has developed.[10] These various initiatives and critical noises might be indicated as Slow Food in a broad sense, just like George Ritzer refers to predecessors and precursors of the process of McDonaldization (Ritzer, 1993: 18–35). I take the same periods as in the previous section to discuss the rise and development of the Dutch

---

[10] The idea of the (earlier) existence of a general category of critical initiatives, countering mainstream modern-industrial food production and consumption is also considered in Goodman and Redclift (this volume) and Guthman (2003).

food chain (1890–1920, 1920–1960, and 1960–1990). These episodes may be considered just as stages in the increasing rationalisation and fastening of its produced food assortment. The focusing on these periods might shed light on some predecessors. Again, the development of the food chain and the social and cultural food context in the Netherlands only serves as a case, reflecting wider developments and touching at examples from elsewhere.

The first period (1890–1920) was characterised by the lengthening and differentiation of the links in the primary and secondary production (see Figure 12.1, Link/Location1). The intensifying process of the production of, for instance, meat consisted of a differentiation in the primary production between arable farming and stock breeding, and thus caused a decline of mixed farming, where cattle breeding only serves the cultivation of crops. According to the historian Theodor Abel, in his *Stufen der Ernährung* (1981), the period of the decline of famines and food instability and the rise of food security in the closing decades of the 19th century, went along with the phase (in the food chain development) in which cattle breeding became a goal in itself. Animals were now fed on vegetable products, which previously made up the main human diet, while humans began to eat meat if they could afford it (Abel, 1981: 65–73). This is the beginning of the process, which agrarian historians have labelled as the 'climbing of the protein ladder' (Goodman and Redclift, this volume). Meat was a scarce article and a high status part of the meal, but its production and consumption rose with industrialisation and the lengthening and differentiation of the chain. In the long run and with the increase of income, it became available to the masses. Since the 1980s, meat consumption in Europe took the shape of Big Macs and other items, following the lead in fast food restaurant development of the United States (Fast Food in Europe, 1990; Levenstein, 1993: 227–237). It might be not too bold to assume that the start of the vegetarian movement, which took place in this very period of the rise of the meat production in the second half of the 19th century, can be seen as a counter-movement against this aspect of food chain development and its various consequences. The vegetarian movement started in middle of the 19th century in England, and it came to the Netherlands in the 1890s (Van Otterloo, 1990: 184–210). It was a minority of the fortunate class of wealthy and educated people (those who could afford to eat meat at the time) that was motivated to deny themselves this luxury product. They legitimated their choice on various principles and do so today, among which animal welfare, health, environmentalism, natural food, humanity and civilisation combine to form a lifestyle. Comparing the 'back to nature'-movements of the 1890s with those of the 1970s and 1980s, continuity in motivations and principles is striking (Van Otterloo, 1990: 189–191).

In the second period (1920–1960) a new technological phase in production (and consumption) of food appeared in the Netherlands. New rational methods of processing and refinement of raw materials for food production became in use, while the fast food cafeterias and the *snack automats* became popular places to eat out in the cities (Van Otterloo, 2000c: 265–267, 308–309; Albert de la Bruhèze and Van Otterloo, 2003). As a counterpart of this development, new 'back to nature' agricultural and food movements emerged. This had been the case in the Netherlands in the 1920s and 1930s. These movements were the Reform Movement and the Movement for Bio-Dynamic Agricultural Methods, which both were imported from Middle European Countries like Germany and Austria (Dr. Rudolf Steiner), and Switzerland (Dr. Bircher Benner), and were diffused also elsewhere in Europe. Whole-food, less industrially processed food and organically grown food belonged to the important material and symbolic artefacts and agricultural methods of such movements. Similar natural and health alternatives to refined factory food (white bread, for instance) had been developed in the United States in the 19th century by religious groups such as the Seventh's Day Adventists, or by physicians like Dr. Allinson, who advocated whole-grain bread.

In the 1950s critical noises in the Dutch Consumer's Union's Guide denounced the widespread use of pesticides in gardening and agriculture (van Otterloo, 1995: 258). This was six years before the appearance of Rachel Carson's famous book *Silent Spring* of 1962, which had a great influence on the success of the environmental movement. White, uniform and spongy factory bread was scorned upon by a member of the Dutch Parliament in this same period, and ever since brown varieties became increasingly in demand.[11] Such limited and relatively polite noises were heralding the renaissance of the movement for health and natural food, building up at the waves of the environmental movement since the 1960s and early '70s. The shock of the Club of Rome report, *Limits to Growth* (1972), formed alongside with *Silent spring* another impetus to a widespread opposition against the ruthless economic and technological growth, exhausting minerals and raw materials, which was to appear in the next decades. The economic growth of the food sector, shaped by the lengthening and differentiation of the food chain in this context was a much contested part of economy and technology (Van Otterloo and Sluijter, 2000: 297–298).

During the third period, 1960–1990, the public distrust of modern food technology with its additives, residues and animal-hostile methods in the bio-industry, increased. In the 1970s the environmental movement arose with a strong 'food wing', advocating with E.F. Schumacher, that 'Small is beautiful' (1973), and that people, animals and nature as a whole do matter. Lengthening and differentiation, leading to large-scale companies and complex links and locations in the food chain, were opposed. Many experimental small-scale and 'self-supporting' communities and communes were founded in the countryside in modern, industrial Western-Europe and the United States. Striving at 'living from the land', these groups were practising short food chains, growing unsprayed crops and dairying from their own free-roaming cows, sheep and goats near their homes. Examples are the *Findhorn Community* in Scotland, since 1962, and *De Kleine Aarde* (The Small Planet) in Boxtel, the Netherlands, since 1972 (van Otterloo, 1990: 192–194). Nowadays, these communities are educational centres, carrying experiments in small-scale technology that might be applied also in less economically developed countries. From the beginning food has occupied a central place in this *ecological* movement, aiming at a life in balance with nature. Similar types of lifestyles were practised at the time in the United States and were oriented towards the organic cultivation of food (Belasco, 1989; Guthman, 2003). *Alternative* ways of living and eating, naturally and healthy remained popular in the 1980s. In the same decade, finally, groups of critical consumers were recurrently going into the details of 'food pollution', diffusing the idea through the media that the increasing complexity of the food chain caused a lowering of food quality. They did build on earlier initiatives, and their message was frightening, implying rising toxic dangers and food un-safety in food production (van Otterloo and Sluijter, 2000: 297–298). In the 1990s discussions over GMF (genetically modified food) and functional

[11] It must be noted that a century before the complaints about low quality factory bread, the new phenomenon of the bread-factory appeared in Amsterdam in the 1850s.

food followed suit. Food scares caused by problems in the production and distribution lines on animal disease like BSE (Bovine Spongiform Encephalopathy or mad cow disease), worsened the situation and gave cause to new protests.

In short, I may conclude from this historical exploration about initiatives, countering the industrialisation and rationalisation of food production and covering a period of hundred years, that Alternative Agricultural Food Networks (AAFN's) are not new. What Goodman and Redclift, this volume, indicate as a matter of a contemporary *quality turn*, is part of a whole *century of debates on food qualities* of various types, passed by in a nutshell in this chapter. Slow Food and its peers also belong to these debates on food, advocated by different societal groups. The contemporary groups go for pure, unspoiled and locally or regionally produced foods and emphasise an artisan or traditional way of preparation. Traditional ways of growing and preparation form, according to this reasoning, part of the culinary cultures of the past, which must be protected and preserved as valuable heritage. If not, these ways and products irrevocably will get lost forever (Van der Meulen, 1998: 3).

These and similar contested aspects of food quality may vary in time and place. In the contemporary case of Slow Food *taste* and the philosophy of pleasure are defended in Italy, while sustainability, green initiatives and bio-diversity are at stake in Germany and Britain. The differences in the dimensions of Slow Food between Mediterranean and Atlantic countries may perhaps be seen as manifestations of long-term culinary traditions (Pujol, this volume; Bruegel and Laurioux, 2002; Mennell, 1985). At the other hand, the frequent virtual or real encounters between so-called *Convivia* (dinner groups) and other units from different countries, open the possibility to an intensive process of blending. National thoughts and practices, expressing feelings of discontent with modern, industrial food production are melted into a complicated mix. The French farmers' protests against McDonald's and its evolution into an alternative Slow Food movement or AAFN, is a comparable case in point. Other contemporary initiatives and trends opposing Fast Food and all it stands for deal with regional agricultural production, linking farmers immediately to consumers (initiatives to adopt a cow, a chicken, or an apple tree and suchlike). Bio-food, or organic food, for instance, shows many characteristics that are analogous to the demands of the Slow Food movement, including health (Guthman, 2003). It even was integrated in a Dutch Fast Food chain, named *Shakie's*, and situated at the Central Station of Amsterdam and Utrecht (Van der Mark, 2003). In Germany consumers mentioned similar food characteristics such as *healthy* and *ecological,* as motivations to buy *regionally produced food*, see Figure 12.3 (Dorandt, 2002: 332).

Higher prices of regionally and biologically produced foods, though, form the most important dimension, impeding the choice of the consumers for this type of slow foods. This is why acceptance is tardy, see Figure 12.4 (Dorandt, 2002: 335). Slow Food gastronomy, at the other hand, does not seem to worry about prices, which historically seen, is not amazing. Slow Food movements, today and in the past, are strongly inclined to be carried by elite groups and consist of people with a rather high level of cultural and economic capital in the sense of Bourdieu (1984).

The figures date from the 1990s. More data are needed for other places (European countries) and times (the 20th century) to get insight in differences and similarities in what Goodman and Redclift have labelled as AAFN's, striving for a *quality turn*. 'In *prosperous* (italics by the author) markets the emphasis is shifting from cheapness to quality, parity and esteem for local artisan methods'; 'Local foods, organic foods,

Figure 12.3 Attributes of 'regional food' from consumers' point of view in 1998–1999

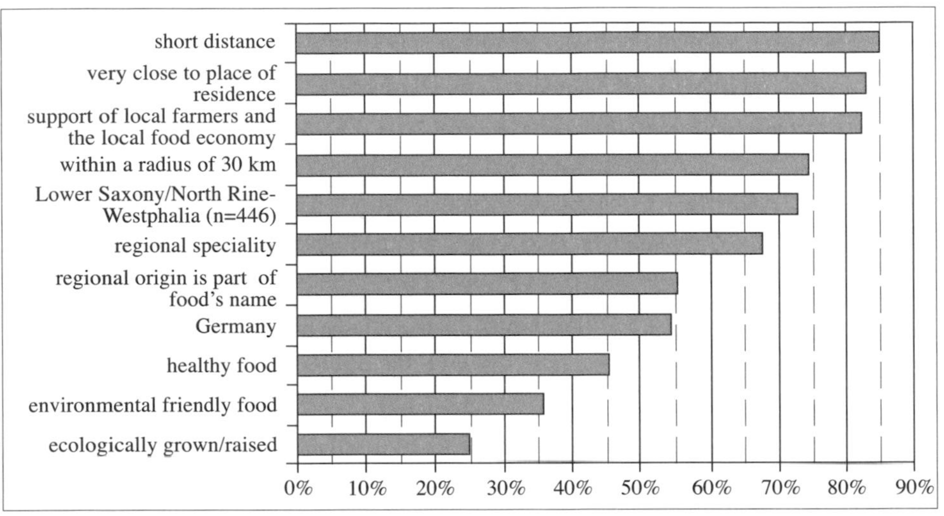

Source: Dorandt, 2002: 332.

Figure 12.4 Reasons for acceptance (dark grey) and disapproval (light grey) of higher prices of regional food in 1998–1999

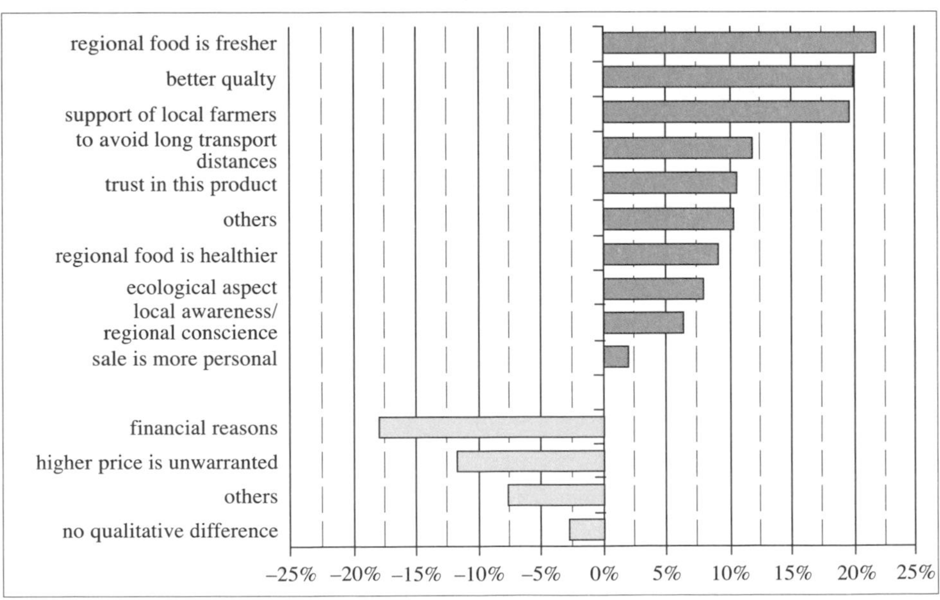

Source: Dorandt, 2002: 335.

traditional foods, GM-free foods, and the like have all become popular in recent years and consumers look for some enhanced security through some re-engagement with 'natural' qualities' (Murdoch and Miele, 2004). In this way, the contemporary AAFN's, consisting of consumers and producers, and/or 'alternative food movements' try to counter the development of extreme *lengthening* of the food chain. As indicated above, this process of contest repeatedly manifested itself in various forms and continued from the beginning up to the end of the 20th century. Actors and actor groups propagate in this way shorter distances between producers and consumers. Doing so at the same time they contribute to and contest the second important development of the food chain: its *differentiation*. Both activities take the shape of simplifying complicated links, while forming food chain links themselves. The same is true for the expansion and diversifying of the assortment of food products by adding, for instance, regional and farmer's products next to high tech industrial fast foods. The condition for maintaining alternative networks, markets and its products is the consumer's willingness to pay more money, which is not always present, as we saw in Figure 12.4 above. Murdoch and Miele appropriately made the restriction of *prosperity* to their assertion, that emphasis in contemporary food choices is shifting from cheapness to quality.

V. Concluding remarks

Critical initiatives and movements, time and again, have tried to slow down the pace of the evolution of the food chain into the direction of efficiency, calculability, predictability and control. The resulting fast food products, eaten out or at home, were thought to have negative consequences for personal health (obesity), the treatment of animals (chicken and pig raising), the environment (reduction of bio-diversity), human labour, fair trade, and finally, taste. This list is not complete. The persons and groups involved try to shorten the distance between locations, to simplify the links and partly to untwine the networks. Fast Food and Slow Food, from a history of technology point of view, are 20th-century trends opposing each other, while both are intimately related to the long term rise of the food chain and the change of its 'slow into fast' food assortment. The analyses of Miele and Murdoch, and of Goodman and Redclift are enlightening, however, they miss a long-term perspective. From this basic assumption it is clear that both Fast Food and Slow Food are corollaries, inherent to the development of the food chain over time. As such the critical noises under the banner of Slow Food in a general sense have come to belong to the *intermediate field* that emerged between production and consumption. This intermediate field contributed to the general process of condensing of the chain, as a counterpart against its lengthening and differentiation.

Both trends are spread throughout Europe, but not at the same time or in the same pace or degree. Both trends, in my opinion, may be seen as simultaneously integrating and differentiating influences in the European food chain and assortment, depending on the level of analysis. They belong to the mutual pervasion of technology and society. Modern society, however, presumably will not abandon its preference for 'fast' food. Therefore, I (we) may expect that the characteristics of fast food will possibly permeate slow food initiatives in so far, that the rational time-saving element must remain available, see for instance Shakie's freshly pressed, but ready to drink natural juices, to be ordered at the railway station. This is in line with Julie Guthman's critical remarks on the binary character of fast food and slow food/organic food. Systematic historical research into these long-term trends appears to be absent or fragmentary.

Some final hypothetical remarks touch at the recurrent alternatives and countertrends over a century, accompanying the rise and lengthening of the chain. Their appearance seems to be bound to periods of affluence, when problems of *quantity* as food shortages are solved and a longing for *quality* appears. Moreover, the AAFN's and other critical food networks do not stay alone to defend their stand for long. Soon they will be kept company by other actors (authorities, scholars, technologists and other experts, the media) to build together the *intermediate field* between the production and consumption of food, that is characteristic for the rise of the modern-industrial food chain in the twentieth century. This intermediate field attunes food supply and demand and has granted the chain in the last decades of the 20th century its *condensing*, reflexive, qualities.

Bibliography

Abel, Th. (1981) *Stufen der Ernahrung. Eine historische Skizze*, Göttingen.

Adam, B., Beck, U. and Loon, J. van (2000) *The risk society and beyond. Critical issues for social theory*, London.

Albert de la Bruhèze, A. and Otterloo, A.H. van (2003) 'Snacks and snack culture in the rise of eating out in the Netherlands in the twentieth century', in M. Jacobs and P. Scholliers, *Eating out in Europe*, Oxford, pp. 317–335.

Alfino, M. (1998) 'Postmodern hamburgers: Taking a postmodern attitude toward McDonald's', in M. Alfino, J.S. Caputo and R. Wynyard (eds), *McDonaldization revisited: critical essays on consumer culture*, Westport, pp. 175–189.

Ariès, P. (1997) *Les fils de McDo. La McDonaldization du Monde*, Paris.

Baltesen, F. (1996) 'De zegetocht van McDonald's', in www.bedr-horeca.nl.

Beck, U. (1992) *Risk society. Towards a new modernity*. London.

Belasco, W.J. (1989) *Appetite for change. How the counterculture took on the food industry*, New York.

Biologische Landbouw (1999) Den Haag, in www.minlnv.nl.

Boas, M. and Chain, S. (1976) *Big Mac: The unauthorised story of McDonald's*, New York.

Bourdieu, P. (1984) *Distinction. A social critique of the judgement of taste*, London–New York.

Bové, J. and Dufour, F. (2002) *De wereld is niet te koop. Boeren tegen junkfood [The World is not for sale. Farmers against junkfood]*, Rotterdam.

Bruegel, M. and Laurioux, B. (2002) 'Introduction', in M. Bruegel and B. Larioux (eds), *Histoire et identités alimentaires en Europe*, Paris, pp. 9–19.

Camaggio, G. and Lagioia, G. (2002) 'Potatoes fast food. Consumption, technology and future trends', in C.A.A. Butijn [et al.] (eds), *Changes at the other end of the chain. Everyday consumption in a multidisciplinary perspective,* Wageningen–Maastricht, pp. 73–81.

Dagevos, J.C. (2002) *Panorama voedingsland: traditie en transitie in discussies over voedsel*, Den Haag.

Dorandt, S. (2002) 'Shopping behaviour of private households towards regional food', in Butijn, C.A.A. [et al.] (eds), *Changes at the other end of the chain. Everyday consumption in a multidisciplinary perspective,* Wageningen–Maastricht, pp. 327–338.

Driel, J. van and Boon, M. (1991) *Private consumption expenditure and price index numbers for the Netherlands, 1921–1939 and 1948–1988*, Voorburg.

Elias, N. and Scotson, J.L. (1965) *The established and the outsiders*, London.

Fast food in Europe (1990) *Quick serving catering in West Germany, United Kingdom, France, Italy, Netherlands and Belgium* (Special report no. 2027), London.

Featherstone, M, Lash S. and Robertson, R. (eds) (1995) *Global modernities*. London.

Ferguson, P.P. (1998) 'A cultural field in the making: Gastronomy in 19th century France', *American Journal of Sociology*, 104, pp. 597–641.

Fischler, C. (1999) 'The ' McDonaldization of culture', in J.L. Flandrin and M. Montanari (eds), *Food, a culinary history from antiquity to the present*, New York, pp. 530–547.

Gabaccia, D.R. (1998) *We are what we eat. Ethnic food and the making of America*, Cambridge (Mass).

Grefe, Ch., Heller, P., Herbst, M. and Pater, P. (1987) *Das Brot des Siegers. Die Hamburger-Konzerne*, Bornheim-Merten.

Gabriel, Y and Lang, T. (1998) *The unmanageable consumer. Contemporary consumption and its fragmentation*. London.

Goodman, D. and Redclift, M. (1991) *Refashioning nature. Food, ecology and culture*, London–New York.

Goody, J.R. (1982) *Cooking, cuisine and class. A study in comparative sociology.* Cambridge.

Guthman, J. (2003) 'Fast Food/Organic Food, reflexive tastes and the making of "yuppie chow"', *Social & Cultural Geography,* vol. 4, 1, pp. 45–58.

Hartog, A.P. den (ed.) (1995) *Food technology, science and marketing: European diet in the twentieth century,* East Lothian.

Jacobs, M. and Scholliers, P. (2003) *Eating out in Europe. Picnics, gourmet dining and snacks since the late eighteenth century*, Oxford–New York.

Jobse-van Putten, J. (1995) *Eenvoudig maar voedzaam. Cultuurgeschiedenis van de dagelijkse maaltijd in Nederland*, Nijmegen.

Kalb, D. [et al.] (eds) (2000) *The ends of globalization. Bringing society back in*, Lanham–Oxford.

Kincheloe, J.L. (2002) *The sign of the burger. McDonald's and the culture of power*, Philadelphia.

Knop, B. and Schmitz, M. (1983) *Currywurst mit Fritten. Von der Kultur der Imbissbude*, Zürich.

Kroc, R. and Anderson, R. (1977) *Grinding it out: The making of McDonald's,* New York.

Leonhäuser, I.-U. (2002) 'Concerning food patterns in a comparative way', in C.A.A. Butijn [et al.] (eds), *Changes at the other end of the chain. Everyday consumption in a multidisciplinary perspective,* Wageningen–Maastricht, pp. 19–30.

Levenstein, H. (1993) *Paradox of plenty. A social history of eating in Modern America*, Oxford–New York.

Love, J.F. (1986) *McDonald's: Behind the Arches,* New York.

Mark, T. van der (2003) 'De lastige weg van bio-food', *Haagsche Courant* , 8 February 2003.

Mennell, S., Murcott, A. and Otterloo, A. van (1992) *The sociology of food: Eating, diet and culture*. London–New Delhi.

Meulen, H. van der (1998) *Traditionele streekproducten, Gastronomisch erfgoed van Nederland*, Doetinchem.

Miele, M. and Murdoch, J. (2002) 'Slow Food', in G. Ritzer (ed.), *McDonaldization. The reader*, London, pp. 250–254.

Miele, M. and Murdoch, J. (2003) 'Fast Food/Slow Food: Standardizing and differentiating Cultures of Food', in R. Almas and G. Lawrence (eds), *Globalization, localization and sustainable livelihoods*, Aldershot, pp. 25–43.

Murdoch, J. and Miele, M. (2004) 'A new aesthetic of food? Relative reflexivity in the "alternative" food movement', in M. Harvey, A. McMeekin and A. Warde (eds), *Qualities of food: Alternative theoretical and empirical approaches*, Manchester, pp. 55–75.

Mintz, S.W. (1985) *Sweetnees and power. The place of sugar in modern history.* New York–Hammondsworth.

Nestle, M. (2002) *Food politics. How the food industry influences nutrition and health*, Berkeley–London.

Oddy, D. J. (2003) 'Eating without effort: the rise of the fast-food industry in Britain, 1920–2000', in M. Jacobs and P. Scholliers (eds), *Eating out in Europe. Picnics, gourmet dining and snacks since the late eighteenth century*, Oxford–New York, pp. 301–315.

Otterloo, A.H. van (1990) *Eten en eetlust in Nederland 1840–1990*, Amsterdam.

Otterloo, A.H. van (1995) 'The development of public distrust of modern food technology in the Netherlands. Professionals, laymen and the consumer's union', in A.P. den Hartog (ed.), *Food, technology, science and marketing. European diet in the twentieth century*, East Linton, pp. 253–268.

Otterloo, A.H. van (2000a) 'Voeding in verandering', in J.W. Schot [et al.] (eds), *Techniek in Nederland in de twintigste eeuw,* III. Landbouw en voeding, Zutphen, pp. 237–247.

Otterloo, A.H. van (2000b) 'Nieuwe produkten, schakels en regimes 1890–1920', in J.W. Schot [et al.] (eds), *Techniek in Nederland in de twintigste eeuw*, III. Landbouw en voeding, Zutphen, pp. 248–261.

Otterloo, A.H. van (2000c) 'Prelude op de consumptiemaatschappij in voor- en tegenspoed 1920–1960', in J.W. Schot [et al.] (eds), *Techniek in Nederland in de twintigste eeuw*, III. Landbouw en voeding, Zutphen, pp. 262–280.

Otterloo, A.H. van (2000d) 'Ingredienten, toevoegingen en transformatie: heil en onheil', in J.W. Schot [et al.] (eds), *Techniek in Nederland in de twintigste eeuw,* III. Landbouw en voeding, Zuthpen, pp. 295–310.

Otterloo, A.H. van and Sluijter, B. (2000) 'Naar variatie en gemak 1960–1990', in J.W. Schot [et al.] (eds), *Techniek in Nederland in de twintigste eeuw,* III. Landbouw en voeding, Zutphen, pp. 280–295.

Ritzer, G. (1996) *The McDonaldization of society: An investigation into the changing character of contemporary social life*, Thousand Oaks.

Ritzer, G. (1998) *The McDonaldization thesis: Explorations and extensions*, London.

Ritzer, G. (1999) *Enchanting a disenchanted world: Revolutionizing the means of consumption*, Thousand Oaks.

Ritzer, G. (2000) *The McDonaldization of society: New century edition*, Thousand Oaks.

Ritzer, G. (ed.) (2002) *McDonalidization. The reader*, Thousand Oaks.

Schlosser, E. (2001) *Fast food nation: The dark side of the all-American meal*, Boston.

Schot, J.W. (2000) *De bouwput van techniek en maatschappij. Uitgangspunten van een nieuwe contextualistische techniekgeschiedenis,* Eindhoven.

Schot, J.W. (2001) *De maakbaarheid van Nederland,* Twente.

Seidel-Pielen, E. (1996) *Aufgespießt. Wie der Döner über die Deutschen kam*, Hamburg.

Smart, B. (ed.) (1999) *Resisting McDonaldization*, London.

Tansey, G. and Worsley, T. (1995) *The food system: A guide*, London.

Thoms, U. (2003) 'Industrial canteens in Germany 1850–1950', in M. Jacobs and P. Scholliers, *Eating out in Europe. Picnics, gourmet dining and snacks since the late eighteenth century*, Oxford–New York, pp. 351–373.

Wagner, C. (1995) *Fast schon Food: die Geschichte des schnellen Essens*, Frankfurt.

Walton, J.K. (1992) *Fish and chips and the British working class, 1870–1940*, Leicester.

Warde, A. and Martens, L. (2000) *Eating out: Social differentiation, consumption and pleasure*, Cambridge.

Watson, J.L. (1997) *Golden Arches East: McDonalds in East Asia,* Berkeley.

13 Industrialising catering. Technological developments and its effects in the twentieth century

Ulrike Thoms, Freie Universität, Berlin

I. Introduction

Food technology has been a topic for the interdisciplinary history of food since a long time. This chapter reports on the state of the art, marks the most important relevant questions and identifies big problems of the historical development of catering. It will first sketch out the increasing role of catering (in the sense of supplying people with prepared meals), and so deal with the middle of the food chain. Then, it will turn to the changes within catering, as to the rationalisation within the canteens in regard to consequences for the upper and lower end of the nutrition chain, and finally it will compare the development throughout Europe. The latter, however, is still difficult because of problems of collecting reliable and comparable data. This chapter hopes to show that, in general, the most important innovations in food technology can be traced back to the specific demands of catering and its conditions, which changed primary production decisively, and then influenced private households.

Especially ethnology and social history have conducted numerous studies on traditional household techniques and the change of kitchen appliances (Schärer and Fenton, 1998; Thoms, 1998; Teuteberg, 1997; von Hausen, 1999; Hessler, 2001). Economic history has studied the rise of new food technologies in brewing, baking and the changes in food industry in general (especially Den Hartog, 1995a), as for example appertising, (Bruegel, 1998), freezing (Oddy and Oddy, 1998; Teuteberg, 1991; Hellmann, 1990), and packaging (Den Hartog, 1995b), new 'artificial' products as margarine (Van Stuyvenberg, 1969: 83–121; Pelzer and Reinhold, 2001) or infant food (Thoms, 1994). All of these studies have been restricted to the industry itself or to per capita consumption of 'new' foodstuffs.

The latter is also true for the meanwhile numerous studies that deal with the feeding of particular groups of people. Especially the Annales-school was convinced that data from 'mass feeding' in institutions reflect a wider consumption level of the lowest social classes (Hamelmann, 1989; Barlösius, 1989). Mass feeding in institutions would thus supply researchers with long and reliable series of consumption data. The *Annalistes* almost never took into consideration specific circumstances, as for example the economic situation, the age structure, and the general aims of the institution, which highly influenced the choice of foodstuffs and their preparation.

Studies that have considered institutional feeding almost never address the role of technology. Only during the last few years, the history of fast food has become a topic of academic work (Allen, 2002; Albert de la Bruhèze and Van Otterloo, 2003; Ritzer, 2000; Oddy, 2003; Seidel-Pielen, 1996; Wagner, 2001; Schlosser, 2001; Grefe et al., 1987; Knop and Schmitz, 1983; Lohof, 1979; Walton, 1992). These newer studies stressed the increasing demand for cheap and fast food, which results from major social and economic developments. It is probably not a coincidence that studies questioning the role of technology came from Dutch researchers. They pinpointed the importance of technological change, as a precondition for the increase in eating out (Den Hartog, 2003, de la Bruhèze and Van Otterloo, 2003, Van Otterloo, this volume). The Netherlands do not have a particular strong restaurant culture and lunch is cold, mostly taken from home to work, and the one warm meal is eaten at home. Starting from some forerunners in the 1920s and 1930s and influenced by the US, the development is closely connected to the intensive commercialisation and rationalisation process of the twentieth century.

II. The increasing role of catering services

There are different types of catering facilities. One type supplies people with all daily meals, running from breakfast to supper, as for example the army, hospitals, prisons, orphanages and senior residences. In this chapter, I will concentrate on canteens of factories, which belong to an 'open type', as they deliver only particular meals, mostly lunch and/or breakfast. These meals are completed by meals at home. In their openness to 'normal' life and their strong relation to the working process, they fully reflect the impact of industrialisation on changing food habits much more than the 'closed types' do. The latter always have special interests, as for example the cure of diseases or re-education of prisoners, which entangle the way in which the feeding is organised.

The separation of home and work has been characterised as one of the most important elements of industrialisation, which forced more and more people to eat at work. In former studies I have sketched out that canteens were the organisational answer to match the bodily needs and functions to the requirements of the industrial process. In the course of time, canteens themselves underwent a rationalisation process similar to the factory production itself (Thoms, 2003). Low cost – high profit criteria were adopted to industrialisation as to nutrition, and in Germany the term *Rationelle Volksernährung* ('rational feeding') precisely illustrated the popularity of this connection (Tanner, 1999). A closer look at the price structure of canteens of the nineteenth and twentieth centuries clearly shows that the workers shared this view. Canteen food was only accepted and bought by them, when it was cheap. Industrialising catering, thus, meant to make food as cheap as possible and this, I would say, is a common European feature.

How important canteens and other facilities for having a warm lunch were for the contemporaries, depended much on the role of the midday meal in the meal system of a society. In nineteenth-century Germany lunch was the main meal of the day and it was almost always a hot one (Sandgruber, 1988). German social reformers as well as nutritionists underlined the importance of this custom even in the 20th century, and the hot midday meal of the whole family remained their ideal, even though the father often had to eat at work (Cremer, 1956: 65). Therefore they stressed the importance of a long ninety-minutes lunch break. This was not so much the case in other European countries. In England for example, children often had and have lunch at school, whereas German children return home from school around midday. Moreover, the short break of only thirty minutes had its offspring in the early and highly industrialised UK, where there was no possibility to have a family meal at all. Even before the Second World War, Dutch and English people preferred a more frugal lunch, having a hot meal after work (Burnett, 1998: 117ff; de la Bruhèze and Van Otterloo, 2003). However, the shortage of time created a further demand for eating facilities. In fact, the short break was introduced to German factories since the 1890s; [1] and in France cheap restaurants, soup kitchens and dairy shops sprang up and offered home cooking at modest prices to serve the need for hot meals (Pitte, 1999: 477). Only in the last decades this pattern is changing again towards a snack in the noon break in Germany.

Time became more important in respect to the increasing wages of the kitchen personnel. Data for canteens are scarce and scattered, but the administration reports of hospitals clearly illustrate the trend (Figure 13.1). According to an inquiry of leading hospital administrations, the annual costs of a kitchen maid was about 254 Marks in 1906, but about 298 Marks in 1911 (*Taschenbuch,* 1910: 8ff, 52ff; Z[eidler], 1912), and the cost for wages per patient's day of stay rose enormously, despite the administration's efforts to rationalise.

Figure 13.1 Expenditure for wages of the kitchen staff per patient's day of stay in the Hospital St. Jacobs, Leipzig 1875–1908 (Reichsmark per patient and day)

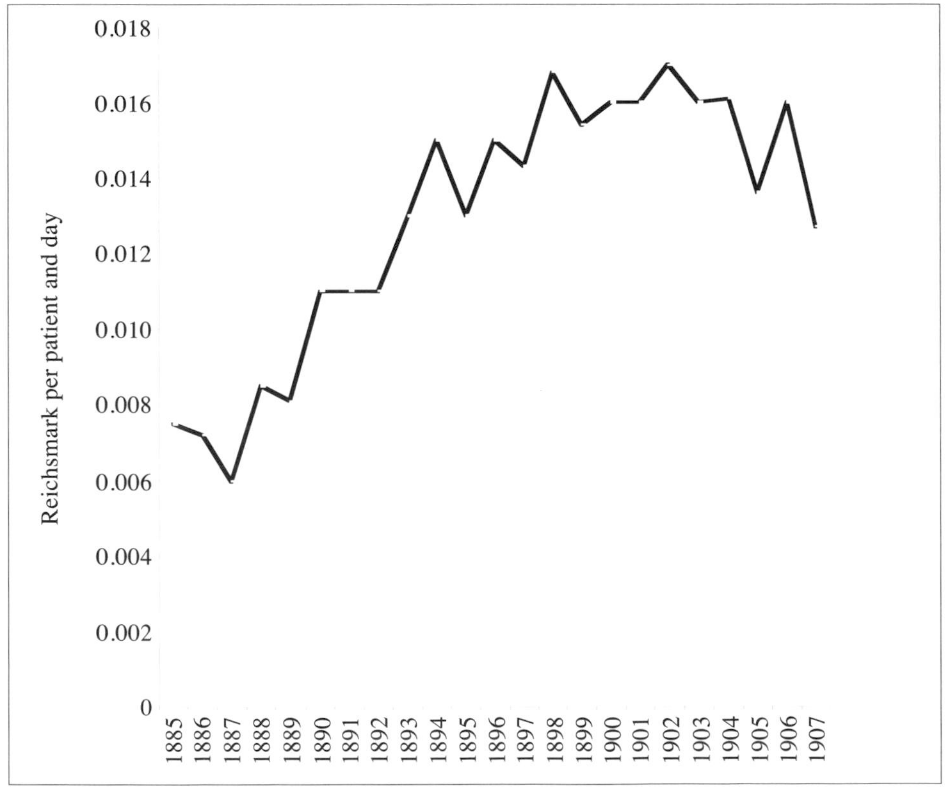

Source: Verwaltungsberichte des Magistrats Leipzig, 1875.

[1] See the example of Siemens at Halske in Berlin, Deutschmann, 1988: 142, 251.

Wages were rising, as the example of a single German canteen illustrates. The cost for kitchen personnel increased by 30 percent between 1875 and 1891. The reduction of its percentage in the total cost, which fell by six percent, goes back to the doubling of prices for the raw material. Nevertheless, both factors underlined the necessity of rationalisation within the kitchen.

Table 13.1 The composition of cost of meals from the canteen of the Emser Plumb and Silverworks in Ems, Germany, 1875–1891 (Pfennig per portion)

	Cost per Portion (Pfennig)			Percentage		
	Wages	Fuel	Raw Material	Wages	Fuel	Raw Materials
1875	2.7	0.6	12.1	17.4	4.2	78.4
1885	2.9	0.5	18.5	13.2	2.6	81.6
1891	3.5	0.8	24.8	11.3	2.7	84.4

Sources: Post and Albrecht, 1893, II: 337; *Amtliche Mittheilungen,* 1885: 216ff*; Einrichtungen,* 1876: 68ff.

The proportion of wages is low compared to today. In more recent times, a higher percentage of the wages had to be spent on personnel. In the 1970s and 1980s, canteens spent wages half of their expenditures on food items. But in 2001, the cost for personnel had decreased to 17 percent, which illustrates the far-reaching consequences of rationalisation within the big kitchens.

III. Rationalisation within the canteens, and cooking technology

Expenditure on kitchen staff did hardly matter until the end of the 19th century. When 'rational feeding' was discussed, quality of food and its composition, especially its protein content, were stressed. But the rise of wages changed this situation, especially as opportunities to compensate higher expenses with higher meal prices were rather limited, as higher prices always resulted in a downfall of participation of the number of diners.

Rationalisation of appliances and food was the answer to the rising wage bill. In regard to appliances, first rationalisation measures – mainly to reduce fuel consumption – were taken at the beginning of the 19th century. This was not special for mass catering, as it started from private households. Rationalisation of the kitchen in private households, however, stopped at a theoretical stage, whereas new techniques were more easily applied in mass cooking (for example, steam cooking). At the end of the 19th century, an industry of mass cooking appliances had developed. Engineers debated on their ideal construction in their journals, and lists of manufacturers and food processing firms were published (Maßregeln, 1894: 386ff)

Rationalising food preparation was most promising. Still in 1985, the preparation of the different foodstuffs, that is the washing, the selection of spoilt parts, the cutting and so on took 70 percent of the whole preparation time of a meal (Loeschke and Höfs, 1985: 11). This opened a broad field to inventive engineers, which combined principles already known from inventions for the private household with powerful machines. Peeling potatoes was a special problem, as mass catering in Germany heavily depended on the cheap and nutritious potato dishes. Potatoes fitted the food preferences of workers as well as the possibilities of kitchen equipment, which allowed only boiling. Peeling potatoes takes a lot of time. At the end of the 19th century, the usual daily portion amounted to 500–1000 g per capita intake. The amount of potatoes, which could be peeled by hand, differed from 2.4 kilo and 12.5 kilo per hour and person, so that automatic peeling seemed rather promising. First potato-peeling machines appeared in the late 1870s (Giedion, 1987: 601),[2] and models from around 1900 were able to peel 25 kilo potatoes per hour.[3]

A major problem was that the new machines could not peel all types of potatoes. Especially those sorts, which were liked most in mass catering, were not suited for machine peeling at all, as the colour of their red skins was mitigated to the fruit itself during peeling and turned into blue during cooking. But people refused to eat blue potatoes.[4] Another point was that a round and regular form of potatoes was most important for rubbing off the skin in the machine. In fact, the manufacturers informed the buyers of the machine, which sorts of potatoes were suited best for their products.[5]

On the long run this had far reaching consequences for agriculture, which adjusted to the needs and concentrated on the growing of particular sorts (I will come back to this point). The same happened with dried vegetables, which have a long tradition in Europe. Since the middle of the nineteenth century they were processed on an industrial scale. They were available all over the year, being easy to be transported and to be stored, they were cheap and allowed commercialising the surplus production in traditional vegetable-growing areas. Finally, dried vegetables need no preparation before cooking at all. So it is no wonder that they became a popular ingredient of hot-pots in mass catering. This differed with canned food, which was a luxury item before the Second World War, still and at best in use for the military war rations (*Eiserne Ration*).[6] But as dried vegetables were more expensive than fresh ones, workers did not buy them at all, because household work of women was unpaid and not calculated in financial terms. Moreover, workers often owned a small piece of land to grow vegetables (Rosenbaum, 1992).

Nevertheless, dried potatoes were not produced on a very large scale before World War I. Once again, the military forces were the motor of change: during the First World War, canning factories dried large amounts of potatoes for the provision of the soldiers and for the wartime bread (Eckart, 1992: 339). People did not like these wartime surrogates at all, and these disappeared from the menu until the Second World War, when they again had become a factor in the soldier's ration: The state requisitioned the production of whole factories for the need of the army (Eckart, 1992: 339). In 1949 powdered potatoes occurred on the market as brand articles, now in beautifully coloured packages. This product was meant to make dumplings or potato pancakes (Eckart, 1992:

---

2 For the construction of these machines see 'Kartoffelschälmaschinen', 1906: 237–238; see also Klaffke, 1910: 395–398.

3 Archives of the Humboldt University, Berlin, Charité-Direktion, Nr. 1835, Bl. 98.

4 Archives of the Humboldt University, Berlin, Charité-Direktion, Nr. 1553, Bl. 223.

5 See the letter of Dreßler 15 April 1903, in: Archives of Humboldt University, Charité-Direktion, Nr. 1835, Bl. 101.

6 This was not the case with the army, where they were part of the emergency rations (*Eiserne Ration*) from early on, because in times of war financial aspects played a minor role. On the consumption of canned and dried vegetables and the high social differentiation of its consumption, Thoms, 2005: 590.

340). Histories of particular enterprises may tell us that this product was made for more comfort for stressed housewife. Yet, I found numerous advertisements for products of this kind in the canteen manager's journals from the 1950s onwards. Since that decade, processed potato products were a booming business. Between 1949 and 1959 the per capita amount of processed potatoes climbed from 2 to 25 kilo.[7] In 1974–1975 a fifth of all eaten potatoes was consumed in form of processed products. In mass catering, these figure was even higher, as 30 to 50 percent of all potato dishes and 90 to 95 percent of mashed potatoes and potatoes dumpling were made from convenience products.

Table 13.2 Per capita consumption of potatoes, West-Germany, 1948/49–1975 (in kg)

	Fresh potatoes	Processed Potatoes
1948/49	224	–
1954/55	157	1
1964/55	114	5
1974/75	70	20

Source: 'Kartoffeln – die beliebte Beilage', *Großküchen + GV* , 28/12, 1976: 8.

This development had important consequences for agriculture: As every cook knows, every potato sort has different qualities and is suited for special purposes. As this is determined genetically, the importance of potato breeding increased ever since. So the factory owners were themselves active in breeding, to fit the raw material to production requirements, especially as some of the founding fathers of the German potato processing industry owned degrees in agriculture (Eckart, 1992: 339; Schnetkamp, 1992: 331). They tried to find out which sorts were the best for certain preparations, and they developed own certificate breeds (Eckart, 1992: 339; Schnetkamp, 1992: 331). The German State supported these industrial efforts by the foundation of the 'Institute for Starch and Potato Technology' (*Institut für Kartoffelforschung*) in 1972, which was to conduct research in the field of the industrial processing of potatoes. This institute made tests to find out those sorts which were suited best for frying, boiling or drying. Moreover those breeds, which fit the demands of industrial processing, were certified.

In the nineteenth century there had been 2,500 different potato sorts, nowadays 180 are certified and thus allowed for growing. Only ten, however, are qualified for industrial processing. Growing them is furthered by the development of contract farming. Factories began to buy all of the harvest of farmers who delivered it at fixed dates and got fixed prices for a fixed quality. For the factories this was a safeguard against shortage of the raw material and a guarantee for a steady working production without any need of storage.

Contract farming was spreading even in the growing of vegetables. One point was the new 'career' of dried vegetables. To avoid the association of wartime food, they were not simply called dried vegetables (*Dörrgemüse*) but dehydrated vegetables (*Dehydro-Gemüse* or *Quellgemüse*) (Beim Quellgemüse, 1975: 23–29). The manufacturers were

[7] For the USA, see 'Kartoffelprodukte', 1950: 160. Most of the amount was manufactured into potato stripes (12%), potato starch (7%), frozen products (2,5%) dried products (2,5%), and other conserved and mixed products (1%).

eager to promote its advantages and told the canteen's managers that one cube meter was sufficient to store 7,000 to 11,000 portions, and that there was no need for special conditions for storage, no need for preparation and no waste. But again, the consumption of these products first was limited to mass catering, save its role as an ingredient in convenience soups. These were exactly the same arguments the producers of the mashed potato and potato-pancake powder used (Der Kartoffel, 1975: 10).

IV. Frozen food

Freezing industry played an increasing role for the growing importance of contract farming, too. Frozen meat was rather popular and half of all meat which was consumed in the UK was frozen meat as early as the 1930s (Fleischmann, 1934: Appendix). But the success with other frozen foods was rather limited. In the USA they were produced from the 1930s onward, first on a more or less experimental stage. From the 1940s onward, scientific journals spread the idea that rationalisation and mechanisation of the kitchen would make cooking unnecessary for the housewife. She would simply open the freezer, take out a meal, which was prepared by master cooks and then frozen, heat it in a high-frequency radio-wave oven, and just serve it some minutes later. In fact, the sales figures of frozen food boosted between 1934 and 1944 from 39 to 600 million pound-packets in the USA (Giedion, 1987: 652).

The development in Europe was much behind the US-experience. The Nazi regime was very well aware of this and of the rationalisation potential of frozen food, especially in regard to the campaign 'Fight against Waste' (*Kampf dem Verderb*) it had launched in 1936 (Ziegelmayer, 1947: 599–603). In the 1940s the regime initiated a research programme to develop the freezing technique for the army's purposes. A special institution, the *Reichsanstalt für Lebensmittelfrischhaltung*, had the assignment to investigate deep freezing. In 1939 Solo-Feinfrost was founded, an enterprise controlled by the leading military institution (*Oberheereskommando*). It undertook serial trials in deep freezing and introduced the freezing of fruit and vegetables. In 1940 it already produced 22,000 tons of frozen food, but the war hindered a further spreading (Hilck and Auf dem Hövel, 1979: 34f, 40).

These experiences explain the rapidity with which the new technology and frozen products spread after the end of the Second World War. Nevertheless, a first campaign of the newly founded 'German Deep Freezing Institute' (*Deutsches Tiefkühlinstitut*), the 'German Commission of Information in Political Economy' (*Bundesausschuß für volkswirtschaftliche Aufklärung*), and the 'German Fish Promotion' (*Deutsche Fischwerbung*) in 1955, which was directed to private households, failed. Most of the household simply could not afford a freezer (Wildt, 1996: 63ff). In 1960 only three percent of all German households owned a deep freezer and even in 1970 only ten percent of all German (as well as English) households had the possibility to store frozen food for a longer time (Teuteberg, 1991: 154; Oddy and Oddy, 1998: 299).

Costs were another point. Pre-cooked dishes were and are relatively expensive, because preparation and storage have to be paid. But still in the 1950s the budgets of private households were rather limited. As kitchen work is 'invisible' and unpaid, the cost factor of work did not count in the private household at that time, but more the convenience aspect (Tiefgekühlte Fertigspeisen, 1959: 125). This is quite different in collective catering, where every minute has to be paid. Higher prices for frozen food are acceptable, thus, if compensated by a minus in (cost of) work. In fact, industry was expanding so much in the 1950s, that workforce was running short. The West-German government signed a contract with Italy in 1955 on the recruitment of Italian workers for German enterprises. At the same time, enterprises expanded their social welfare systems to bind the workers to their company, and to take care for those without own households. So they invested large sums in new canteens, too. There were 4,480 canteens in 1953, 5,080 in 1954 and 5,612 in 1955 (Krohn, 1956: 10–12; 10.000 Kantinen, 1956: 6).

This opened a new market for the German food industry, which was very well aware of the chances and conquered the expanding market according to a plan (Qualitätsanspruch, 1956: 8–10). Food producers had been keen on selling food to canteens, hospitals, the army and so on since long,[8] as they consumed large quantities at once, so that huge parts of the production could be sold to a low number of clients. This meant low administration costs. The enterprises were willing to adjust to the customer's wishes in regard to special packing or portion size or the production of special articles (Hein, 1966, H4: 8–12).

There is indeed evidence that the participants in mass catering were the first who consumed frozen food in big amounts, and paved the way for a broader consumption of frozen food. In the Netherlands canteens have been a test market for pre-cooked frozen food (de la Bruhèze and Van Otterloo, 2003: 14). Sales numbers for Switzerland from 1964 to 1975 do not only show that collective kitchens were the first to consume the new products, but that their consumption grew much faster than overall consumption. I can find similar results for Germany in recent years, where nearly half of all frozen food is bought in big gastronomy-packages and their sales figures grow faster, still.

Table 13.3 Increase in the consumption of frozen food in private and collective households in Switzerland 1964–1970

	Total consumption (in tons)		Consumption in collective households (in tons)		Change from 1964–1970 (in percent)	
					Total consumption	Collective households
	1964	*1970*	*1964*	*1970*		
Vegetables	3011	7284	752	2393	242	318
Fruit	417	729	107	292	175	273
Fruit juices	197	680	56	292	345	521
Fish and sea food	4298	9342	627	1846	217	294

Source: Neidhart, 1971/72: 6.

[8] In case of sales crisis, agricultural associations had appealed to the government that the army and the state institutions, like Hospitals and so on, should buy their products, see Thoms, 2005.

Figure 13.2 Number of canteens in West-Germany and their turnover per employee, 1959–1999

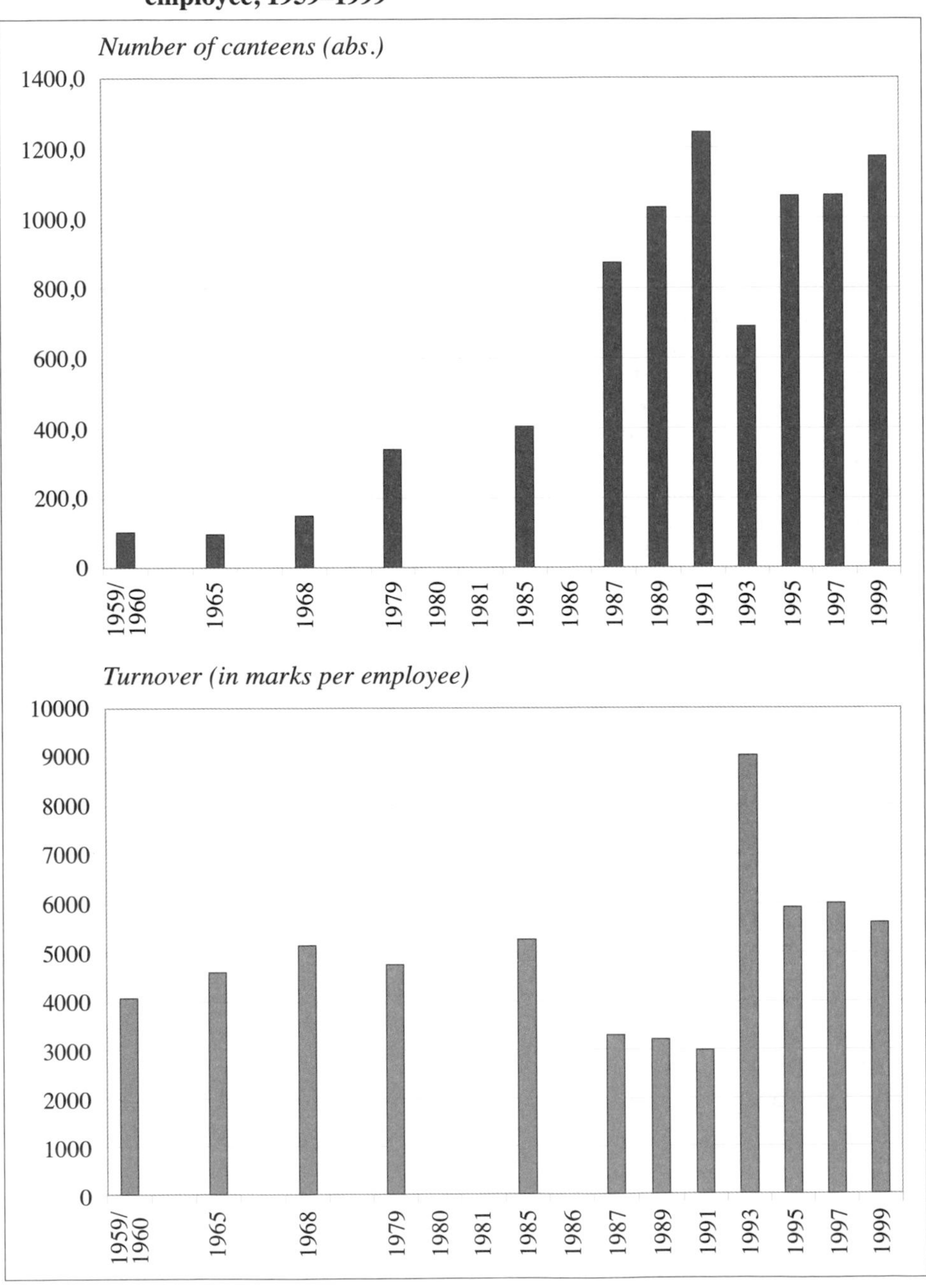

Source: Statistisches Jahrbuch der BRD, 1959ff.

Table 13.4 Sales figures of frozen food 1991 and 2001 in Germany (in tons)

	Household packages			Gastronomy packages		
	1991	2001	Change (%)	1991	2001	Change (%)
Vegetables	200,459	267,504	52.1	116,680	165,150	2.1
Pre-cooked dishes	70,010	222,669	218.1	82,248	215,010	671.5
Meat, Game, Poultry	228,400	198,783	-13.0	337,767	344,778	161.4
Potato products	167,469	182,401	8.9	172,910	180,210	4.2
Bakery	75,278	169,631	125.3	40,239	310,450	41.5
Pizza	74,939	159,571	112.9	4,200	10,690	77.4
Fish	88,403	152,383	72.4	51,732	91,779	289.4
Baguettes, Snacks	37,547	60,499	61.1	13,972	14,308	2.4
Fruit, Fruit-juices	8,749	9,862	12.7	14,580	56780	154.5
Total	951,254	1,423,303	49.6	834,328	1,389,155	66.5

Source: Cool facts, 2002.

Rationalisation was not simply limited to the outsourcing of preparation as one step in the nutrition chain. Forced by the very small kitchen, the aeroplanes were the first to substitute the cold snack food or pre-cooked fresh meal by complete frozen meals on their long flights. Since the late 1950s, Maxim's in Paris delivered 200,000 single meals per year to PanAm (Hilck and Auf dem Hövel, 1979: 79).

Meanwhile engineers had developed special convector ovens to reheat the frozen meals in a short time (Hilck and Auf dem Hövel, 1979: 81). The availability of this technology was an ideal condition for the further development of industrial production of frozen meals for collective kitchens. In 1958 *Apetito in Rheine* was the first German enterprise in producing pre-cooked frozen meals on an industrial scale, and a number of others followed in the 1960 (Hilck and Auf dem Hövel, 1979: 85). There were different establishments specialised in the development, construction and installation of the technical appliances (Firmenspiegel, 1966: 12–18).

The combination of frozen meals and convector ovens allowed to provide even small groups for which it had not been profitable to cook before. Whereas 70 per cent of all hot meals were served in canteens of enterprises with over 500 employees, half of all convectors where to be found in firms with less than 500 employees. Especially for smaller enterprises, which provide most of the working places in Germany, re-heating kitchens were rather interesting, because they allowed the reduction of the large investment sums of kitchen equipment to less expensive re-heating and freezing facilities. This meant a reduction in building and running cost, a point that was heavily checked in advertisements of frozen dishes.

Rationalisation of meal production was only one point, another was the rationalisation of meal composition in regard to nutritional guidelines. In the first German nutrition report (1969), the 'German Society for Nutrition Research' (*Deutsche Ernährungsgesellschaft*) recommended the consumption of frozen food in collective kitchens. The second report (1972) again stressed that losses in vitamin content and the change of taste were rather small (Hein, 1974: 87f). Industry called upon this opinion. Moreover, it published the

Table 13.5 Costs per menu in a defrosting and in a normal kitchen in 1964 (DM)

Defrosting kitchen Expenditure for	300 Menus	500 Menus	Normal kitchen Expenditure for	300 Menus	500 Menus
Raw material (food)	690.00	1125.00	Raw material (food)	360.00	575.00
Personel	206.00	312.00	Personel	624.00	816.00
Energy	25.00	45.00	Energy	70.00	145.00
Administration	15.00	25.00	Administration	33.00	55.00
Depreciation	35.00	62.50	Depreciation	110.00	208.00
Interests on capital	17.50	31.25	Interests on capital	55.00	104.00
Rooms	25.00	40.00	Rooms	50.00	80.00
Total	1013.50	1640.75	Total	1302.00	1983.00
Cost per menu	3.38	3.28	Cost per menu	4.34	3.97

Source: Düsterberg, 1971–1972: 2, 18.

nutrient content of frozen meals on the packages to meet the demands of professional nutritionists to control the calorie intake. On the other hand, this helped to replace the trained cook and dietician not only as a manual worker in the production process itself, but even in the planning of meal composition. In other words, this helped to replace qualified personnel by unskilled workers. The food industry was very well aware of this aspect and interpreted it as a chance for the sale of defined diets for different physical needs and age groups (Hein, 1974: 84). It reached its goal, as enterprises handed responsibility for what the workers or institution inmates found on their plate over to the food industry.

Rationalisation went further on. The assembly line entered the kitchen. Based on studies on movements of the kitchen staff, preparation steps were arranged in the most economic way. Some steps were fully automatic, for example by cooking automats, which were constructed for nearly all different foodstuffs and preparation techniques. With frying automats only one person is needed to put the pieces of meat or fish on a kind of conveyor band, which then transports them through boiling fat. Up to 1,000 pieces per hour can be fried by these machines. Other automats were constructed for automatically cooking pasta or steam-cook vegetables. Also there were the dishwashers. Nowadays, the equipment of canteen and hospital kitchens does not differ much from menu factories, save the latter wrap the ready meals for sale.

This completed the industrialisation of a highly mechanised and rationalised food production, which first had only mechanised and outsourced single steps of the preparation and cooking process. On the other hand the growing importance of technical equipment and knowledge made financial capital more and more important and furthered the well-known concentration process, which later on spread to the contract caterers, too.

V. The European Perspective: Differences

To what extent these technological possibilities were used, how canteens were organised, and especially, to what extent they were a commercial or a more charitable affair, differs largely within Europe. Especially the UK had a fair number of cheap commodities for eating out and in regard to mechanisation and commercialisation of the catering trade, that is to say to fast food, it lay ahead. From the nineteenth century onward, the fish and chips shops offered a relatively cheap opportunity for eating out (Walton 1992). In other European countries, philanthropic organisations took over the practical organisation of canteens, as for example the 'Swiss Association of People's Service' (*Schweizer Verband Volksdienst*), which ran a huge number of canteens throughout the country (Tanner, 1999). In Berlin the *Volksküchen- und Speisehallengesellschaft*, founded in 1866 (Allen, 2002), run public kitchens, which addressed workers too. But since the number of cheap restaurants was established, the number of visitors went down. For Berlin it is shown that some of the formerly philanthropic organisations turned into commercial enterprises. They meant a severe concurrence to the canteens. As provision was secured the enterprises did not see a necessity to build canteens. The employees mostly ate outside, if they had the choice and if they did not prefer to bring something from home. Generally, visitors were reluctant to visit canteens and the frequency depended much on the meal price (Thoms, 2005). In times of need, the number went up, as for example during the First World War.

The existence of both German States after 1945 shows that the development of catering is not necessarily bound to free economic forces. From the early nineteenth century onward, socialists have underlined the role of community kitchens and their mechanisation for women's liberation from household work. They developed an idea of family life which differed clearly from the concept propagated by the bourgeoisie and especially by paternalistic employers. During the long 19th century, the hot family lunch around midday was a symbol of 'healthy' family life and a measure for the housewife's virtues. After the end of the Nazi regime, which heavily promoted *Gemeinschaftsverpflegung* as a way of creating and spending an 'esprit de corps', both German states underwent quite different developments. In the former GDR, collective eating was part of the socialistic way of living. Canteens and school meals were seen as a social contribution of the state, which financed half of the costs, organised it in most cases, and did a lot of research to improve the quality of food. In 1988–1989 nearly 80 per cent of all workers and 86 percent of school children were supplied with lunch from such commodities (Schilling, 1996: 156). More than half of the whole food consumption of the GDR took place in canteens and other catering enterprises of the GDR. This resulted from high state subsidies for the meals, a high percentage of working women, which should not be burdened by household work and by a deficient supply of convenience food in the shops (Donat, 1996: 8). This contrasts sharply with the situation in the FDR, where this percentage was at only 10 percent, where schools meals were uncommon at all, and only 46 percent of all employees had access to canteens, of which 57 made use of this opportunity (Köhler, 1996: 267). But after reunification, when the subsidies in eastern Germany were cut down, participation reduced there at once.

It seems that the mingling of ideological and moral aspects has had strong influences on the development of canteens, but it seems worth noting that the idea of outsourcing, of buying food or ready-prepared dishes or complete meals has spread all over Europe since the end of Second World War. Especially French catering enterprises, like SODEXHO, are working on a global scale and are still expanding their business. Against the declining trend of the catering industry's turnover in recent years, the turnover of contract caterers is still growing. In the UK it rose from 2,257 to 3,514 million pounds between 1995 and 2001. Half of this was spent in canteens of business and industry (www.caterer.com/facts), whereas only three percent of all German canteens were run by caterers (Köhler, 1996: 281). But the development in Germany is similar. In contrast to the turnover of the catering industry turnover and profits of canteen caterers are rising. In 1997 the German caterers had a profit of 30 Mrd. DM, from which 61 per cent came from canteens. That means that the turnover increased by 8.3 per cent in comparison to 1996 (gv-praxis 1998, ch. 5, p. 42).

VI. Summary and outlook

There is a lot of work in bringing together the rather diverging numbers and figures on the development of catering. I have mainly used the example of German canteens to demonstrate some general developments in the twentieth century and have dealt with some general problems in the possibility of comparing developments and figures throughout Europe, which result from cultural food differences.

Invented primarily for the use in the private household, frozen foods were accepted in the catering trade much earlier, because they saved work and thus money. Only the changes on this larger scale drew forward the changes in food technology and finally resulted in changes of agriculture. It turns out that the explanation of changes in the food market and food consumption can thus not be explained by developments in the private household alone. Figures support this finding: up to half of all consumed food, and in some cases even all of certain items, are sold to catering enterprises. In Germany, people spend 74.6 Mrd. Euro on eating out in 2001 (www.bve-online.de/zahlen/index.html). As 70 percent of all expenditures in gastronomy goes to the raw material (www.destatis.de/basis/d/insol/kostltab2.htm), that is food, gastronomy spends 52 Mrd. Euro on food and drinks, whereas food industry has had a total turnover of 126.6 Mrd in 2001 (bve-online.de/zahlen/index.html). These figures do matter, and we cannot ignore them by explaining changes in consumption by changes within the private household alone.

Table 13.6 Food sold to the catering industry as percentage of total food sales in 1986

Pasta	44	Frozen Fish	40	Vegetable oil	42
Rice	30–40	Sausages	45	Mayonnaise	35
Frozen Potatoes	50	Tinned soups	28	Salad dressing	40
Cooled Potatoes	100	Soup powder	26	Ketchup	40
Tinned Vegetables	25	Smoked/Salted fish	30	Frozen fruit	80
Frozen Vegetables	41	Gravy powder	30	Frozen bakery	35
Pickled Vegetables	25	Tomato puree	60	Frozen bread	30
Frozen Meat	75	Vegetable fats	75	Instant Coffee	30
Frozen meat meals	45	Animal fat	70		

Source: Gira, 1986, cited after *Ernährungsbericht* 1988: 242ff.

Instead, we should have a closer look and ask where food is bought, where, how and under what circumstances it is consumed. We then should find out how the shifts are related to the changing role of eating out and catering, and what follows from this change for the production as a precondition of processing. Evidence from household education shows that an increasing number of children in Germany does not know basic cooking techniques anymore, because they seldom see how the food is being prepared. This poses the question of how and how far the technological changes of catering influence eating behaviour, nutritional knowledge and cooking skills. Finally, we have to ask how far catered food does indeed influence the development of taste. Experiences with children show that they get used to the food they eat in the *kindergarden* or at school, so that they finally love things they never get at home. The problem is that what goes on in the catering trade is hidden for the 'normal' consumer. It is a world of its own, only known to the experts and insiders who are simply not interested to allow a closer look at their daily practice, because it might destroy some of the customers' illusions.

Bibliography

'10.000 Kantinen und Großküchen. Ständiges Anwachsen der Kantinen und Kasinos/ Industrie zieht Nutzanwendungen' (1956) *Deutscher Kantinen-Anzeiger*, 31, 4, p. 6.

Albert de la Bruhèze, A. and Otterloo, A. van (2003) 'The rise of eating out in the Netherlands in the Twentieth Century. Snacks, meal-patterns and the food-chain', in M. Jacobs and P. Scholliers, P. (eds), *Eating out in Europe. Picnics, gourmet dining and snacks since the late eighteenth century*, Oxford–New York, pp. 317–336.

Allen, R. (2002) *Hungrige Metropole. Essen, Wohlfahrt und Kommerz in Berlin*, Hamburg.

Amtliche Mittheilungen, aus den Jahres-Berichten der mit Beaufsichtigung der Fabriken betrauten Beamten (1885) Berlin.

Barlösius, E. (1989) 'The history of diet as a part of the vie matérielle in France', in H.-J. Teuteberg (ed.), *European food history. A research review*, Leicester, pp. 90–108.

'Beim Quellgemüse summieren sich Vorteile' (1975) *Großküchen + GV*, 27: 9, pp. 23–29.

Bruegel, M. (1998) 'From the shop floor to the home: Appertising and food preservation in households in rural France, 1810–1930', in M. Schärer and A. Fenton (eds), *Food and material culture*, East Linton, pp. 203–247.

Burnett, J. (1998) 'Time, place and content: The changing structure of meals in Britain in the nineteenth and twentieth centuries', in M. Schärer and A. Fenton (eds), *Food and material culture*, East Linton, pp. 116–132.

Cool facts (2002) Tiefkühlkost, Deutsches Tiefkühlinstitut, Cologne.

Cremer, H.D. (1956) 'Die ernährungsphysiologischen Probleme der Großverpflegung', in *Probleme der vollwertigen Ernährung in Haushalts- und Großverpflegung. Die Vorträge auf der Mainzer Tagung 1956 der Deutschen Gesellschaft für Ernährung*, Frankfurt am Main.

Deutschmann, C. (1988) *Der Weg zum Normalarbeitstag. Die Entwicklung des Arbeitstages in der deutschen Industrie bis 1918*, Frankfurt am Main.

Donat, P. (1996) 'Die Entwicklung des Ernährungsverhaltens der DDR-Bevölkerung vor und nach der Wende', in T. Kutsch and S. Weggemann (eds), *Ernährung in Deutschland nach der Wende: Veränderungen in Haushalt, Beruf und Gemeinschaftsverpflegung*, Witterschlick, Bonn, pp. 1–20.

Düsterberg, K. (1971–1972) 'Tiefgekühlte Fertiggerichte – Möglichkeiten, Grenzen und Problematik', in F. Forster (ed.), *Die Tiefkühlung in der Großküche*, Zürich, pp. 15–19. (Berichtsband IGEHO 71).

Eckart, O. (1992) 'Das andere Wort für Kartoffel. Ein Unternehmen schreibt Kartoffelgeschichte', in O. Helmut and K.Z. Ziessow (eds), *Die Kartoffel. Geschichte und Zukunft einer Kartoffel*, Cloppenburg, pp. 339–346.

Die Einrichtungen für die Wohlfahrt der Arbeiter der größeren gewerblichen Anlagen im Preußischen Staate, bearb. im Auftrage des Ministers für Handel, Gewerbe und öffentliche Arbeiten (1876), 3 vols, Berlin.

Ernährungsbericht (1988) ed. by the Deutsche Gesellschaft für Ernährung e.V. im Auftrag des Bundesministers für Jugend, Familie, Frauen und Gesundheit und des Bundesministers für Ernährung, Landwirtschaft und Forsten, Frankfurt am Main.

'Firmenspiegel des Tiefkühlmarktes' (1966) *Deutscher Kantinen-Anzeiger*, 5, pp. 12–18.

Fleischmann, O. (1934) *Der Weltmarkt in Gefrier- und Kühlfleisch. Eine Untersuchung über Erzeugung, Handel und Verbrauch des Gefrier- und Kühlfleisches*, Diss. Cologne.

Giedion, S. (1987) *Die Herrschaft der Mechanisierung. Ein Beitrag zur anonymen Geschichte*, Frankfurt am Main.

Gira (1986) *A Selective Appreciation of Relevant Features of the West German Cätering Market*, Geneva.

Grefe, C., Heller, P., Herbst, M. and Pater, S. (1987) *Das Brot des Siegers. Die Hamburger-Konzerne*, Bornheim-Merten.

gv-praxis (1998) *Die Wirtschaftsfachzeitschrift für Grossverpflegung*, Bielefeld (www.gv-praxis.de).

Hamelmann, T. (1989) *Die historische Nahrungsforschung in Frankreich: Probleme, Methoden, Resultate* (Hausarbeit im Rahmen der Ersten Staatsprüfung für das Lehramt an der Sekundarstuffe II, Ms.), Münster.

Hartog, A.P. den (ed.) (1995a) *Food technology, science and marketing: European diet in the twentieth century*, East Linton.

Hartog, A.P. den (1995b) 'Serving the urban consumer: The development of modern packaging with special reference to the milk bottle', in A.P. den Hartog (ed.) (1995a), *Food technology, science and marketing: European diet in the twentieth century*, East Linton, pp. 248–267.

Hartog, A.P. den (2003) 'Technological innovations and eating out as a mass phenomenon in Europe: a preamble', in M. Jacobs and P. Scholliers (eds), *Eating out in Europe. Picnics, gourmet dining and snacks since the late eighteenth century*, Oxford–New York, pp. 263–280.

Hausen, K. von (1999) 'Häuslicher Herd und Wissenschaft. Zur frühneuzeitlichen Debatte über Holznot und Holzsparkunst in Deutschland', in M. Grüttner, R. Hachtmann and H.G. Haupt (eds), *Geschichte und Emanzipation*, Frankfurt am Main–New York, pp. 700–727.

Hein, G. (1966) 'Tiefkühlkost – für jede Großküche', *Deutscher Kantinen-Anzeiger,* 48, H4, pp. 8–12.

Hein, G. (1974) 'Der Markt für tiefgefrorene Fertiggerichte – gegenwärtiger Stand und Entwicklungstendenzen', in *Tiefgefrorene Gerichte in der Gemeinschaftsverpflegung. Vortragstagung der Deutschen Gesellschaft für Ernährung am 1.2.1974 in der Kongreßhalle Berlin, Internationale grüne Woche*, Berlin, pp. 82–88.

Hellmann, U. (1990) *Künstliche Kälte. Die Geschichte der Kühlung im Haushalt*, Giessen.

Hessler, M. (2001) *Mrs. Modern Woman. Zur Sozial- und Kulturgeschichte der Haushaltstechnisierung*, Frankfurt am Main.

Hilck, E. and Auf dem Hövel, R. (1979) *Jenseits von minus Null. Die Geschichte der deutschen Tiefkühlindustrie*, Hamburg.

'Der Kartoffel auf der Spur' (1975) *Großküchen + GV* , 27, 1, p. 10.

'Kartoffeln – die beliebte Beilage' (1976) *Großküchen + GV* , 28, 12, p. 8.

'Kartoffelprodukte – beliebte Nahrungsmittel' (1959) *Ernährungs-Umschau*, p. 160.

'Kartoffelschälmaschinen' (1906) *Zeitschrift für Krankenanstalten*, 2, p. 237.

Klaffke (1910) 'Die neue Kartoffelschälmaschine "Oceana". Vortrag, gehalten auf der 14. Hauptversammlung der Vereinigung der leitenden Verwaltungsbeamten der Kranken-Anstalten von Rheinland und Westfalen in Essen a.d. Ruhr', *Zeitschrift für Krankenanstalten*, 6, pp. 395–398.

Knop, B. and Schmitz, M. (1983) *Currywurst mit Fritten. Von der Kultur der Imbissbude*, Zürich.

Köhler, B. (1996) 'Die Ernährung der Beschäftigten – Kompromisse zwischen Kantine und eigenem Herd', in T. Kutsch and S. Weggemann (eds), *Ernährung in Deutschland nach der Wende: Veränderungen in Haushalt, Beruf und Gemeinschaftsverpflegung*, Witterschlick–Bonn, pp. 263–282.

Köhler, T. (2002) *Sie werden plaziert. Die Geschichte der Mitropa*, Berlin.

Krohn, M. (1956) 'Warum Tiefkühlware? Ein Großversuch mit tiefgefrorenen Lebensmitteln', *Deutscher Kantinen Anzeiger,* 31, 4, pp. 10–12.

Loeschke, G. and Höfs, J. (1985) *Großküchen. Grundriß und Ausstattungsplanung für Küchen zur Gemeinschaftsverpflegung*, Wiesbaden–Berlin.

Lohof, B.A. (1979) 'Hamburger stand: Industrialization and the American fast-food phenomenon', *Journal of American Culture*, 12, pp. 519–533.

'Maßregeln zur Erzielung einer besseren Ernährung der Fabrikarbeiter' (1894) *Arbeiterfreund,* 3, pp. 386–387.

Mühl, A. (1992) *75 Jahre Mitropa: die Geschichte der Mitteleuropäischen Schlafwagen- und Speisewagen-Aktiengesellschaft*, Freiburg im Breisgau.

Neidhart, T. (1971–1972) 'Die Entwicklung der Tiefkühlung in der Großküche', in Forster, F. (ed.), *Die Tiefkühlung in der Großküche*, Zürich, pp. 6–11. (Berichtsband IGEHO; 71).

Oddy, D.J. (2003) 'Eating without effort: the rise of the fast-food industry in Britain in the twentieth century', in M. Jacobs and P. Scholliers (eds), *Eating out in Europe. Picnics, gourmet dining and snacks since the late eighteenth century*, Oxford–New York, pp. 301–316.

Oddy, D.J. and Oddy, J.R. (1998) 'The iceman cometh: the effect of low-temperature technology on the British diet', in M. Schärer and A. Fenton (eds), *Food and material culture*, East Linton, pp. 287–307.

Pelzer, B. and Reinhold, R. (2001) *Margarine. Die Karriere der Kunstbutter*, Berlin.

Pitte, J.-R. (1999) 'The rise of the restaurant', in J.-L. Flandrin and M. Montanari (eds), *Food. A culinary history from the ancient time to the present*, New York, pp. 471–480.

Post, J. and Albrecht, A. (1893) *Musterstätten persönlicher Fürsorge von Abeitgebern für ihre Geschäftsangehörigen*, 2 vols, Berlin.

'Der Qualitätsanspruch der Großküchenbetriebe und die deutsche Nahrungsmittel-Industrie' (1956), *Deutscher Kantinen-Anzeiger,* 31, 7, pp. 8–10.

Ritzer, G. (2000) *The McDonaldization of Society*, Pine Forge.

Rosenbaum, H. (1992) *Proletarische Familien*, Frankfurt.

Sandgruber, R. (1988) 'Zeit der Mahlzeit. Veränderungen in Tagesablauf und Mahlzeiteneinteilung in Österreich im 18. und 19. Jahrhundert', in B. Nils-Arvid [et al.] (eds), *Wandel der Volkskultur in Europa. Festschrift für Günter Wiegelmann zum 60. Geburtstag*, Münster, pp. 459–472.

Schärer, M. and Fenton, A. (eds) (1998) *Food and material culture*, East Linton.

Schilling, D. (1996) 'Erscheinungen und Tendenzen des Konsumverhaltens im Gaststättenwesen und in der Gemeinschaftsverpflegung', in T. Kutsch and S. Weggemann (eds), *Ernährung in Deutschland nach der Wende: Veränderungen in Haushalt, Beruf und Gemeinschaftsverpflegung*, Witterschlick - Bonn, pp. 155–170.

Schlosser, E. (2001) *Fast food nation. What the all-American meal is doing to the world*, London.

Schnetkamp, E. (1992) 'Die Entwicklung der Kartoffel-Veredelung zu tiefgefrorenen Nahrungsmitteln", in O. Helmut and K.Z. Ziessow (eds), *Die Kartoffel. Geschichte und Zukunft einer Kartoffel*, Cloppenburg, pp. 331–338.

Seidel-Pielen, E. (1996) *Aufgespießt. Wie der Döner über die Deutschen kam*, Hamburg.

Stuyvenberg, J.H. van (ed.) (1969) *Margarine. An economic, social and scientific history 1869–1969*, Liverpool.

Tanner, J. (1999) *Fabrikmahlzeit. Ernährungswissenschaft, Industriearbeit und Volksernährung in der Schweiz 1890–1950*, Zürich.

Taschenbuch für den wirtschaftlichen und verwaltungstechnischen Krankenanstalts-Betrieb (1910) *Ein Auskunfts- und Nachschlagebuch über Bau, Einrichtung, wirtschaftlichen Betrieb, Organisation und Verwaltung der Krankenhäuser, Hospitäler, Lazarette, Kliniken, Irren- und Pflegeanstalten, Heilstätten, Sanatorien etc.*, Leipzig.

Teuteberg, H.-J. (1991) 'Zur Geschichte der Kühlkost und des Tiefgefrierens', *Zeitschrift für Unternehmensgeschichte*, 36, pp. 139–155.

Teuteberg, H.J. (1997) 'Die Rationalisierung der Küche am Beispiel des Elektroherdes seit dem späten 19. Jahrhundert', in H.J. Gerhard (ed.), *Struktur und Dimension. Festschrift für Karl Heinrich Kaufhold zum 65. Geburtstag, II: Neunzehntes und zwanzigstes Jahrhundert*, Stuttgart, pp. 456–476.

Thoms, U. (1994) '"Der Tod aus der Milchflasche". Säuglingssterblichkeit und Säuglingsernährung im 19. und 20. Jahrhundert', in *Kein Kinderspiel. Das erste Lebensjahr. Eine Ausstellung des Westfälischen Museumsamtes Münster*, Münster, pp. 58–70.

Thoms, U. (1998) 'Changes in the kitchen range and changes in food preparation techniques in Germany, 1850–1950', in M. Schärer and A. Fenton (eds), *Food and material culture,* East Linton, pp. 48–76.

Thoms, U. (2003) 'Industrial canteens in Germany 1850–1950', in M. Jacobs and P. Scholliers (eds), *Eating out in Europe. Picnics, gourmet dining and snacks since the late eighteenth century*, Oxford–New York, pp. 351–372.

Thoms, U. (2005) *Rationalisierung der Anstaltskost. Gefängnis- und Krankenhausverpflegung im 18. und 19. Jahrhundert*, Stüttgart.

'Tiefgekühlte Fertigspeisen' (1959) *Ernährungs-Umschau*, 6, p. 125.

Wagner, C. (2001) *Fast schon Food. Die Geschichte des schnellen Essens*, Bergisch Gladbach.

Walton, J.K. (1992) *Fish and Chips and the British Working Class*, Leicester.

Wildt, M. (1996) *Vom kleinen Wohlstand. Eine Konsumgeschichte der fünfziger Jahre*, Frankfurt am Main.

Z[eidler, B], (1912) 'Vergleichende Übersicht über die Betriebsergebnisse und Betriebsverhältnisse von 58 Krankenanstalten und Heilstätten', *Zeitschrift für Krankenanstalten*, 8, pp. 345–368, 481–492 and 641–646.

Ziegelmayer, W. (1947) *Die Ernährung des deutschen Volkes. Ein Beitrag zur Erhöhung der deutschen Nahrungsmittelproduktion*, Dresden–Leipzig.